GEOMAGNETISM OF BAKED CLAYS AND RECENT SEDIMENTS

Edited by

K.M. CREER, P. TUCHOLKA
Department of Geophysics, University of Edinburgh, Edinburgh (United Kingdom)

and

C.E. BARTON
Graduate School of Oceanography, University of Rhode Island, Kingston, Rhode Island (U.S.A.)

ELSEVIER   Amsterdam — Oxford — New York — Tokyo 1983

ELSEVIER SCIENCE PUBLISHERS B.V.
Molenwerf 1
P.O. Box 211, 1000 AE Amsterdam, The Netherlands

Distributors for the United States and Canada:

ELSEVIER SCIENCE PUBLISHING COMPANY INC.
52, Vanderbilt Avenue
New York, NY 10017

ISBN 0-444-42231-5

Printed in The Netherlands

Contributors

M. Aitken — Research Laboratory for Archaeology and the History of Art, Oxford University, Oxford OX1 3QJ, UK.

P. A. Alcock — Research Laboratory for Archaeology and the History of Art, Oxford University, Oxford OX1 3QJ, UK.

S. K. Banerjee — Department of Geology and Geophysics, University of Minnesota, 108 Pillsbury Hall, Minneapolis, Minnesota 55455, USA.

M. Barbetti — Department of Physics, The University of Adelaide, Adelaide, South Australia, Australia.

C. E. Barton — Graduate School of Oceanography, University of Rhode Island, Kingston, RI 02881, USA. (formerly at Department of Geophysics, University of Edinburgh.)

R. Bradshaw — School of Botany, Trinity College, Dublin 2, Eire. (formerly at Department of Geophysics, University of Edinburgh.)

S. P. Burlatskaya — Institute of Physics of the Earth, B. Gruzinskaya 10, Moscow G-242, USSR.

G. D. Bussel — Research Laboratory for Archaeology and the History of Art, Oxford University, Oxford OX1 3QJ, UK.

W. S. Chang — Institute of Geophysics, Academia Sinica, Beijing, China.

C. Y. Chao — Institute of Geophysics, Academia Sinica, Beijing, China.

R. M. Clark — Mathematics Department, Monash University, Clayton, Victoria 3168, Australia.

K. M. Creer — University of Edinburgh, Department of Geophysics, King's Buildings, Mayfield Road, Edinburgh EH9 3JZ, UK.

K. P. Games — Sub-Department of Geophysics, University of Liverpool, Oliver Lodge Laboratory, PO Box 147, Liverpool L69 3BX, UK.

R. E. W. Hedges — Research Laboratory for Archaeology and the History of Art, Radiocarbon Accelerator Unit, Oxford University, Oxford OX1 3QJ, UK.

K. Hirooka — Geological Laboratory, Faculty of Education, Fukui University, Fukui, Japan.

M. Kovacheva — Geophysical Institute, Academician Bonchev Street, 1113 Sofia, Bulgaria.

T. C. Li — Institute of Geophysics, Academia Sinica, Beijing, China.

S. P. Lund — Department of Geological Sciences, University of Southern California, Los Angeles, CA 90007, USA.

M. W. McElhinny — Division of Geophysics, Bureau of Mineral Resources, PO Box 378, Canberra City, ACT 2601, Australia.

J.S. Mothersill — Faculty of Science, Lakehead University, Thunder Bay, Ontario P7B 5EL, Canada.

F.O. Oldfield — Department of Geography, The University, Liverpool L69 3BX, UK.

S. Papamarinopoulos — 40 Byron Street, Argyroupolos, Athens, Greece. (formerly at Department of Geophysics, University of Edinburgh.)

D.J. Schove — St. David's College, 29 South Eden Park Road, Beckenham, Kent, BR3 3BQ, UK.

C.J Shaw — Research Laboratory for Archaeology and the History of Art, Oxford University, Oxford OX1 3QJ, UK.

V.P. Shcherbakov — Geophysical Observatory Borok, Nekouzskiy Region, Yaroslavskaya Oblast, USSR 152742.

V.V. Shcherbakova — Geophysical Observatory Borok, Nekouzskiy Region, Yaroslavskaya Oblast, USSR 152742.

A.M. Sinito — Departamento de Ciencias Geologicas, Facultad de Ciencias Exactas y Naturales, Pabellon 2, Ciudad Universitaria, Buenos Aires, Argentina.

R. Sternberg — Department of Geosciences, University of Arizona, Tuscon, Arizona 85721, USA.

R.C. Thomas — Department of Engineering Science, Oxford University, Parks Road, Oxford, UK. (formerly at Department of Geophysics, University of Edinburgh.)

P. Tucholka — University of Edinburgh, Department of Geophysics, King's Buildings, Mayfield Road, Edinburgh EH9 3JZ, UK. (on leave from Institute of Geophysics, Polish Academy of Sciences, Warsaw, Poland)

P. Tucker — Warren Spring Laboratory, Department of Industry, Stevenage, UK. (formerly at Department of Geophysics, of Edinburgh.)

G.M. Turner — Department of Physics, Victoria University, Private Bag, Wellington, New Zealand. (formerly at Department of Geophysics, University of Edinburgh.)

D.A. Valencio — Departamento de Ciencias Geologicas, Facultad de Ciencias Exactas y Naturales, Pabellon 2, Ciudad Universitaria, Buenos Aires, Argentina.

J.F. Vilas — Departamento de Ciencias Geologicas, Facultad de Ciencias Exactas y Naturales, Pabellon 2, Ciudad Universitaria, Buenos Aires, Argentina.

D. Walton — Department of Physics, McMaster University, Hamilton, Ontario L8S 4M1, Canada.

S.P. Wang — Institute of Geophysics, Academia Sinica, Beijing, China.

S.F. Wei — Institute of Geophysics, Academia Sinica, Beijing, China.

Q.Y. Wei — Institute of Geophysics, Academia Sinica, Beijing, China.

A. Wintle — University of Cambridge, Sub-Department of Quaternary Research, The Godwin Laboratory, Free School Lane, Cambridge CB2 3RS, UK.

D. Wolfman — Arkansas Archaeological Survey, Box 1356, Arkansas Technical University, Russellville, Arkansas 72801, USA.

Preface

Our understanding of the long term behaviour of the geomagnetic field is severely limited by the short span of time, approximately four centuries, for which instrumental records are available. This book is about ways and means by which the geomagnetic record may be extended back in time.   Basically we depend on two sources of information: on evidence inadvertently left behind by ancient civilizations and on evidence left as a result of natural geological processes. It is the job of research workers in the fields of archaeomagnetism and palaeomagnetism respectively to recognize these clues and from them to deduce past directions and magnitudes of the field.

The generally accepted usage of these names is not entirely logical: 'archaeomagnetism' is widely used to include studies of the remanent magnetism of archaeological relicts while 'palaeomagnetism' is used to include studies of the remanent magnetism of rocks. These definitions are somewhat misleading, in that the Greek prefix 'archaeo-' means 'oldest' while 'palaeo-' means merely 'ancient'. Thus strictly, archaeomagnetism should refer to the oldest geomagnetic studies (say to results derived from Pre-Cambrian or Archaean rocks) while palaeomagnetism should refer to studies of rocks which are not quite so old (say to results derived from Phanerozoic rocks, including of course the Quaternary). However the usage of these two terms is now deeply rooted in the literature and we realize that any attempt to redefine them logically would not gain general acceptance.   Thus, throughout this book we have used the term archaeomagnetism to refer exclusively to results derived from archaeological materials and we have used the term palaeomagnetism to describe results derived from naturally occurring materials.

It follows that the range of time accessible to archaeomagnetic studies will always be limited to that during which human beings have existed: at present the vast majority of results have been derived from existence left behind by settled civilizations but we should expect that it should be possible, in future, to extend the archaeomagnetic record if methods for extracting information from archaeological features left behind by semi-nomadic tribes and by cave dwellers can be devised. On the other hand, the discipline of palaeomagnetism has already been developed to the stage where we have a rudimentary knowledge of the behaviour of the geomagnetic field over the last 4000 million years and a much more detailed knowledge over Mesozoic and more recent time. This research has revealed a very broad spectrum of variations ranging from $\sim10^6$ to $\sim10^7$ years which is characteristic of changes in polarity bias of the main dipole through $\sim10^5$ to $\sim10^6$ years which is characteristic of the frequency of polarity reversals to $\sim10^2$ to $\sim10^3$ years which is characteristic of secular variations (sensu stricto). In this book we are concerned exclusively with the last of these three ranges of characteristic times.

The concept of creating this book arose from conversations during a symposium on "Time Scales of Geomagnetic Secular Variations" held at the 4th Assembly of the International Association of Geomagnetism and Aeronomy held at Edinburgh in August 1981. As often happens during international symposia, enquiries were made by some participants as to whether we had any plans for publication of the papers read. We hold mixed views about the publication of symposium proceedings because very often, the best of the presented papers either have already been published or are just about to be published elsewhere or the research has not reached the stage at which publication is justified. Thus volumes in which symposium papers are presented are very often unsatisfactory in that the coverage of the subject is unbalanced or incomplete. Hence their value to the research worker or to the student is limited.

Nevertheless it seemed to us that the time was ripe for a survey of archaeomagnetic and palaeomagnetic results, first because of the large volume of new data which have recently become available and second because results from the two types of source, namely human artefacts and geological deposits have rarely been discussed in the same context although they clearly complement one another. Therefore we decided to use the best of the papers presented at the IAGA Symposium as the core of the book and to ask the authors of these papers to upgrade them so as to give them a more lasting value. In addition we decided to invite other research workers, who have made a notable contribution to these fields but who were unable to attend the IAGA meeting, to contribute to the book so as to improve the coverage and overall balance. Finally we decided to edit the contributions with the aim of producing a uniform style of presentation as far as possible. At the beginning we did not fully appreciate the extent of editorial work we were involving ourselves with. The amount and character of editorial modifications we considered necessary varied widely: some of the contributions had to be substantially shortened which required extensive rewriting on our part while some few required hardly any modification at all. Wherever substantial modifications were made to the text or wherever figures had to be redrawn, the approval of the authors was requested and we appreciate their ready agreement.

The contributions have been collected together into four chapters and the references listed in each one have been assembled at the end of each chapter, our intention being that the volume should be read as a book rather than as a collection of papers.

The division of work between the editors has essentially been as follows: KMC has carried out most of the editorial changes to the submitted manuscripts; PWT planned and supervized the preparation of the camera ready copy which we submitted to the publishers and CEB did much of the early planning work including persuading many of the contributors to write for us before he moved from the University of Edinburgh to the University of Rhode Island.

We would like to express our appreciation of the facilities placed at our disposal by the University of Edinburgh, in particular to the Department of Computer Science for the use of their word processing and printing equipment.

K. M. Creer
P. Tucholka
C. E. Barton

Edinburgh, May 1983

Units

In 1973 the International Association of Geomagnetism and Aeronomy passed a resolution specifically recommending the use of SI units in the field of geomagnetism. It was anticipated, in IAGA, that a smooth and orderly transition from the centimetre-gram-second (cgs) based emu system to the metre-kilogram-second (MKS) based SI system would follow. The other associations affiliated to the International Union of Geodesy and Geophysics (IUGG) passed similar resolutions at about the same time.

However, a decade later, the SI system of units has still not won general acceptance among geophysicists and in particular many palaeomagnetists and rock magnetists see little merit in it. The broad based resistance to the adoption of SI units is reflected in the policies adopted by the editorial boards of journals covering the various fields of geophysics: most recommend but few insist upon the use of SI units. Current attitudes have recently (February 1982) been briefly reviewed by Gregg Forte, Editorial Services Manager of the American Geophysical Union in EOS (Vol. 63, pp. 140-141). Since emu are still fairly widely used and since those who continue to use them feel strongly about it we opted to accept the use of either emu or SI units in the contributions to this volume according to the individual preferences of the respective authors.

In the field of electromagnetism, the adoption of the SI system was complicated first, because it involved rationalization so that the 4π appear in different places in equations written in the old and new systems, as well as the change-over from a c.g.s. base to MKS base and second, because there was a protracted discussion about whether magnetic moment should be defined as having the same units as H (Am^{-1}) or B (tesla). Therefore in order to specify the particular form of SI units recommended by IAGA we reproduce the relevant resolution from IAGA News, No. 12, September, 1973, p4.

"Resolution 3

IAGA, considering that SI units are achieving international recognition as a single standard for worldwide use, recommends adoption of SI Units in the field of geomagnetism. Specifically IAGA recommends that:

– (a) Values of the geomagnetic 'field' be expressed in terms of the magnetic induction B (SI Unit tesla = Weber/metre) (b) If it is desired to express values in gamma, a note should be added stating that "one gamma is equal to one nanotesla".

– (a) Values of 'intensity of magnetization' be expressed in terms of magnetization M (SI Unit ampere/metre). (b) If it is desired to express values in emu, a note should be added stating that 'one emu is equal to 10 ampere/metre'.

– (a) Values of susceptibility be expressed as the ratio between magnetization M and the magnetic field H. (b) If, during the transitional period, it is desired to use values of susceptibility in emu, a note should be added stating that 'χ_{SI}' is equal to '$4\pi\chi_{emu}$'

We are still at the 'transitional period' but we expect that the SI system will ultimately gain general acceptance, if only because the younger generations of research workers have been educated only in the SI system and are therefore, ignorant of the elegant simplicity of the traditional em system of units.

Abbreviations

1. Types of remanent magnetization

ARM	anhysteretic remanent magnetization
CRM	chemical remanent magnetization
DRM	depositional remanent magnetization
GRM	gyroremanent magnetization
IRM	isothermal remanent magnetization
MD	multi-domain (grain)
PDRM	post-depositional remanent magnetization
PNRM	partial natural remanent magnetization
PSD	pseudo single domain (grain)
PTRM	partial thermoremanent magnetization
RRM	rotational remanent magnetization
SD	single domain (grain)
SIRM	saturation isothermal remanent magnetization
SRM	shear remanent magnetization
TRM	thermoremanent magnetization
VRM	viscous remanent magnetization

2. Geomagnetic field parameters

ADF	axial dipole (geomagnetic) field
GAD	geocentric axial dipole
GED	geocentric equatorial dipole
NDF	non-dipole field
RD	radial dipole
RDM	reduced dipole moment
SV	secular variation
VADM	virtual axial dipole moment

VDM   virtual dipole moment

3. Measurement techniques

AC   alternating current
ADR   annual dose rate
AF   alternating field (demagnetization)
CF: CS   constant flux : constant sedimentation rate
CIC   costant initial concentration
CRS   constant rate of sedimentation
CVMSE   cross-validation mean square error
D   declination
DC   direct current
ED   equivalent dose (thermoluminescence)
F_A or F^A ancient field magnitude
F_D or F^D dipole field magnitude
F_P or F^P present field magnitude
I   inclination
J   intensity of remanent magnetization
J_S saturation magnetization
k, χ susceptibility
M^A moment acquired on cooling in ancient field
M^L moment acquired on cooling in laboratory field
MDF   median destructive field
Q   Q-ratio (J/k)
T_B blocking temperature
TL   thermoluminescence

4. Units

A   amp
AD   years anno domini
BC   years before Christ
BP   calendar years before present
bp   radiocarbon years before present
c. g. s.   centimetre-gram-second (system of units)
e. m. u.   electromagnetic unit
G   gauss

MKS	metre–kilogram–second (system of units)
Oe	oersted
SI	systeme international (of units)
T	tesla

Table of Contents

1.
Magnetization
Processes

1.1. INTRODUCTORY COMMENTS

By the Editors

In this chapter the processes by which materials may naturally acquire and retain records of past geomagnetic secular variations are discussed from both theoretical and experimental viewpoints. Essentially three kinds of material are used for archaeomagnetic and palaeomagnetic investigations. Since they differ substantially in their physical and chemical properties it is not surprising to find that the processes by which they become magnetized depend on quite distinct physical phenomena. Pottery and the walls of the kilns in which they are baked become magnetized in the ambient magnetic field when they cool from high temperature. Natural sediments become magnetized by rotation of spontaneously magnetized grains (which most sediments contain to a degree) towards the direction of the ambient geomagnetic field both during and after deposition while the water content is high. Adobe bricks become magnetized by rotation of magnetized grains towards the ambient field direction due to the shock of impact when the moist mud is thrown into the mould. The acquisition of TRM and PTRM by baked clays is discussed first in section 1.2. Then the acquisition of DRM and PDRM by unconsolidated sediments is dealt with in sections 1.3 and 1.4 and finally the acquisition of the SRM by adobe bricks is described in section 1.5. The processes described in sections 1.3 and 1.4 occur naturally whereas those described in sections 1.2 and 1.5 require the intervention of human beings. The availability of three such different types of material as recorders of the ancient geomagnetic field allows the possibility of the detection of systematic errors inherent in any one of them and it would be rather difficult to identify such errors were only a single method of investigation available. Furthermore, our confidence in the results provided by any one type of material is strengthened whenever a given result is confirmed by agreement with another result obtained from investigations using one of the other types of material of the same age. In this respect we are still not able to make such comparisons to the extent we would wish because as yet no proven method has been developed to obtain ancient field magnitudes from lake sediments, though magnitude results from baked clays can be compared with those from adobe bricks. It is however, possible to compare secular variations in direction as recorded by all three kinds of material.

The results derived from lake sediments should be regarded as complementary to those derived from baked clays or adobe bricks. This is because lake sediments are capable of providing a continuous time sequence of data while baked clays and adobe bricks never can provide more than, at best, a patchy record. Also, the amplitudes of directional changes and of intensities recorded by lake sediments are reduced because of smoothing over times of the order of a few centuries whereas baked clays and adobe bricks provide data pertaining to a particular 'instant' of time related respectively to the rate of cooling and to the time of impact of the mud in the brickmakers mould.

1.2. MAGNETIZATION PROCESSES IN BAKED CLAYS

By Peter Tucker and Ruth C. Thomas,
University of Edinburgh, UK

1.2.1. Introduction

During cooling, baked clay can acquire and retain a permanent record of the Earth's magnetic field. The discipline of archaeomagnetism is concerned with the interpretation of such records carried by baked clay artefacts such as pottery and the walls of the kilns in which they were fired. The magnetization of the clays is of thermal origin, termed thermoremanent magnetization or TRM. As with the other forms of magnetization, TRM arises when a direct field (here the geomagnetic field) acts in conjunction with an energy perturbation (in this case heat). After firing, the baked clay cools through its Curie temperature and cooperative magnetization is established. With further cooling through a given temperature, known as the blocking temperature, this magnetization becomes fixed and a record of the Earth's field, at the time of firing, is preserved. The record contains information on both the direction and the intensity of the field. It is a relatively easy process to decipher the archaeodirection record but a much more lengthy and complex process to determine the archaeointensity. A considerable amount of research (see Chapter 3 of this volume) has been concerned with deciphering the archaeointensity record. As a preliminary to the later discussion, it is instructive at this point, to examine in greater detail, how the baked clays acquire their magnetic records.

1.2.2. Magnetic mineralogy

The TRM characteristics are strongly influenced by the physical and chemical properties of the magnetic grains contained within the sample. The most common magnetic minerals found in baked clays are the iron oxides magnetite and haematite. Which of these two minerals predominates depends not only on the source material but also on the ambient oxidation conditions during the firing of the sample.

The specific thermoremanence for haematite-bearing samples will be approximately two hundred times lower than that for samples containing an equivalent assemblage of magnetite grains. The stability of the thermoremanence against subsequent change or overprinting of the record will, however, be considerably higher in the case of haematite. It is this second criterion that combined with the chemical stability of haematite when refired in the laboratory, has led many workers to attempt to select only haematite-rich samples for analysis.

The size of the magnetic grains also has considerable influence on the specific thermoremanence (see Day, 1977). Over a critical size range (of order 0.05 to 0.5μm for magnetite) the grains each consist of a single domain (SD). Above these sizes the grains are multi-domain (MD). The boundary in magnetic properties may not be so clear cut.

There is evidence of a transition zone where, although several domains may be present the magnetic behaviour still retains some single domain like characteristics. Such grains are termed pseudo-single domain (PSD). The specific thermoremanence is high for single domain material, but falls rapidly with grain size over the PSD range, finally becoming independent of size for truly multi-domain material. The stability of the thermoremanence shows roughly the same trend with a higher stability usually recorded for single domain than for multi-domain material, though at the lower end of the single domain range the stability falls away.

For many of the pottery samples studied by the earlier workers, it was assumed that the magnetic grains were predominantly in the single domain range. Evidence is now accumulating that a substantial fraction of the carriers of thermoremanence, in such samples, are more probably in the form of small multi-domain grains (e.g. Thomas 1981). To fully characterize the magnetization processes one must therefore consider both single domain and multi-domain cases.

1.2.3. Single domain TRM

Neel (1949, 1955) developed a theory of TRM for single domain grains, based on a non-interacting dispersion of aligned uniaxial particles. In the absence of an applied field, there would be two equally probable stable directions of magnetization separated by an energy barrier of height $E = vH_cM_s/2$, v is the grain volume; H_c, the coercive force and M_s the spontaneous magnetization. The moment changes from one orientation to the other if an externally applied field (H_A) is sufficient to drive it across this energy barrier. In an applied field, the moment orientated parallel to the field would have a lower energy than that antiparallel to the field. If the dispersion were heated close to the Curie temperature, the height of the barrier would tend towards zero and in consequence, even the weakest applied field should be sufficient to produce saturation alignment of the magnetic moments. Thus on cooling, the resulting TRM should always equal the saturation magnetization and be independent of the inducing field. This is, however, contrary to all experimental evidence. To overcome the problem, Neel introduced the concept of thermal agitation, where, even in the absence of an applied field the probability (dp) of a change in orientation occuring in a time dt is finite, given by

$$dp = C \exp[-E/kT] \, dt \qquad (1)$$

where T is the absolute temperature, k = Boltzmann's constant and C a frequency factor. This immediately leads to the concept of a relaxation time (τ) for the process given by:

$$1/\tau = dp/dt = C \exp[-E/kT] \qquad (2)$$

In an applied field, Neel showed by similar arguments that there would be two relaxation times

$$1/\tau = C \, (1 \pm H_A/H_C) \, [1 - H_A^2/H_C^2]^{0.5} \exp(-E'/kT) \qquad (3)$$

with $E' = - v \, M_s(H_C \pm H_A)^2/H_C$ for changes into and out of the field direction respectively.

It follows that the relaxation time is a very strong function of temperature. Indeed τ can vary from a fraction of a second to thousands or millions of years with, as little as a few degrees drop in temperature. Near to the Curie temperature, τ is small so the grains rapidly relax towards an equilibrium. In an applied field, the difference in grain energy between the two orientations (given by $\Delta E = vM_sH_A/kT$) causes a net alignment of grain moments (vM_s) parallel to the inducing field. The total magnetization M(T) is thus given by:

$$M(T) = vM_s(T) \, [\exp(\Delta E/2) - \exp(-\Delta E/2)] / [\exp(\Delta E/2 + \exp(-\Delta E/2)] \qquad (4)$$

or

$$M(T) = vM_s(T) \tanh(vM_s(T)H_A/kT) \qquad (5)$$

On cooling, the relaxation times rapidly increase so preserving this alignment. A blocking temperature (T_B) may be defined for which the relaxation time is of order the time of measurement. At room temperature the resulting TRM becomes:

$$M_{TRM} = vM_{so} \tanh(vM_{so}H_A/kT) \qquad (6)$$

which is the fundamental equation of single domain TRM. For very weak inducing fields, equation 6 reduces to $M_{TRM} \propto H_A$, that is, the weak field TRM is linear with applied field. This concept forms the basis to all current (comparative) methods of palaeointensity determination.

A word should be said here about pseudo-single domain TRM, that is, where grains at sizes slightly above the single domain threshold may, like SD grains, carry a permanent magnetic moment. Stacey and Banerjee (1974) show that the TRM of such grains also follow a tanh dependence on applied field.

The $\tanh(H_A)$ dependence of the SD and PSD models, however, cannot in most cases, be made to fit experimental data; the theoretical curves being too steep and saturating at far too low a field. Dunlop and West (1969) showed that this discrepancy could be resolved assuming magnetic interactions to be present; i.e. an interaction field would modify the effective inducing field. An alternative explanation could be that the assumed single domain carriers in the experimental samples were in reality two or three domain grains, in which case one might look instead, to the multi-domain models. Day (1977) highlights that all measured TRM induction curves (whether they be obtained from SD, PSD or MD material) are usually best described by MD rather than SD or PSD theory.

1.2.4. Multi-domain TRM

Multi-domain TRM results from the motion of domain walls subject to the internal demagnetizing field, and to lattice defects in an externally applied field. The process is thermally activated. Thus, the domain walls themselves, can be thought to have blocking temperatures. The blocking process is complex, because each wall is also dependent on the relative positions of neighbouring walls. It is the total configuration of all the walls that defines the resultant remanent magnetization.

Theories of multi-domain TRM have been developed by Neel (1955), Everitt (1962), Dunlop and Waddington (1975), Stacey (1958) and Schmidt (1973). It should be noted that with the exception of Stacey, the theories are derived for a system comprising two domains separated by a single wall. A treatment of truly multi-domain material with many domains is too complex to solve analytically. Schmidt (1973) however has attempted to extend the theory to a three domain system and

Tucker and O'Reilly (1980) have extended the basic two domain models to include assemblages of non-interacting domain walls.

Neel (1955) developed the theory of multi-domain TRM in terms of the expansion, with decreasing temperature, of a 'square' hysteresis loop (implicit for a two domain system). The temperature dependence of the wall-defect interaction and the field dependence of the blocking temperature are not considered explicitly but are contained in the temperature dependence of the coercive force (H_c).   Neel assumed that H varies as (T_c-T) near to the Curie temperature (T_c) and that the saturation magnetization varies as $(T_c-T)^{0.5}$ near to T_c. Thus :

$$H_C/H_{C0} = (M_S/M_{S0})^2 \qquad\qquad (7)$$

Where $H_{C0} = H_c$ at T_0 and $M_{S0} = M_S$ at T_0

The form of the hysteresis loop is shown in Figure 1-1a, the demagnetizing field causing the loop to be sheared with slope $1/N$; N being the demagnetizing factor. At $T = T_c$ the loop will be infinitessimally small and an applied field (H_A) will give the magnetization defined by the point Q, as shown. As the temperature decreases, the loop will expand, the magnetization following the locus of Q. Q will move back along the descending branch of the hysteresis loop, as long as the ratio $r = AQ/AB = (H_A +H_C)/M_S$ continues to decrease.   At the minimum, $dr/dT = 0$ giving:

$$dH_C/dT = [(H_A + H_C)/M_S].dM_S/dT \qquad\qquad (8)$$

which from equation 7 gives the blocking condition $H_A=H_C(T_B)$ and $r_{min}=2H_A[NM_S(T_B)]$.   From this point, r increases as the temperature falls but Q, the point representing the magnetization, cannot move up the descending branch of the loop and is effectively fixed in value. At room temperature, the specimen therefore has a magnetization given by :

$$M=r_{min}M_{S0} = 2H_A/NM_S(T_B) \qquad\qquad (9)$$

It is not altered when H_A is removed, hence from equation 7:

$$M_{TRM} = (2/N) (H_A \cdot H_{C0})^{0.5} \qquad\qquad (10)$$

That is, the TRM varies as the square root of the applied field.   However, the experimental evidence is that the TRM acquired in weak fields, is in fact, proportional to the applied field; so to take account of this, Neel extended the model to include the effect of thermal fluctuations.   The thermal agitation, at high enough temperatures, may provide sufficient energy, so that the domain wall is unable to localize on an energy well and will fluctuate around the position of minimum energy. Neel invoked a fluctuation field H_f such that for $H_A< H_f$ the blocking condition becomes $H_C(T_B)= H_f$ with:

$$H_f/H_C(T_B) = (M_s(T_B)/M_{S0})^2 \qquad\qquad (11)$$

The magnetization at T_B is simply H_A/N therefore at room temperature

$$M_{TRM} = (H_A/N)(M_{S0}/M_s(T_B)) = (H_A/N)(H_{C0}/H_f)^{0.5} \tag{12}$$

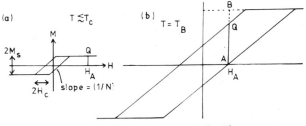

Sheared 'square' hysteresis loop of Neel's 1955 model.

(For explanation see text)

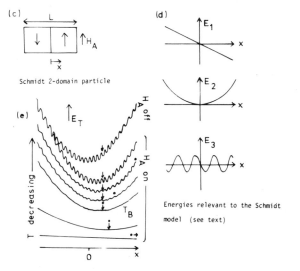

Schmidt 2-domain particle

Energies relevant to the Schmidt model (see text)

• = maximum stable wall displacement

↓ = domain wall location

Figure 1-1: Schematic representation of the wall blocking process (after Schmidt 1973).

Schmidt (1973) gave a pictorial description of the thermal blocking process. His model was developed in terms of the temperature dependence of the energies of interaction between a 180° domain wall and the applied field, demagnetizing field and structural defects. The interaction energies; magnetostatic energy (E_1), self demagnetizing energy (E_2), and assumed wall defect interaction energy (E_3), are shown in Figure 1-1d. On minimizing the total energy, Schmidt found that the TRM is given by

$$M_{TRM} \propto H_A^{(1-[1/(p-1)])}$$ (13)

with p given by $H_C \propto M_S^{p-1}$; $2 \leqslant p \leqslant 10$. It should be noted, that on setting p=3, the theory gives the same result as the Neel model. As Neel had done, Schmidt extended his model to include thermal agitation blocking at low fields. A schematic representation of the blocking process is shown in Figure 1-1e.

1.2.5. Partial TRM's and the law of additivity

On inspection of the TRM models, it is readily seen that an assemblage of grains will very rarely be described by a single blocking temperature.   A spread in coercivities or grain volumes will inevitably produce a range of blocking temperatures. Thus, if a magnetizable sample is field-cooled through a limited temperature range, only those grains whose blocking temperatures lie within this range acquire a thermoremanence.   Thellier (1946) established a law of additivity for these partial thermoremanences (PTRMs). The total TRM acquired on cooling can thus be interpreted as a sum of PTRMs acquired independently over several discrete temperature intervals (summing to the total cooling range).   The law of additivity has been well verified both theoretically (Neel, 1949) and experimentally for single domain grains. Stacey (1963) showed that the law of additivity might also apply to multi-domain material under certain conditions. Stacey and Banerjee (1974) cited evidence in support of a law of additivity for multi-domain TRM. Tucker and O'Reilly (1980) however, have shown that for large multi-domain grains, the law of additivity does in fact break down, as the TRM also includes contributions from the high field isothermal remanent magnetization (IRM), acquired after each cooling stage.   This final reservation may, however rarely, be applicable when considering the relatively fine grains normally found in baked clays. A general law of PTRM additivity can usually be assumed for such material. Indeed, this assumption is implicit in Thelliers (Thellier and Thellier, 1959) method of palaeointensity determination.

It is often assumed, a priori, that the magnetization gained on cooling through a temperature interval, is totally destroyed on heating through that same temperature interval.   In fact, the unblocking temperature is always higher than the corresponding blocking temperature. This feature is a direct prediction of Neel's (1949) model for single domain grains. In the single-domain case, the differences between the two temperatures will, in many instances, be very small.   For multi-domain grains however the unblocking temperature may often be considerably higher than the blocking temperature (Tucker and O'Reilly, 1980).

1.2.6. Some possible modifications to primary remanence.

It should be noted, that the NRM measured in the laboratory, will frequently not represent the total TRM acquired during the original firing. Processes which can partially or totally destroy the TRM record, such as the production of VRM, CRM or subsequent refiring, are discussed more fully in subsequent chapters.

1.3. MAGNETIZATION OF UNCONSOLIDATED SEDIMENTS AND THEORIES OF DRM

By Peter Tucker,
University of Edinburgh, UK

1.3.1. Introduction

In order to assess the fidelity of the geomagnetic record carried by unconsolidated sediments, it is necessary to understand the recording process itself, and to ascertain the degree to which the original record may subsequently be modified. The primary magnetization of sediments arises through the statistical alignment of detrital grains. The process has been termed detrital remanent magnetization or DRM. Alteration of the original DRM can occur through misalignment effects, viscous remagnetization of the detrital grains or by chemical growth of a new magnetic phase. This section is primarily concerned with the physical aspects of DRM.

DRM, in common with other types of remanent magnetization, results when magnetizable material is subjected to an energy perturbation in the presence of a direct magnetic field. The net magnetic moment is driven towards the applied field direction when, as the result of the perturbation, the constraints on such alignment are temporarily reduced or removed. For example, in the case of thermoremanent magnetization (TRM) the application of heat lowers the barriers to domain-wall movement or domain rotation. On cooling, the barriers are re-established, the relaxation time increases and the grain moment becomes effectively blocked in. With DRM, the detrital grains already carry a permanent magnetic moment. These individual grain moments are not altered during DRM acquisition. During deposition, the whole grains rotate under the influence of the torque developed between their own respective magnetic moments and the applied (Earth's) magnetic field. The remanent magnetization is blocked in (i.e. free rotation is inhibited) by the physical constraints imposed when the grain is finally trapped within the sediment. A mechanical perturbation of the sediment may subsequently cause a temporary reduction in the magnitude of the constraining forces by, for example, 'loosening' the packing structure whereupon free rotation again becomes possible. The magnetization resulting from such a realignment within the wet sediment is usually referred to as a post-depositional remanent magnetization or PDRM. A schematic representation of the entire deposition process is shown in Figure 1-2. Each aspect will be treated in more detail below.

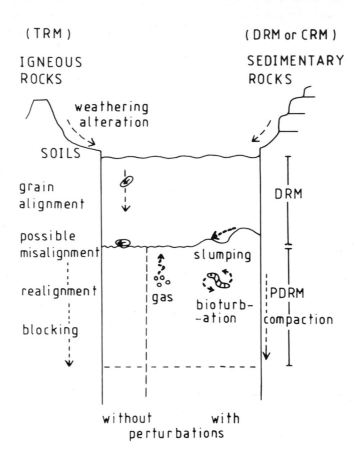

Figure 1-2:   Schematic representation of the deposition process.

1.3.2. The time-scale for grain alignment and the concept of blocking

To elucidate the concept of magnetization by grain alignment, it is instructive to first consider the most simple case, that is the rotation of an isolated spherical magnetized grain, moving freely under the influence of a magnetic torque. The equation of motion for the grain is:

$$I\Theta + \lambda\Theta + m B \sin\Theta + L = 0 \qquad (14)$$

where Θ is the angle between the grain moment (m) and the external field (B), I is the moment of inertia and $\lambda\Theta$ is the viscous drag. L represents any mechanical torque, perhaps time dependent, which acts in conjunction with the magnetic torque (mB sin Θ). To solve equation 14, it is appropriate to make certain simplifying assumptions. In general, for small detrital grains, the inertial term is negligible compared with the other terms, and may be neglected (Collinson, 1965). Taking the simple case L=0, the solution to equation 14 is:

$$\tan[\Theta(t)/2] = \tan[\Theta_0/2] \exp[-mBt/\lambda] \qquad (15)$$

As we are most concerned with the component of remanence in the applied field direction (i.e. m cosΘ), equation 15 is best rearranged into the form:

$$\cos[\Theta(t)] = (1-A)/(1+A) \qquad (16)$$

where $A = \tan(\Theta_0/2)\exp(-2KT)$

which further reduces to: $\cos[\Theta(t)] = \tanh(Kt)$ for special case $\Theta_0 = \pi/2$.

$K = (mB/\lambda)$ and Θ_0 is the original grain orientation.

For a spherical particle of radius r, density ρ, and moment per unit mass σ, rotating in a fluid of viscosity $|\eta|$:

$$m = (4/3)\pi r^3 \rho\sigma \qquad (17)$$

and

$$\tau = 8\pi r^3 \eta \qquad (18)$$

By taking reasonable values for the parameters, we can calculate the time τ necessary for alignment. With $\rho = 5 \times 10^3 Kg/m^3$, $\sigma = 0.05$ to $50 Am^2/Kg$, $r = 0.5$ to $50\mu m$, $B = 20$ to $40\mu T$ and $\eta = 10^{-3} Pa.s$ for water (note 1Pa.s=10 poise), the characteristic time for apparent (say 99%) saturation is $\tau_{99} = 0.002$ to 3 seconds. As can be seen, the alignment time is very short indeed.

A region of laminar flow is usually observed stretching for a few centimetres above the sediment/water interface. Thus, even if the detrital grains were completely randomized in turbulent flow above the interface, they could almost certainly regain a high degree of alignment during their final free fall through the laminar region.

In the sediment itself, the proximity of adjacent grains would act to increase the effective viscosity. For a spherical grain moving in a spherical void (radius R), the viscosity is increased by the factor $1/[1-(r/R)^3]$ (Landau and Lifshitz, 1959) with a corresponding increase in alignment time. Tucker (1980a) measured the alignment times in a magnetite/quartz sediment, of grain size 1 to 38μm, and found that a tanh(t) dependence provided an adequate description of the build up of remanence. The alignment time τ_{99}, was given approximately by the empirical relationship $B\tau_{99}= 3 \times 10^{-3}$. That is, for geophysically realistic fields, $\tau_{99}= 75s$ to 150s.

Hamanno (1980), in an extended treatment, considered theoretically an assemblage of identically shaped spheroidal grains of aspect ratio (a/c) with a distribution of grain sizes. He derived an expression for the relaxation time for alignment in terms of the void ratio (e) of the sediment. His results are shown pictorially in Figure 1-3.

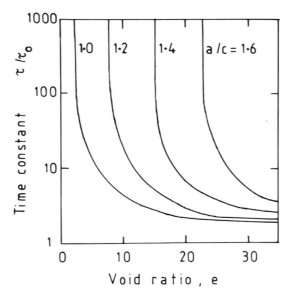

Figure 1-3:   Relaxation time for alignment as a function of void
ratio e for an assemblage of spheroidal grains of different
aspect ratios (a/c). After Hamanno (1980).

where it will be seen that with a decrease in void ratio, a critical void ratio is approached where τ increases dramatically and the grain is blocked in. This is, of course, analogous to the blocking condition for TRM. The critical void ratio is largest for the most elongated grains (ie. highest aspect ratios). Thus, it is seen that this form of blocking, brought about by a fall in effective void size usually as the result of compaction, occurs at a finite depth below the sediment surface. A range of blocking depths is predicted because of the distribution of grain and void geometries encountered in any real sediment. Lovlie (1974) experimentally demonstrated the phenomenon of a finite blocking depth. He found a depth of a few centimetres for a sediment slowly deposited in the laboratory.

1.3.3. Grain energy and alignment in the absence of energy perturbations

The rotation of grains in water-filled voids was first proposed by Irving (1957), in order to account for the observed post-deformational remagnetization of certain slumped Torridonian sandstones. Subsequent laboratory experiments (e.g. Irving and Major, 1964) have demonstrated the feasibility of such a process. Tucker (1980a), however has shown experimentally that even under the most favourable conditions less than 20% of the remanent magnetization is susceptible to realignment in this way. Even near the surface there may be sufficient physical constraints to prevent realignment. Field alignment only becomes possible when these constraints are broken.

A pictorial approach, describing the acquisition of TRM and isothermal remanent magnetization (IRM) in multi-domain grains, has been given by Schmidt (1973). A similar treatment is given here for DRM. We can imagine the grains as being trapped in energy wells of various depths. The total energy of the grain is given by the sum of the field energy ($-mB\cos\theta$) and the energies associated with the constraints. These terms are shown schematically in Figure 1-4 a and b. At a critical applied field (Figure 16c), the field energy overcomes the barrier energy and the grain becomes able to rotate towards the field direction ($\theta=0$). When the field is subsequently relaxed the grain becomes trapped in this new position.   In contrast, rotation against viscous (or frictional) torques (ie. free rotation in water-filled voids) takes place in the absence of any constraints. Here, field alignment occurs even in the lowest applied fields (Figure 1-4d). The grains, of course, remain free for subsequent realignment. A PDRM will only be produced when, or if, constraints are eventually built up.

The depth of the energy wells depends on the form of constraint.   Tucker (1980a) identified two types of constraining force, each overcome at a critical activation field. He interpreted these as rotations, involving local displacements of the surrounding grains (relevant to non-cohesive sediments in near-surface environments) and rotation against surface tension effects associated with entrapped gas bubbles. The results are portrayed schematically in Figure 1-5.

1.3.4. Energy perturbations

At this point, it is necessary to reconsider the solution to equation 14, taking into account perturbations from any mechanical torques (L). On their own, these torques would act to destroy or reduce the grain alignment. For small free-falling grains, or small grains freely rotating within voids, Brownian motion becomes significant

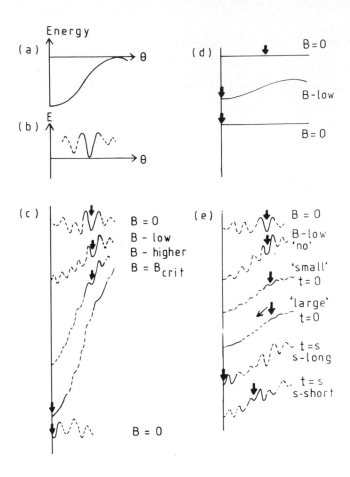

Figure 1-4: A conceptual energy representation of the blocking process (a) the field energy; (b) the energy associated with the physical restraints; (c) DRM acquisition through an increase in the applied field – arrows show stable grain position; (d) DRM acquisition for a freely falling grain; (e) Alignment after mechanical disturbance occurring at the time t=0. The grain rotates from t=0 to when the constraints are reestablished at t=s.

(Collinson, 1965; Stacey, 1972). If the grains are not spherical then gravitational forces may prevent complete alignment. This effect is most apparent at the sediment/water interface, where elongated grains would tend to come to rest with their long axes statistically aligned towards the horizontal (King, 1955). The result would show as a flattening of the magnetization inclinations. Similarly, the rolling of grains into surface hollows can give rise to inclination errors (King and Rees, 1966). The relative importance of these two effects depends critically on the nature of the

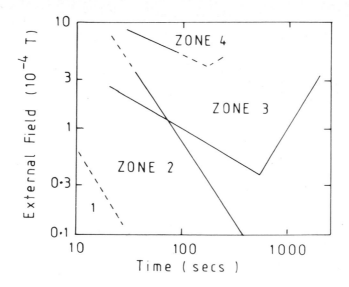

Figure 1-5: A schematic representation of the times and fields over
which various modes of grain alignment occur in an
undisturbed sediment. Zone 1 represents small reversible
movements within an energy well. Zone 2 represents
free-rotation in water-filled voids. Zone 3 and 4
alignments occur above a critical field where the
magnetic torque overcomes the constraining torques.
The energy wells associated with Zone 3 are shallower
than those associated with Zone 4. The data refer to
near-surface conditions in a synthetic sediment.
(After Tucker, 1980a)

sediment. It can be readily seen that these effects would be most prevalent in the
case of large grain non-cohesive sediments. Indeed, it is only with such sediments
that inclination errors have ever been definitely identified. That is not to say that
surface misalignment does not occur with other types of sediment: it may well be that
subsequent   post-depositional   realignment   can   correct   for   the   depositional
misalignment errors.

In the bulk of the sediment, perturbations still remain important. For instance, the
movement of gas bubbles can temporarily disturb the grain alignment (Noel, 1980;
Tucker, 1980a). Other local perturbations can arise from bioturbation or slumping.
The perturbations will generally occur at random time intervals and their net effects will
compound with time (Otofuji and Sasajima, 1981). Deeper within the sediment, where
the magnetic grains are fully blocked, compactions may cause a flattening of the grain
alignment, so giving rise to further inclination errors (Blow and Hamilton, 1978).

In the upper levels of the sediment, perturbations may not only give rise to
misalignment, they may also act to temporarily lower the barriers against field-aided
grain rotation. These barriers essentially arise through the packing of non-magnetic

grain rotation. These barriers essentially arise through the packing of non-magnetic matrix grains around each magnetic particle. Localized shear forces may act to separate adjacent grains, and this would result in a corresponding fall in the effective viscosity. In this way, the magnetic grains are temporarily freed and are able to rotate towards the ambient field direction. After the shear is removed, the sediment structure progressively reasserts itself, and after a characteristic time (s) the constraints have reformed sufficiently to block the grain in its current position. This process is visualized on the energy diagrams of Figure 1-4e. It is analogous to TRM acquisition. The degree of alignment achieved after each perturbation is controlled by the critical time (s). Conceptually, s can be expected to increase with the magnitude of the disturbance, and be largest for loosely packed non-cohesive sediments of high water content.

Slumping is perhaps the most dramatic example where such processes operate. Tucker (1982) gives an example, where the remanent magnetization of a slumped deep-sea carbonate ooze was totally reset during the slumping event, whilst for a nearby more cohesive clay-rich sediment, there was no appreciable PDRM associated with the slumping event. It seems likely that the remagnetization of Irving's (1957) sediments was contemporaneous with the slumping (i.e. perturbation assisted field alignment) rather than is often interpreted, the subsequent free rotation in water-filled voids. Bioturbation which affects the majority of sediments, and even the impact of detritus on surface particles, form other effective perturbation mechanisms.

As with alignment in the absence of perturbations, the above effects require progressively more energy as the sediment becomes more rigid and compacted. In consequence, this mode of realignment, also, is attenuated with depth.

1.3.5. The intensity of DRM and PDRM

Many authors have attempted to derive a theoretical expression for the field dependences of DRM. Collinson (1965) pointed out that an assemblage of small magnetized particles, subject to Brownian motion at temperature T, would behave like a classical paramagnetic gas, obeying the Langevin formula for the fraction of saturation alignment:

$$M/M_o = \coth a - 1/a \qquad (19)$$

where $a = kT/mH$ and k is Boltzmann's constant. Stacey (1972) extended Collinson's theory for a collection of particles with grain moments distributed over all values up to a maximum (m_{max}). On integration of equation 19 he derived the result:

$$M = (M_o/x) \ln[(\sinh x)/x] \qquad (20)$$

where $x = m_{max}H/kT$

For small x, equation 20 reduces to :

$$M = M_o x/6 = M_o m_{max} H/6kT \qquad\qquad (21)$$

Thus, the total moment (M) is linear with applied field for small fields. Whilst the above theory was originally derived for DRM, it should equally apply to grain rotation in water-filled voids. Many experimental studies of both DRM and PDRM have been interpreted on such a model, a linear or smooth curve being fitted to the data. Barton et al (1980) and Tucker (1980a) have questioned whether a smooth increase of intensity with field does indeed completely describe DRM (or PDRM) acquisition data. They both noticed low field 'kinks' in the acquisition curve. Tucker (1980a) correlated these kinks with the onset of the different modes of grain alignment (ie. zones 3 and 4 in Figure 1-5). Reassessment of the earlier data (eg. Johnson et al, 1948; Khramov, 1968) has revealed the possible presence of the low field 'kinks'. Thus, whilst the Stacey (1972) theory may well describe the results of free grain alignment, it appears, on its own, insufficient to describe the intricacies of the complete magnetization process. The theoretical models are compared with experimental data in Figure 1-6.

If the primary magnetization is fixed after a disturbance of the sediment, the intensity would be limited by the characteristic time (s) over which the grains were free to realign. Tucker (1980b) considered this case. He solved equation 14 assuming that L served initially to randomize the grain orientations then freed them for a time, s. The resulting PDRM is given by equation 16 summed over all Θ. Thus, for an assemblage of identical, spherical grains the resulting PDRM intensity (M) is given by:

$$M = m \sum_{\Theta_0} (1-b)/(1+b) \qquad\qquad (22)$$

where $b = \tan(\Theta_0/2) \exp(-2Ks)$

Computation of equation 22 for small disturbances (ie. small s) yields a linear dependence on field (B) for small B. For larger magnitude disturbances this linearity breaks down. The computed curves are shown with a selection of relevant experimental data in Figure 1-6.

1.3.6. Interpretation of the geomagnetic record

Every sediment sample taken for palaeomagnetic analysis must span a definite time-scale, equivalent to the sample thickness divided by the accumulation rate. A range of blocking times can be associated with each horizon within the sample, causing a further spread in apparent magnetic age. Any change in the ambient field during the effective blocking time-scale would produce a palaeodirection that is a weighted mean of the field directions, and a remanent intensity of lower magnitude

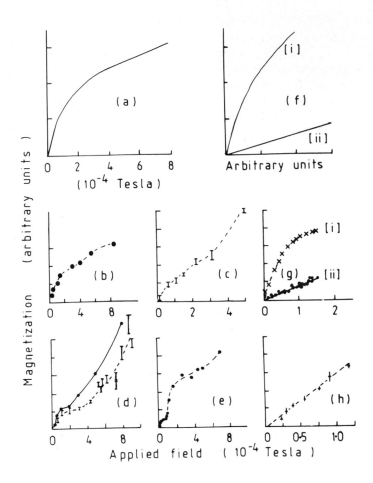

Figure 1-6:   The field dependence of detrital remanent magnetization:
(I) DRM intensities for free-grain rotation.
(a) theory (Stacey, 1972); (b) experimental curves
for redeposited varves (Johnson et al. 1948);
(c) red clays (Khramov, 1968); (d) concentrated
dispersions from lake sediments (Barton et al. 1980);
(e) synthetic sediments (Tucker, 1980a).
(II) DRM intensity attained after a disturbance to the sediment;
(f) theory for a large magnitude disturbance (i)
and a small disturbance (ii),   (Tucker, 1980b);
(g) experimental curves for synthetic sediments (Tucker, 1980b);
(h) deep-sea sediments (Kent, 1973).

than if all the blockings took place in the same field, because the measured
remanence vector is the vector sum of all the individual magnetic moments.

The stable magnetization direction is usually established through progressive AF demagnetization. The removed (low-coercivity) component includes contributions from the larger grains, possibly those which were subjected to the largest gravitational misalignment torques during deposition, and from any viscous material present. Now consider the effects of the possible delayed blocking of some of the magnetic grains: the grains most likely to be blocked last, are those which find themselves in voids which are large with respect to their own dimensions. Statistically, the smaller grains should be those most significantly affected.   The smaller grains are generally those with the highest coercivities.   Two situations arise: (i) if they remain unblocked at the time of measurement, then they would be readily disturbed and randomized by even the lowest alternating fields, (ii) if they were blocked in a later field direction, and this includes 'free' grains blocked in the laboratory field as the result of partial drying, they would impart a high-coercivity 'physical' overprint on the NRM. Such an overprint may not be apparent with the customary 'one-step' demagnetizations practised in routine palaeomagnetic analyses. Drying can also cause a distortion of the sediment structure, with a corresponding distortion of the stable NRM vector. Mechanisms are described by Henshaw and Merrill (1979) and Tucker (1980a).

Despite the gloomy prognosis there are many instances where reliable palaeodirections have been obtained. These will be discussed in the later chapters. One problem which has not been solved however, is 'how do we get reliable palaeointensities from unconsolidated sediments?'. Palaeointensity methods rely on normalizing the stable NRM by a laboratory induced magnetization. Normalization by the initial susceptibility and by a saturation isothermal remanence have been tried without success. These parameters fail since they over emphasize the role of the larger grains (which contribute little to the NRM). Levi and Banerjee (1976) showed that an anhysteretic magnetization (ARM) excited roughly the same coercivity fraction as the NRM. They went on to establish a relative palaeointensity method based on ARM normalization. ARM normalization does not however always give reliable results. This is perhaps, not surprising.

A DRM is acquired when previously magnetized grains rotate in a deposition field. An ARM not only destroys the initial grain remanence, but also, in no way simulates the field alignment of the individual grains. Further, the magnitude of the ARM is often very different from the NRM. If the sediment was redispersed and then deposited in a laboratory field, the laboratory remanence could provide a 'better' normalization parameter. Tucker (1981) reproduced the NRM of a deep-sea turbidite sequence, both in direction and in absolute magnitude, simply by stirring the redispersed material in a controlled laboratory field. At the time of writing, this technique has not been applied to recently deposited sediments where we have independent data on the strength of the Earth's field. It remains to be seen whether it can be developed into a technique that is applicable to the wide range of sediment types that are encountered in palaeomagnetic analyses.

1.4. GRAIN EFFECTS DURING DEPOSITION

By V.P. Shcherbakov and V.V. Shcherbakova,
Institute of Physics of the Earth, Moscow, USSR.

1.4.1. Thermal effects

Depositional remanent magnetization is produced by the alignment of particles carrying an intrinsic magnetic moment (m), into the direction of magnetic field (H) during deposition (Nagata, 1961). If grain alignment were opposed simply by viscous friction it can be readily demonstrated that the sediments should acquire a saturation DRM on settling in the Earth's field, the characteristic time of alignment being given by:

$$\tau \sim 6\eta / J_n H \tag{23}$$

where η is the viscosity of water ($\sim 10^{-2}$ poise), J_n is the intensity of magnetization (~ 1 gauss for magnetite) and H is the ambient field strength (~ 0.5 Oe). Thus, the time of alignment is ~ 0.1s, which is essentially instantaneous.

Experimental results contradict this conclusion as pointed out by Collinson (1965), King and Rees (1966), Khramov (1968) and Stacey (1972) who suggested that thermal fluctuations should be taken into account in order to explain the observed field dependence of DRM. In this case, the magnetization is described by the Langevin expression

$$J_r = mn(\cot a - 1/a) \tag{24}$$

where J_r is the intensity of remanent magnetization, m the magnetic moment of each grain, n the number of grains per unit volume and a=mH/kT where k is Boltzmann's constant and T the absolute temperature. Stacey (1972) showed that for single domain magnetite particles, thermal fluctuations disorientate particles with radii (r) $\leqslant 0.3\mu$m. For grains of haematite, the critical grain size is an order of magnitude larger. The thermal fluctuations affect only the the finest grains. The alignment of grains (including those with r > 0.3μm) can, however, also be affected by other factors, as discussed below.

1.4.2. Coagulation

Settling particles will for various reasons, collide and stick together. In the general case, sticking is caused by Van der Waals' forces (Fuks, 1955). Particles carrying a magnetic moment will also interact. Diffusional and kinematic coagulations should thus be distinguished. The particles collide with each other due to Brownian motion in the first case, and due to directed movement in the second case. One type of kinematic coagulation is 'gravitational', in that 'large' particles 'pick up' smaller ones settling with lower velocities. For diffusional coagulation of non-magnetic particles, it can be shown that intensive sticking will take place for grains smaller than about $1\mu m$ in radius, and that sticking is insignificant for grains larger than this, since these larger particles would normally reach the bottom before touching one another. This holds true for both lake and marine sediments.

Magnetic particles would of course, stick more quickly. However, coagulation of magnetic particles becomes significant only at ferromagnetic concentrations exceeding several percent. Such high concentrations are not normally found in natural sediments, and therefore, coagulation of magnetic particles can be disregarded in natural conditions.

Now consider gravitational coagulation: when a particle of size r settles through a layer of smaller particles, it can be shown that kinematic coagulation prevails over diffusional for $r > 1\mu m$ and under these circumstances small grains will be adsorbed by large ones during deposition.

It thus appears, that even under the most unfavourable conditions, namely deposition in still water, coagulation prevents particles $< 1\mu m$ in size from reaching the bottom. Magnetic particles become stuck to non-magnetic ones, so sharply decreasing the magnetization per unit volume of microaggregates.

1.4.3. Shape irregularity

In general, ellipsoids and discs don't fall down vertically (Happel and Brenner, 1965): a preferred orientation is observed only where the Reynolds number is large (Re > 10), (Fuks, 1955), i.e. for large particles with grain size > $100\mu m$. Small particles of diameter < 50 μm can rotate while falling if they have some shape asymmetry. A pictorial model of such a body is a 'propeller'.

Chernous and Shcherbakov (1980) showed that for an assymetric ferromagnetic particle, if the angle between the deposition field and the gravitational field is $90°$, then a stable direction for the magnetic moment (m) arises where all trajectories converge for all initial conditions. The particle may rotate around this axis. The stable direction is a function of the field inclination, and is more steeply inclined than the deposition field.

The propeller effect, thus causes the magnetic inclination to increase. The declination however remains unchanged and the intensity of magnetization can be shown to decrease smoothly, as the angle between the magnetic moment and the vertical increases from $0°$ to $90°$ (i.e. the magnetization of the suspension depends on the field inclination). The increase in inclination is rather unexpected as other theoretical models and the majority of experimental results show the inclination to be decreased. Increased inclination was, however, observed by Barton and McElhinny (1979) during the early stages of settling. This may well have been due to the effect of hydrodynamic forces.

A full theoretical discussion of the effect of shape irregularity is given by Chernous and Shcherbakov (1980).

1.4.4. General

For a fuller account of the theory of depositional remanent magnetization the reader is referred to Shcherbakov and Shcherbakova (1983)

1.5. MAGNETIZATION OF ADOBE BRICKS

By Ken Games,
University of Liverpool, UK

1.5.1. Introduction

All around the world, whenever human beings have inhabited dry regions, they have made buildings out of the most readily available material: the soil and clay upon which they walk. The process is simple: water is mixed with the clay to give it the right consistency. Some of the mixture is then scooped up and thrown into a wooden mould. The brick or adobe so formed is then turned out to dry in the sun. This basic method has remained largely unaltered through many millennia, in places such as Peru and Egypt.

This contribution is concerned with the discovery that adobe bricks become magnetized in the ambient geomagnetic field at the time of manufacture; at the moment when the mud is thrown into the mould. The NRM so acquired may be called a Shear Remanent Magnetization (SRM), since it arises from the large but short-lived stresses to which the mud is subjected, as it is forced to conform to the shape of the mould. Adobe bricks, being relatively valueless archaeologically, are widely available. Since they can be fairly accurately dated, it occured to us, that they might provide a most useful additional source of archaeomagnetic data.

1.5.2. The magnetic properties of adobe bricks

Adobe bricks start their lives as moist mud. Hence, it is natural to start by asking whether the NRM which they carry was acquired by post-depositional orientation of magnetic carrier grains. This cannot be so, however, because their normal water content, typically ~25% to 30%, is much lower than that at which PDRM becomes 'locked' into unconsolidated lake or marine sediments (~70%) (see section 1.3 of this chapter). A series of experiments in which mud was thrown into containers under controlled conditions was carried out in order to investigate the magnetization process, (Games 1977, 1978). The results show that a remanent magnetization is created at the moment when the mud is thrown into the container – see Figure 1-7.

In addition to the SRM acquired during manufacture, adobe bricks typically possess a softer, secondary, viscous component of magnetization. This has been demonstrated in the laboratory, by throwing mud into a cube-shaped mould aligned with its X-axis parallel to the ambient field, and then immediately turning the mould, (for example, through 90° to bring its Y-axis into the direction of the horizontal component of the ambient field) and then subsequently allowing the bricks to dry out for about 15 days in the new orientation. The result in all cases indicated that a viscous moment, of up to 15% of the total NRM, was built up along the direction of the ambient field in which the brick had dried out. This viscous component could be removed by AF demagnetization in 60mT or more. Thus the two components of remanent magnetization are physically distinct.

The magnitude of SRM acquired under laboratory conditions was found to be proportional to the strength of the applied field, over the range $20\mu T$ to $50\mu T$. Hence, ancient adobe bricks can be used to determine the magnitude of the ancient geomagnetic field, and the results of such studies are described in Chapter 3 of this volume. Although the direction of magnetization within a given brick is uniform to within about 2° (r.m.s. deviation from the average direction obtained from six specimens from a single brick), specimens taken from the corners have been found to deviate from the average direction by ~16°. This is presumably due to the extreme deformation necessary for the mud to fill the sharp corners of the mould. Therefore, samples should not be taken from the corners of bricks.

1.5.3. The magnetic constituents of adobe bricks

A comprehensive study was undertaken to determine the magnetic constituents of four adobe bricks. The investigation was in two parts, the first being a study of the magnetic properties of the bricks as a whole, and the second a study of the dependence of coercivity of remanence and saturation magnetization upon the grain size of the constituent magnetic minerals. The results indicated the presence of a

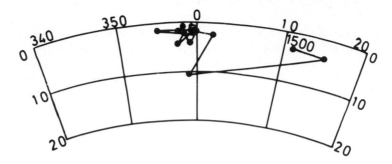

P1p : SRM (Set at 0,0)

Figure 1-7:   A polar plot of the direction of an SRM
given by 'throwing' the mud into a mould under
laboratory conditions (Games, 1978). The direction of the
applied field is at the top centre of the plot.
Each point represents the direction of magnetization
of the sample at a particular maximum AF value.
The direction at 1500 Oe (150mT) is indicated.

stable component of magnetization in the bricks and were consistent with the behaviour
of an assemblage of single domain and multi-domain grains of titanomagnetite. The
presence of titanomagnetite was confirmed by: X-ray diffraction, thermomagnetic
analysis and optical microscopy in reflected light, on the magnetic extract. Exsolved
ilmenite lamellae within titanomagnetite grains, suggested a class 2 oxidation.

1.5.4. Mechanism of SRM acquisition

In order to understand how adobe bricks become magnetized by the throwing
process, we must turn to the science of clay rheology. The following information was
obtained from a chapter on Clay Products by Macey, which appears in a general work
on building materials edited by Reiner (1954) – actual quotations are presented within
inverted commas.

As clays dry, they shrink and harden in such a way that the change in volume is
equal to the volume of water lost. This process continues until the shrinkage stops
with comparative suddenness at a definite moisture content. This moisture content
varies from clay to clay, and at this point the condition of the clay is called 'leather
hard'. It is more difficult to define the consistency of a clay, but for our purposes, the
consistency required for hand moulding or its mechanical equivalent (as is used in
brick-making) is determined by two factors: first the clay must be soft enough to be

mouldable, and second, it must be stiff enough to hold its shape when it is removed from the mould.

One of the theories put forward to explain shrinkage behaviour, is based on a comparison of the lattice structures of ice molecules and clay molecules. The clay particles 'form an arrangement in which each is in a position of minimum potential energy, as far away from its neighbours as the limited amount of water will permit, and so appear to be separated by water films. The forces of repulsion are ultimately balanced by surface tension at the surface of the mixture'.

'Repulsion' between opposing clay particle surfaces is originally brought about by the fact that the ice lattice fits exactly on that of the oxygen in the clay lattice. A monomolecular layer of ice is thus formed (at T>0°C) which would be continued outwards indefinitely, were it not gradually destroyed by thermal agitation and the interference of adsorbed ions. Instead of a simple elastic repulsion, we have a mechanism wherein work is required to bring two opposing surfaces closer together. This will appear externally as a repulsion, but does not imply immediate reversibility, as indeed is the case. Water may be removed from plastic clay by compression between porous pistons, but it will not return immediately or completely upon release of the pressure.

This structure of the clay leads to a mathematical formulation which is derived by assuming that if a pressure, p, is applied to a sample, this pressure is balanced by the force of repulsion. The equation $p = a\ e^{-bw}$ has been found to holdtrue experimentally, where a and b are constants and w is the water content of a particular clay. Next, let us to consider the effect of applying a small constant shearing stress to a sample of a clay. Since there is a large range of particle sizes in the clay, the resulting deformation due to the stress presents a fairly complex picture. However, for small stresses, the system as a whole will be distorted in a pseudo-elastic manner in such a way that it returns to its original state upon removal of the stress. With mouldable clay, however some of the particles will undergo re-arrangement, thus producing a permanent deformation. Since the force of repulsion of the clay particles varies exponentially with their distance apart, one does not expect either of these types of deformation (elastic and plasteic) to be linear with stress.

The total amount of deformation possible under conditions of small stress, such as the weight of the clay itself, is less than 0.1% in a moderately soft clay, and once the particle re-arrangement described above has taken place, the system then becomes very stable. No flow or creep takes place, and the particles are resistant to thermal agitation, consequently such a moulded article will retain its shape even if prevented from drying.

Having discussed small stresses and the corresponding deformation that they cause, the obvious point to consider is the effect of larger stresses. In particular, we will consider the amount of deformation which can take place before failure occurs. Clay is a unique substance in the way that it behaves under both small and large stresses. When the rate of loading is small, as described above, the clay can slowly adjust itself to the applied stress, in such a way that no particle energy barriers are broken, and so there are no new arrangements of particles. Hence, failure under these conditions can occur with very little deformation. When the rate of loading is large, however, there is a hysteresis effect which prevents the surfaces separating once they have been brought together. Since there is no chance of slow adjustment taking place, the deformation resulting from such a large loading rate is much greater than that for a small loading rate. This means that the clay can be plastically deformed by impact, as some metals may be, but in the case of clays the size of deformation is much larger.

'A bar of soft plastic clay in such condition that it can be readily moulded will break almost undeformed with a brittle fracture if the rate of loading is sufficiently small. This is why clay particles crack during drying although still plastic. Once the effect of rate of loading is realized, its influence can be appreciated in the traditional methods of hand moulding. Clay is always thrown violently into a mould and is shaped or punched, not pushed or kneaded. The required deformation is then obtained with the minimum expenditure of energy and with the least risk of cracking'.

1.5.5. Summary

Adobe bricks become magnetized parallel to the ambient field, at the moment when the moist mud is thrown into a mould before they dry out. They may subsequently acquire a VRM both during and after drying out, but it has been shown that this viscous component may be removed easily by AF demagnetization. The intensity of the SRM acquired does not seem to depend on the force with which the mud is thrown into the mould, only upon the strength of the ambient field. Therefore, the throwing process used in brickmaking must achieve sufficiently large shear deformation in the clay, to produce a saturation SRM in the ambient field. This is probably because the throwing force has to be strong enough for the mud to fill the corners of the mould. The drying process would not be expected to cause remagnetization of the SRM, since it involves only slow rates of change of stress, which are less likely to cause deformation than the faster rates of stress change caused by throwing.

The discovery of SRM in adobe bricks suggests that this process may occur when unconsolidated sediments are sampled in the field by hammering tubes into an exposed section. For example, Symons et al. (1980) report a large r.m.s. scatter of ~13° for samples collected by this technique whereas samples drilled in the laboratory in the usual manner yielded directions with a much smaller scatter of ~3°.

REFERENCES

Barton, C.E., McElhinny, M.W. and Edwards, D.J. Laboratory studies of depositional DRM. *Geophys. J. R. astr. Soc.*, 1980, *61*, 355–377.

Barton, C.E. and McElhinny, M.W. Detrital remanent magnetization in five slowly redeposited long cores of sediment, *Geophys. Res. Lett.*, 1979, *6*, 229–232.

Blow, R.A. and Hamilton, N. Effect of compaction on the acquisition of a detrital remanent magnetization in fine-grained sediments. *Geophys. J. R. astr. Soc.*, 1978, *52*, 13–23.

Chernous, M.A. and Shcherbakov, V.P. Fluid-dynamics implications for the acquisition of sedimentary magnetization. *IZVAN*, 1980, *1*, 120–124.

Collinson, D.W. Depositional remanent magnetization in sediments. *J. geophys. Res*, 1965, *70*, 4663–4668.

Day, R. TRM and its variation with grain size. *J. Geomag. Geoelectr.*, 1977, *29*, 233–265.

Dunlop, D.J. and Waddington, E.D. The field dependence of thermoremanent magnetization of igneous rocks. *Earth Planet. Sci. Lett.*, 1975, *25*, 11–25.

Dunlop, D.J. and West, G.F. An experimental evaluation of single domain theories. *Rev. Geophys. Space Phys.*, 1969, *7*, 709–757.

Everitt, C.W.F. Thermoremanent magnetization, III: Theory of multi-domain grains. *Phil. Mag.*, 1962, *7*, 599–616.

Fuks, N.A. *Mechanics of Aerosols*. Acad. Nauk SSSR, 1955. 351 pp. in Russian.

Games, K.P. The magnitude of the palaeomagnetic field: a new, non-thermal, non-detrital method using sun-dried bricks. *Geophys. J. R. astr. Soc.*, 1977, *48*, 315–329.

Games, K.P. *A new method for determining the magnitude of the archaeomagnetic field*. Doctoral dissertation, University of Liverpool, 1978.

Hamanno, Y. An experiment on the post-depositional remanent magnetization in artificial and natural sediments. *Earth Planet. Sci. Lett.*, 1980, *51*, 221–232.

Happel, J. and Brenner, H. *Low Reynolds Number Hydrodynamics*. Prentice-Hall, 1965. 630 pp.

Henshaw, P.C. and Merril, R.T. Characteristics of drying remanent magnetization in sediments. *Earth Planet. Sci. Lett.*, 1979, *43*, 315–320.

Irving, E. Origin of the palaeomagnetism of the Torridonian sandstones of north-west Scotland. *Phil. Trans. R. Soc.*, 1957, *250*, 100–110.

Irving, E. and Major, A. Post-depositional detrital remanent magnetization in a synthetic sediment. *Sedimentology*, 1964, *3*, 135–143.

Johnson, G.L., Murphy, T. and Torreson, O.W. Pre-history of the Earth's magnetic field. *Terr. Magn. Atmos. Elec.*, 1948, *53*, 349–372.

Kent, D.V. Post-depositional remanent magnetization in a deep-sea sediment. *Nature*, 1973, *246*, 32–34.

Khramov, A,N. Orientation magnetization of finely dispersed sediments. *Izv. Acad. Sci. USSR Phys. Earth*, 1968, *1*, 63–66.

King, R.F. Remanent magnetism of artificially-deposited sediments. *Mon. Not. R. astr. Soc. Geophys. Suppl.*, 1955, *7*, 115–134.

King R.F. and Rees A.I. Detrital magnetism in sediments: an examination of some theoretical models. *J. geophys. Res*, 1966, *71*, 561–571.

Landau, L.D. and Lifshitz, E.M. Course in Theoretical Physics. In . Pergamon Press, Oxford, 1959.

Levi, S and Banerjee, S.K. On the possibility of obtaining relative paleointensities from lake sediments. *Earth Planet. Sci. Lett.*, 1976, *29*, 219–226.

Lovlie, R. Post-depositional remanent magnetization in re-deposited deep-sea sediments. *Earth Planet. Sci. Lett.*, 1974, *21*, 315–320.

Nagata T. *Rock Magnetism*. Mazuren Company Ltd., Tokyo, 1961. 346 pp.

Neel, L. Theorie du trainage magnetique des ferromagnetiques au grains fins avec applications aux terres cuites. *Annales de Geophysique*, 1949, *5*, 99-136.

Neel, L. Some theoretical aspects of rock magnetism. *Advances in Physics, a Quarterly Supplement of the Philosophical Magazine*, 1955, *4*, 191-243.

Noel, M. Surface tension phenomena in the magnetization of sediments. *Geophys J.R. astr Soc.*, 1980, *62*, 15-25.

Otofuji, Y. and Sasajima, S. A magnetization process of sediments: laboratory experiments on a post-depositional remanent magnetization. *Geophys. J.R. astr. Soc.*, 1981, *66*, 241-260.

Reiner, M. *Building Materials, their Elasticity and Inelasticity*. North Holland Publishing Co., 1954.

Schmidt, V.A. A multidomain model of thermoremanence. *Earth Planet. Sci. Lett.*, 1973, *20*, 440-446.

Shcherbakov, VP. and Shcherbakova M. On the theory of depositional remanent magnetization. *Geophysical Surveys*, 1983, Vol. *5*. in press.

Stacey, F.D. Thermoremanent magnetization (TRM) of multidomain grains in igneous rocks. *Phil. Mag.*, 1958, *3*, 1391-1401.

Stacey, F.D. The Physical Theory of Rock Magnetism. *Adv. Phys*, 1963, *12*, 45-133.

Stacey, F.D. On the role of Brownian motion on the control of detrital remanent magnetization of sediments. *Pure Appl. Geophys.*, 1972, *98*, 139-145.

Stacey, F.D. and Banerjee, S.K. *The Physical Principles of Rock Magnetism*. Elsevier Scientific Publishing Co., 1974.

Symons, D.T.A., Stupavsky, M. and Gravenor, C.P. Remanence resetting by shock-induced thixotropy in the Seminary Till, Scarborough, Ontario, Canada. *Geological Society of America Bulletin*, 1980, *91*, 593-598.

Thellier, E. Sur la thermoremanence et theorie du metamagnetism. *Comptes Rendus de l'Academie des Sciences*, 1946, *223*, 319-321.

Thellier, E. and Thellier, O. Sur l'intensite du champ magnetique terrestre dans le passe historique et geologique. *Annales de Geophysique*, 1959, *15*, 295-376.

Thomas, R.C. *Archaeomagnetism of Greek Pottery and Cretan Kilns*. Doctoral dissertation, University of Edinburgh, 1981.

Tucker, P. A grain mobility model of post-depositional realignment. *Geophys. J. R. astr. Soc.*, 1980, *63*, 149-163.

Tucker, P. Stirred remanent magnetization: A laboratory analogue of post-depositional realignment. *J. Geophys.*, 1980, *48*, 153-157.

Tucker, P. Palaeointensities from sediments: normalization by laboratory redeposition. *Earth Planet. Sci. Lett.*, 1981, *56*, 298-404.

Tucker, P. Magnetic remanence acquisition in IPOD Leg 73 sediments. *Init. Rep. Deep-Sea Drilling Project*, 1983, Vol. *73*.

Tucker, P. and O'Reilly, W. The acquisition of thermoremanent magnetization by multidomain single crystal titanomagnetite. *Geophys. J. R. astr. Soc.*, 1980, *60*, 21-36.

2.
Dating
and Related Problems

2.1. INTRODUCTORY COMMENTS

By the Editors

The establishment of well based time-scales is crucial for the evaluation of archaeomagnetic and palaeomagnetic secular variation results. Errors or deficiencies in the adopted time-scales can result in the determination of false rates of change of the geomagnetic field elements and this is particularly important where maximum rates of change are of interest. Accurate time-scales are essential for reliable determinations of the form of the geomagnetic spectrum, and here we are interested particularly in long periodicities ($\sim 10^2$ to $\sim 10^4$ years). And if we are to construct the form of the spatial variations of the field at particular epochs of time by combining data from different geographical regions, the prior construction of a reliable time-scale for each region is essential.

Methods of dating may be sub-divided into two classes, relative and absolute. This chapter starts with section (2.2) in which Dr Bradshaw describes the principles involved in the most important method of relative dating, that of palynological correlation, on a regional basis, with radiocarbon dated type-sections.

The most important method of absolute dating is the radiocarbon ('Libby' half life of $^{14}C=5568yr$) method which is discussed, particularly with applications to dating sediments in mind, by Dr Hedges in section 2.3. Also, the exciting possibility of being able to extend the range of time over which the radiocarbon method may be applied (possibly to about 60000bp), using accelerator methods of detection, is discussed.

Even when highly satisfactory radiocarbon ages have been determined, the problem that the radiocarbon and calendar year time-scales differ from one another has to be faced. The relationship between the two time-scales has been the subject of much research and constitutes an intriguing problem which is discussed by Dr Schove in section 2.4. In order to distinguish clearly between ages expressed in the two time-scales in this volume, calendar years before present are identified by capital letters BP while radiocarbon years before present (and other isotope based ages) are identified by lower case letters bp.

The ^{210}Pb method (half life 22–26yr) which has been used for dating very young sediments is described by Professor Oldfield in section 2.5. ^{32}Si (half life ~300yr) based methods are potentially of use in providing a link between the ^{210}Pb and ^{14}C methods.

Two methods are available for dating sediments of up to ~10^5 years of age, which is beyond the reach of the radiocarbon method. Professor Oldfield discusses chronologies based on the decay of ^{230}Th in section 2.5 and the thermoluminescence (TL) method is described by Dr Wintle in section 2.6.   The TL method has been applied to dating archaeological kilns (e.g. Liritzis and Thomas 1980) but its application to dating sediments is still at the research stage. It has also been applied to dating sediments of Brunhes age and thus may help solve outstanding problems of importance in both practical and theoretical areas. For example the development of a more quantitative Brunhes magnetostratigraphy would help with the dating of tectonic events, a factor of some considerable practical importance in the siting of industrial plant and it would allow us to establish whether geomagnetic 'excursions' are global or regional phenomena which is a factor of some theoretical interest.

2.2. PALYNOLOGY AS A RELATIVE DATING TOOL

By Richard Bradshaw,
Trinity College, Dublin, Ireland

2.2.1. Introduction

Changes in the composition of floras as reflected in the palynological record have been used by palaeomagnetists and others as a method for the relative dating of sediments. Absolute dates cannot be derived from pollen data, but when the timing of certain vegetational changes has been established in a region, by radiocarbon dating for example, pollen data can be a convenient way to correlate sedimentary sequences with the established time-scale. This section investigates the problems associated with the use of palynology as a dating tool, identifies the type of vegetational change that is synchronous over large areas, and lays down guidelines for the palaeomagnetist who wishes to use pollen data for dating purposes.

2.2.2. Pollen data and the sub-division of pollen diagrams

Pollen grains and fern spores are often found in great abundance in Quaternary lacustrine sediments. These palynomorphs can be extracted from sequential sub-samples of sediment cores and identified under a microscope. The data are commonly displayed on pollen diagrams as the percentage values for individual taxa of total pollen counted, plotted against depth in the sediment core or against radiocarbon years (Figure 2-1).

Pollen diagrams may show pollen accumulation rates (number of grains accumulated per cm^2 of sediment per yr) which avoid the interdependence of values of the individual pollen types in percentage diagrams. Pollen accumulation rates are greatly influenced by sedimentary processes and are not as useful for comparing data from different sites as are percentage diagrams. Faegri and Iversen (1975), Moore and Webb (1978) and Birks and Birks (1980) have written texts which cover the principles of pollen analysis.

Pollen diagrams covering the same time period from comparable sites within one geographical area are often strikingly similar in form. Lake Mary and Wood Lake lie 100 km apart in Wisconsin, USA and their pollen diagrams share many common features (Heide, 1981) (Figure 2-1). Both sediment cores cover at least the last 10000 yr as shown on the time-scale, which was derived from radiocarbon dates. The diagrams have been divided into local pollen assemblage zones (LM 1-4 and WDL 1-4) by using a stratigraphically-constrained clustering routine (Gordon and

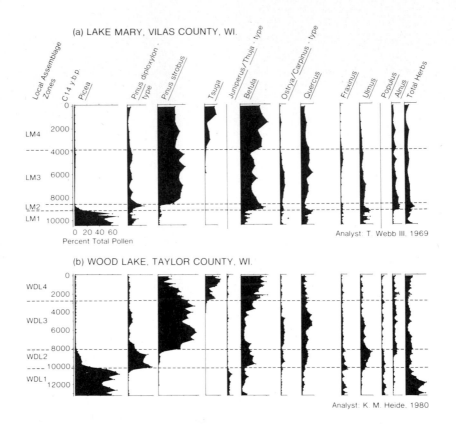

(a) LAKE MARY, VILAS COUNTY, WI.

Analyst: T. Webb III, 1969

(b) WOOD LAKE, TAYLOR COUNTY, WI.

Analyst: K. M. Heide, 1980

Figure 2-1:   Percentage pollen diagrams from two lakes 100km apart in
Wisconsin, USA. All terrestrial pollen were included in the
calculation sum. The vertical scales are ^{14}C years bp.
(after Heide, 1981)

Birks, 1972). Both LM-3 and WDL-3, for example, are characterized by high
percentages of Pinus strobus pollen with low percentages of Picea and Pinus
banksiana/resinosa pollen. Dividing the pollen diagrams into zones has established a
biostratigraphy for the two sites, and in this example the four local pollen assemblage
zones correlate well with each other: the description of WDL-1 would also describe
LM-1. Examination of the radiocarbon dates shows that the zone boundaries occurred
at different times in each site. Transition 3-4 dates from 3800bp at Lake Mary but at
2800bp at Wood Lake. If the Lake Mary core were to be dated purely by correlation
with Wood Lake, a dating error of 1000 radiocarbon years would be introduced,
assuming that the radiocarbon dates are reliable.

The asynchroneity of most pollen zone boundaries has been demonstrated several
times for data from the Holocene (Smith and Pilcher, 1973). Many zone boundaries

are defined by the first arrival of large numbers of pollen of a tree species at a site, and when these first arrivals are mapped over wide areas, migration maps for the taxon can be produced. Isochrones in radiocarbon years bp, mapping the advancing front of significant numbers of <u>Quercus</u> pollen northwards through Britain, show a consistent pattern (Figure 2-2). The first significant arrival of <u>Quercus</u> pollen at a site in Britain cannot be used as a time-marker of accuracy greater than ±3000 yr unless published isochrone and isopoll maps are consulted (Davis, 1976; Huntley and Birks, 1982; Birks, 1982). There are certain palynological events in the Holocene which appear to be largely synchronous over wide areas, but in general, a major change in a pollen diagram will only yield a very approximate date by correlating the episode with the same dated episode in another pollen diagram from the area.

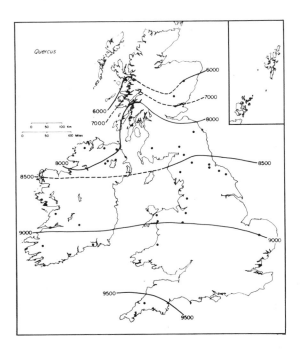

Figure 2-2:   Isochrone map in ^{14}C years bp for the British Isles of the first sustained values of <u>Quercus</u> pollen at the sites marked with dots, (after Birks, unpublished).

Radiocarbon dating provides a good chronology for the Holocene, but the dates are often difficult to interpret for ages greater than about 10000bp. For earlier inter-stadials and inter-glacials, the assumption that a pollen-derived biostratigraphic correlation can be directly transformed into a chronostratigraphic correlation is more frequently made. In the absence of an absolute dating system, only relative dates can be applied; but the demonstration that two sedimentary sequences probably cover the same time period can be of value to the palaeomagnetist. Walker and Lowe (1981) have used pollen zones for time correlation in preference to radiocarbon dates for early Holocene deposits. Turner and West (1968) and West (1981) discuss the establishment of characteristic biostratigraphies for earlier inter-glacials. Watson and Wright (1980) and Birks (1982) discuss in detail the relationship between biostratigraphy and chronostratigraphy.

2.2.3. Synchronous vegetational events in the Holocene

While many pollen zone boundaries are demonstrably asynchronous during the last 10000yr, there are certain events which appear to be synchronous over large areas. Changes in pollen diagrams are attributable to climatic change, migration, human activity and other agencies and it would be surprising if some of these changes did not affect a large geographical area during a short period of time. The marked decline in Ulmus pollen percentages in NW Europe has been so frequently dated to around 5000bp that this event can now be used as a satisfactory time marker. There have been many hypotheses proposed about the cause of the Ulmus decline, but perhaps the most satisfactory is that of a fatal disease which spread throughout the area at a rate which radiocarbon dates cannot resolve (Rackham, 1980). A similar decline has been observed in Tsuga pollen in NE United States, which dates to around 4700bp and again has been ascribed to disease (Davis, 1981). Webb (1982) has collated dates for the Tsuga decline and prepared a histogram showing the distribution of radiocarbon dates and interpolated radiocarbon dates for the depth interval of the steepest decline in Tsuga percentages. The samples were weighted according to the available dating control, and the ease of detection of the steepest decline. The resulting histogram shows a distribution of dates suggesting that this was a synchronous event throughout the area studied in NE United States (Figure 2-3).

Chronologically well-defined palynological events are rare, but if the local pollen chronology is established, as it is wherever palynologists have been active, pollen analysis can quickly establish an approximate date for a sample. Throughout much of lowland Britain there is a massive rise in numbers of Corylus pollen around 9000bp and sediment samples can easily be sorted into pre- and post 9000bp groups. Similarly, the arrival times of major pollen producers in the Holocene can be estimated for any European site from dated isopoll maps (Huntley and Birks, 1983) and used to place a sediment sample in a defined time interval. Certain declines in pollen numbers are probably under climatic control, and again can be used as dating horizons in conjunction with published maps. Caution is always necessary as the Tilia decline was probably caused by human activity and has a wide range of dates associated with it (Turner, 1962).

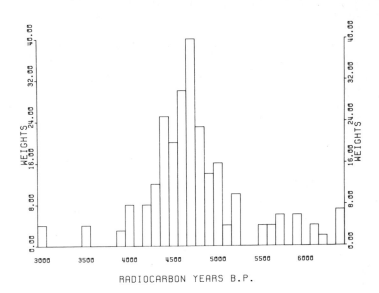

RADIOCARBON YEARS B.P.

Figure 2-3:   Weighted histogram of the dates for the <u>Tsuga</u> decline
in which the weights for all dates in each 100-year interval were
added together. Weights were   assigned as follows: 5 = ^{14}C date
in interval of <u>Tsuga</u> decline and <u>Tsuga</u> decline more than 10%.
4 = bracketing ^{14}C dates about the <u>Tsuga</u> decline. Points
were subtracted if <u>Tsuga</u> pollen decreased by less than 10%,
if more than one <u>Tsuga</u> decline occurred in a core, or if dating
errors were identified in the core. (after Webb, 1982)

This discussion has focussed on pollen studies from the Holocene in NW Europe
and NE United States reflecting when and where most palynological research has taken
place. Palynology has been studied in several tropical areas and is reviewed by
Flenley (1979). The low pollen productivity of the chiefly insect-pollinated tropical
trees, and the difficulty with identification of pollen types below the family level pose
special problems. Palynological studies are not restricted to forested areas, and
pollen diagrams are published from deserts, tundras and grasslands.

2.2.4. Dating strategies

Palynology can be used as a dating tool in two distinct ways: -

a. Biostratigraphic correlation of the sediments under study with well-
   dated sediments at a nearby site. On the basis of this correlation the
   dates can be tentatively transferred from one site to the other.
   Alternatively, dates may be estimated from published isochrone or
   isopoll maps.

b. The biostratigraphic correlation of two or more undated sediments from an area, and the subsequent inference of a chronostratigraphic correlation of the sediments. The 'true' chronostratigraphies, were they to be established by an independent means, may not match the biostratigraphies, but in the absence of any other dating method this system of relative dating with its inherent inaccuracies may be acceptable.

For either method to be successful, the following conditions must hold: –

i. The sites must collect their pollen from areas of similar vegetation. The majority of pollen sampled at the sites to be correlated must derive from the same plant taxa. A sediment core from a grassland community cannot be accurately correlated with one from a forest community, so the past and present patterns in vegetation dictate the distance over which two cores can be correlated. This distance is variable. Similar pollen assemblage zones can be described from forested areas in both Wisconsin and Michigan USA but different zones are obtained from either side of the narrow prairie-forest ecotone in Minnesota.

ii. Major vegetational changes must occur simultaneously around each site. As discussed earlier, vegetational changes vary from the broadly synchronous to the distinctly asynchronous. Specialist knowledge is required to select the synchronous changes in any given study area.

iii. The sites must have similar properties as pollen traps. The influence of the physical properties of a sampling basin on its contained pollen record has received considerable study which is reviewed by Jacobson and Bradshaw (1981). The size of the basin and the presence or absence of inflowing streams are two of the important variables. Some plant taxa produce much more pollen than others, and there is great variation in the dispersal properties of pollen types. Large basins tend to collect more of the well-dispersed pollen than do small basins.

2.2.5. Conclusions

Lake sediments from all over the world have been analysed by palynologists and in many cases the inferred vegetational changes have been dated by independent methods. Biostratigraphic correlation using pollen assemblage zones can permit these dates to be transferred to other lake sediments which are of interest to the palaeomagnetist. Palynology can also provide a useful system of relative dating for sediments laid down beyond the range of the radiocarbon dating method.

2.2.6. Acknowledgements

I thank Dr. K. Heide for permission to use unpublished material. I had useful discussions with Drs. H. J. B. Birks and P. Gibbard. Dr. H. J. B. Birks and A. J. Alexander critically read the manuscript.

2.3. RADIOCARBON DATING OF SEDIMENTS

By R.E.M.Hedges,
Oxford University, UK

This section is not intended to be a critical review of the dates obtained on sediments by the radiocarbon method, but rather an account of how the method may be applied to sediment dating in general. Some features of radiocarbon dating become particularly important when applied to sediments, and careful consideration of these should help in the assessment of the dates obtained.

2.3.1. The principles of radiocarbon dating

While these are well known, it is the extent to which they can be known to apply precisely to the particular context which determines the reliability of the result.

(i) All the ^{14}C measured in a sample was originally formed in the stratosphere from the reaction of cosmic ray produced neutrons with the abundant ^{14}N nuclei. (Some rather exceptional cases can be cited; for example the production of 'bomb' ^{14}C from nuclear weapons testing; also the in situ production of ^{14}C in such nitrogen rich materials as the protein in bone by direct reaction with the very low neutron flux at ground level or from neutrons from absorbed uranium. The contribution of ^{14}C by this process is only significant for material more than 8 half-lives old, however.

While measurements on other radio nuclides and on meteorites suggest that the cosmic ray flux has not drastically changed in the last 5 Myr (Lal and Peters, 1972; Schaeffer et.al., 1963), there is good experimental evidence to support the contention that changes in the Earth's magnetic field can modulate the rate of production of ^{14}C, giving changes in production rate by at least 10% (Bucha, 1970). Short term changes (less than 100 years) can be much greater, but their effect is diminished by the relatively slow equilibration of ^{14}C throughout the biosphere.

(ii) The ^{14}C produced reacts with the atmosphere to give carbon dioxide Unlike most other cosmogenic nuclides which do not form stable gaseous products and are therefore precipitated after attachment to aerosols, carbon dioxide is rapidly distributed throughout the atmosphere, mixing with the stable isotopes $^{12}CO_2$ and $^{13}CO_2$ It then gradually equilibrates with the major reservoir of CO_2, that dissolved in the ocean. Equilibration with the surface water is much faster than with the deep ocean (containing about 90% of the total CO_2), and the apparent 'age' of the CO_2 in the deep ocean is found to be approximately 2000 years. There is considerable interest in, and fair success at, developing quantitative models (Siegenthaler et al., 1980) to account for the atmospheric equilibration of CO_2 because of recent significant changes in CO_2 levels brought about by human actions. Since a radiocarbon date depends upon relating the measurements of $^{14}C/^{12}C$ in the specimen to that which prevailed in the reservoir which the specimen 'sampled', the level of $^{12}CO_2$ prevailing is also important. There is some evidence that the total CO_2 content in the atmosphere was much lower during the last glaciation (Berner et al., 1980; Oeschger, 1983). Since atmospheric CO_2 is such a small fraction of the total available CO_2, this presumably represents a shift in equilibration and could apply equally to the reservoir ^{14}C. However, while equilibrium is being established, variations in the reservoir $^{14}C/^{12}C$ value could come from changes in the mixing rates between deep and surface ocean waters, for example from major changes in circulation patterns during glaciations.

(iii) The carbon atoms of any specimen being measured were able to exchange freely with some suitable reservoir until the event being dated. Usually this event was the death of the organism, although carbonates may equilibrate with ground water CO_2. The most suitable reservoir is the atmosphere, and plants remain the most clear cut samples of atmospheric CO_2 composition. There is negligible uptake of CO_2 derived from the carbonate of calcareous soils, for instance (Tauber, 1983). Some complications are brought in by the rather considerable degree of isotopic fractionation as the plant 'fixes' the CO_2, but this can be corrected for by measuring the fractionation change in the $^{13}C/^{12}C$ ratio.

(iv) There is no further chemical exchange of carbon atoms between specimen and environment until measurement. This is clearly only an ideal situation. A major problem in most radiocarbon dating is knowing the extent to which contamination, either as a result of biological, geochemical or laboratory processes, has changed the isotopic composition. This is discussed in some detail later.

(v) Measurements of the remaining $^{14}C/^{12}C$ ratio allows the age to be determined once the half life is known and the initial $^{14}C/^{12}C$ of the sampled reservoir is known. Problems in knowing the original reservoir composition have been mentioned. A 'Radiocarbon Date', expressed as years bp, is simply based on the assumption that the original level was the same as nowadays, and that the half-life is the original Libby value of 5568 years. Corrections are applied for isotopic fractionation from the reservoir to the specimen, on the assumption that that $^{13}C/^{12}C$ content was the same as it is today. More precisely, a Radiocarbon Date is defined as:

$$T = 8033 \ln[1/(1 + D^{14}C \cdot 10^{-3})] \text{ years bp}$$

where $D = d(^{14}C) - 2[\delta^{13}C + 25] [1 + d(^{14}C)/1000]$

and

$$d(^{14}C) = 1000 \times [(14/12)_{sample}]/[(14/12)_{standard}] - 1$$
$$\delta(^{13}C) = 1000 \times [(13/12)_{sample}]/[(13/12)_{standard}] - 1$$

The (14/12) standard is 95% of the activity of NBS oxalic acid and the (13/12 standard is PDB (a marine carbonate). In practice, the Radiocarbon Date requires calibration against some other time-scale where this is possible; the difference between the Radiocarbon and Calendrical time-scales can be significant (~1000years). The features of the calibration curve will now be described more fully.

2.3.2. Specific application to terrestrial sediments

The main questions with sediments are: what reservoir is being sampled, which are the events whereby carbon atom exchange with the reservoir ceases and to what extent does the system remain free from further incorporation of extraneous carbon atoms?

Most sediments dated contain 1-10% of carbon, excluding carbonate. Where the material is wholly derived from biological activity within the water column above the sediment (i.e. no exogenous sources), carbon is introduced through photosynthetic activity fixing the CO_2 at the surface. This will be in equilibrium with atmospheric CO_2, and also may contain dissolved bicarbonate. Some CO_2 may therefore be derived from carbonate rocks, depending on the local geochemistry. Since modern groundwaters in carbonate districts seldom have ^{14}C 'ages' greater than 1200 years or so, this represents a likely extreme error in the reservoir age (Mook, 1980). However, this still leaves room for significant error. There are difficulties preventing the estimaton of the hard water correction from purely internal evidence. The $\delta^{13}C$ values are very similar for terrestrial carbonates (about -5 o/oo with respect to PDB) and for atmospheric CO_2 (about -7 o/oo). An equilibrium level of bicarbonate ions in solution implies a continuous input of Ca^{2+}, which should be observable as deposition in the sediment.

Where there is a substantial terrigenous input, the effect of 'hard water' diminishes. A possible problem of allochthonous material is that it may be reworked, and therefore not be contemporary with the formation of the sediment. Careful study of the sediment microfossils, as well as the local geomorphology, should provide information on this, so that additional caution can be applied to the interpretation of the date.

It is generally assumed that the cessation of carbon atom exchange occurs with the deposition of the sediment. There is nevertheless extensive biological and chemical activity in the sediment for a very considerable time after deposition. Much of this may not involve exchange with the reservoir, for example bioturbation, but relatively little is known about the extent to which materials dissolved in the pore waters in loosely compacted sediment can exchange, either through biological or chemical processes. Most biological activity will be devoted to the extraction of energy from the precipitating detritus, and this will not in itself alter the ^{14}C content. However, some CO_2 can be formed by oxidation, and this will be transported. The bulk of the carbon content is present as high molecular weight, highly cross-linked polyphenolic material, representing the evolution of humic acids towards kerogen with time.   Since this material has a strong absorptive capacity for more mobile organics such as amino acids and lipids, it is possible that the chemical degradative processes that take place can give rise to a net directional displacement of material. The expulsion of pore water during compaction is another possible source of relative movement. It is also possible to envisage the downward transport of selective material through bacterial action, where trophic levels, lower in the sediment, consume material arising from the actions of a different level above.

Again, there is very little evidence as to the degree of error in carbon-dating sediments brought about by these processes, since their effects are not easily observable except by implication. An outline of the various sources of error in dating is presented in section 2.3.5.

2.3.3. Archaeological dating

Most archaeological dating is based on measurements of $^{14}C/^{12}C$ in either charcoal or bone. The general points of radiocarbon dating still apply, but with the additional factor of interpreting the ^{14}C date as an archaeological event. Thus a date of a large piece of charcoal may be considerably earlier than the felling date of the tree, which in any case may be re-used. However, since the sample is much more specifically defined, most archaeological samples present fewer problems, apart from those of stratigraphic context and archaeological interpretation.

2.3.4. Laboratory measurement of $^{14}C/^{12}C$

Living material has a $^{14}C/^{12}/C$ ratio of about 10^{-12}. Since the ^{14}C content decays by about 1% in 80 years, the measurement must be made to at least this accuracy. However, with material several half-lives old, a less accurate measurement is permissible since an error of 80 years in, say, 20000 years, is unnecessarily small, and probably much less than other errors.

The measurement of such a small concentration of ^{14}C to high accuracy presents problems quite specific to radiocarbon laboratories. Until very recently the only approach has been to detect the radioactivity of ^{14}C itself. Since this is at a relatively low energy, it is not easy to discriminate against the background from cosmic rays and other radioactive components in the apparatus. Thus the subtraction of the background becomes more and more serious with older dates, and a limit is reached at 30000-40000 years. A second limit is the very low count rate for such samples. But for samples within the usual range of 1-2 half lives, provided there is adequate material, and it is well free from contamination, modern methods can achieve accuracies of better than ± 0.5%. Two different methods are employed: Gas counting, in which the sample is converted to a gas, usually CO_2. This must be carefully purified, and is then counted in a well shielded enclosure along with other counters containing standards and other samples. Alternatively, the sample is converted to benzene, diluted to a standard volume, and the Liquid scintillation observed photometrically. In this way each sample or standard is placed repetitively in front of the counter over a long period of time. Since a modern sample has about 15 disintegrations per minute per gram of carbon, a 5 gram sample will take about 2 hours to accumulate 10000 counts. In practice, the counting is considerably longer, partly to establish an adequate background subtraction. However, during the counting time, only a very small proportion of the ^{14}C atoms present are actually observed. Recently, much smaller counters have been developed. These are stable for long periods (Otlet et. al., 1983), and should enable counting times of several months to be employed in the measurement of samples of a few tens of milligrams.

A more radical new development is the measurement of $^{14}C/^{12}C$ by mass spectrometry rather than by radioactivity (Gove, 1978; Kutschera, 1981). In this way about 1% of the ^{14}C atoms can be sampled, and the quantity of carbon required for dating is only 1 to 5mg. This method has yet to approach the measurement accuracy of the 'conventional' method, but has many advantages. Thus the background is virtually eliminated, or rather seems to be set by machine and laboratory contamination, and is in any case in the region of 50000 – 60000 years at present with the prospect of being further reduced. Thus much older samples can be dated provided they are free from contamination by modern material. Also the method holds out the capability to be much more selective in exactly what material is being dated (Hedges, 1981). Since this is a major difficulty especially with sediments, the possibility of dating specific extracts, be they individual fossils or certain chemical compounds, holds out promise for a very worthwhile improvement. However, the technique is still in its infancy: it also requires considerable resources, both financial and technical. Nevertheless, the Laboratory at Oxford is specifically interested in exploiting the technique for the dating of sediments.

Having measured the $^{14}C/^{12}C$ content, it is also necessary to know the extent of fractionation between the reservoir and the material being measured. In general this fractionation comes about through chemical processes obeying thermodynamic laws which imply that the fractionation will, to first order, be proportional to the difference in mass. Thus a measurement of the change in $^{13}C/^{12}C$ ratio, which can be made very accurately on a relatively small sample, can be used to estimate the change in $^{14}C/^{12}C$ ratio (White, 1981). Some fractionation can occur in the laboratory, but this can be minimized by employing chemical reactions for which there is a complete yield of product.

2. 3. 5. Sources of error in radiocarbon dates

Errors can arise at a number of different levels in the measurement of a radiocarbon date, and consideration of their origin should help to make the process more reliable. There is a tendency to take the estimated laboratory error as representing the overall reliability of the date, particularly when quoting a date out of context.

(a) Laboratory measurement errors

The most easily quantified error is that from the statistics of sampling the population of ^{14}C atoms. Since sample size and counting time are operational limitations, counting statistics are often the major source of error. As greater precision is sought, however, other sources of inaccuracy become more important. At present there is a division between 3 or 4 laboratories in the world specializing in high accuracy and precision (quoting measurement errors of better than ±0.5%), and most laboratories, often doing routine dating, with errors 2 to 3 times as large. In all cases, it would seem from intercomparisons of results from similar samples sent to the major laboratories, that the actual error can be somewhat larger than the quoted error, even when the quoted error tries to take into account sources of error beyond those of counting statistics (International Study Group, 1982). Where the quoted laboratory error is crucial, the reasons for its value should be scrutinized. The reliability of the error estimate seems to get worse, the more sample preparation is necessary.

(b) Fractionation effects

There is no evidence to indicate that the correction from the fractionation measurement of ^{13}C is not entirely adequate at present levels of accuracy. (The total fractionation of ^{14}C can amount to 5%, so that first order corrections might not be sufficient. However, most samples of similar type have very similar fractionations.) However, if the specimen to be dated has components from several sources (e.g. the CO_2 from carbonate as well as atmospheric CO_2 as mentioned above), it would be wrong to interpret the δ^{13}C measurement as a fractionation effect and such cases would require individual consideration.

(c) Reservoir effects

Differences in carbon equilibration between the major reservoirs are accounted for in calibration (q.v.), but more local changes would have to be specially accounted for. For example, in exchange of the water between marine and lacustrine environments could change the apparent ^{14}C age of the water by about 400 years.

(d) Calibration curve

The difference between the Radiocarbon Age and the Calendrical Age for tree rings has now been measured for nearly the last 9000 years. This difference applies accurately throughout the Northern Hemisphere and takes account of the global variations in ^{14}C production rate and reservoir equilibration. (The Southern Hemisphere is believed to be about 4-5 o/oo lower in ^{14}C content, because of its larger ocean area). The effect of the curve is to provide systematic corrections both as long term trends (millennia), thought to be due to the changing value of the Earth's dipole field, and shorter term changes (a few centuries), thought to be due to solar activity). Shorter variations may also occur, but these are comparable with the noise level of the measurements. The systematic corrections, plus the short term corrections, can occasionally make for ambiguous dates, that is, the ^{14}C level was higher at some earlier epoch by more than the amount by which it has decayed to a later epoch. More often, the error in the calibrated date may be increased (although sometimes the reverse is true). Furthermore, even if the error in the radiocarbon date has a Gaussian distribution, the error in the calendrical date will not necessarily be Gaussian (for an ambiguous date, it is obviously bimodal).

The present situation, with regard to calibration, is that the fullest use of the existing data, up to but not including the high precision measurements, is published in Klein et al., (1982) where curves and tables for calibrating the mean and the error for a given radiocarbon date mean and error are presented. The curve itself, which has a fair degree of statistical noise associated with it, reproduces fairly satisfactorily the same pattern as the more recent high precision results. The high precision data are by no means complete, but extensive lengths of the calibration curve have been measured (i.e. 0 -2000bp, Stuiver (1982); 3700bp - 5700bp, Pearson (1980), and will eventually replace the less accurate curve. For periods beyond 6000bp - 7000bp, there is so far only the one measurement series made by Stuiver (1970) on laminated sediments from the Minnesota Lake of the Clouds. Further back than this, there is nothing at all.

(e) Sample origin

Mistaking the origin of the sample or some of its components can lead to substantial errors that are virtually impossible to quantify. In sediments, it is easier to imagine how older material can be incorporated by reworking of terrigenous detritus, by metabolism of carbonate-derived carbon, or through the presence of geologically old carbon such as graphite from very ancient organic sediments. A good example where the effect of carbon metabolized from carbonate is observed, is given by Shotton (1972). With the prospect of dating milligram sized samples, a great deal of rejection of undesirable material, or hand-picking of delicate and non-transportable organic structures can be made. On the chemical side it would be interesting to see to what extent specific lipids, or amino acids, differ in date from the humic acids, and to what extent these correlate with identifiable plant fossils.

(f) Sample contamination

Some potential contaminating processes which may occur while the deposited sediment is consolidating have already been mentioned. These can conceivably make the sample appear too young. If the rate of deposition varies abruptly, bioturbation can lead to apparent errors rather than just averaging the age. On the whole, sediments are better protected against contamination by modern carbon than most samples. At 10000 years, a less than 1% contribution of modern (post bomb) carbon will reduce the date by 400 years. Whether the pore water is a significant source of modern contamination is not known, and this would be worth careful investigation. The effect would be most serious for sediments poor in carbon content deposited in environments which are now rich in dissolved organic material.

2.3.6. Summary

The principles of radiocarbon dating apply directly to the dating of sediments. However, the reliability of the date depends crucially on an understanding of the origin and composition of the sample. A conventional radiocarbon measurement requires several grams of carbon as sample, and the associated laboratory error in a radiocarbon date is ±100 years or so. Samples up to, and sometimes beyond, 30000bp can be measured, generally with an increasing error in the date. New developments with accelerator mass-spectrometry promise to make dates on milligram-sized samples, with an accuracy approaching that of the conventional method, and certainly better in the range 20000bp – 50000bp. The main improvement in accuracy, though, should come about from the greater specificity of the material that has been selected for dating. The effect of calibration on the dating error is shown to be fairly complex, and a brief account of the present state of calibration curve construction is given. I thank Richard Gillespie for reading this manuscript and for useful discussions.

2.4. TREE-RINGS AND VARVES

By D. J. Schove,
St David's College, Beckenham, Kent, UK

2.4.1. Magnetism of tree-rings and varves

Remanent magnetism is usually impossible to measure in tree-rings but easy to measure in varves. Tree-rings, on the other hand, can easily be counted with precision whereas the accuracy of varve counts is almost always uncertain (glacial varves excepted). The cross-dating of tree-ring and varve chronologies is therefore necessary before Holocene varves can be placed precisely in the absolute time scale. This is best done locally in the first instance, but the principles of teleconnection can be applied to the cross-dating of different time-series in different continents. Varves can thus be dated in the absolute Bristlecone time-scale back to 6700BC and in cross-dated tree-ring scales from the Alps and the Danube well before that.

In iron-rich soils in England and in North America, some attempts have been made to measure the remanent magnetization of tree-rings directly. So far, no satisfactory results have been published. On the other hand, varves which form at the bottom of deep lakes or estuaries often provide ideal material for analysis. As indicated in Figure 2-4, there are well-counted series available back to before 12000bp involving varve-counts of some 13000 years. Pioneer work by Johnson et al. (1948) in New England was completed in the 1940s when the glacial varved sequences were still plainly exposed at the edges of clay pits – they have since been covered by vegetation. In the 1950's Holocene varves in Sweden were studied by Griffiths (1955) at Birmingham, and in 1970's glacial varves in Sweden were studied by Noel and Tarling(1975) and by Noel (1975). Further studies being made in both Canada and Sweden by Morner. Interglacial varves have been studied by Tucholka (1980) and others, but for high-precision dating the present need is for further studies of the long sequence of Late Glacial and Holocene varves that are available in Europe. The magnetic chronology of NW England can be expressed precisely in radiocarbon chronology and we now need to have an exact chronology to correspond. As our Figure 2-4 indicates, the gaps are at present longer than the period investigated.

2.4.2. Time-Scales

The radiocarbon (bp) time scale has naturally been used in the past in dating palaeomagnetic features. However, from 8300bp (=7200BC or 9150BP) it is now possible to derive calibrated radiocarbon dates (BP or BC), thanks to new ^{14}C measurements made at La Jolla and Heidelberg. The upper case initials BP have

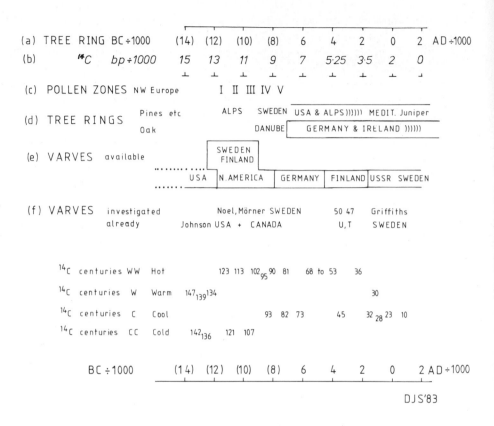

Figure 2-4:   Varves studied palaeomagnetically and their position in the absolute
tree–ring and radiocarbon time–scales:

(a) <u>absolute</u> dates based up to 8000BC on an arbitrary
radiocarbon deviation of 950yr, (b) <u>radiocarbon</u> dates bp (uncalibrated),
(c) <u>pollen-zones</u> in NW Europe, (d) <u>tree-ring</u> <u>chronologies</u> useful
for dating varves, (e) <u>varves</u> available for magnetic investigations,
(f) <u>palaeomagnetic</u> work on varves with names of authors, U and T
correspond to <u>Ulmus</u> and <u>Tsuga</u> declines (see section 2.6.3)
(g) <u>radiocarbon</u> <u>centuries</u> outstanding climatically,
(h) <u>absolute</u> <u>chronology</u> <u>BC.</u>
The Ulmus decline began in radiocarbon century no. 53 (5399–5300bp)
which corresponds to the 43rd century BC.

often been used by authors who have not specified whether their dates are calibrated
or not: the use of lower–case letters for uncalibrated dates, suggested by Schove
(1967) to avoid ambiguity, is now becoming generally accepted and has been adopted
throughout this book.   The differences are discussed in section 2.4.5, and also by
Klein et al. (1982) and Stuiver (1982).

The absolute tree-ring scale has not yet been confirmed for ages older than 7200BC when the radiocarbon deviation was 850yr (see Bruns et al., 1983), but pines from the Alps or Scandinavia, and oaks from the Danube and the Middle East, provide floating chronologies that may ultimately be attached to the Becker and Ferguson tree-ring scales from Central Europe and California respectively. Our main hope, however, rests with varve-scales for the preceding 'Missing Millennium' (cf. Schove, 1979, Figure 2).

Varve-counts are usually unreliable as very narrow varves are easily missed. At 6800BC for instance, bristlecone tree-ring determinations now enable us to confirm that this date, 8750BP, is 8000bp so that the deviation is about 750yr : varve counts, both at Saarnich Inlet, British Columbia, Canada and at the Lake of the Clouds, USA, had previously suggested a deviation of ~ 550yr so that some 200 varves must have been missed.

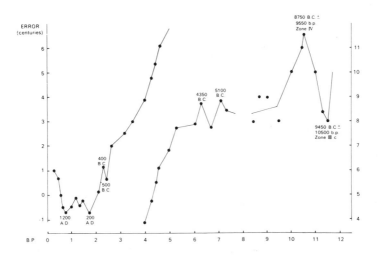

Figure 2-5: A highly simplified ^{14}C error curve: the difference between the conventional radiocarbon date (bp) and the true or tree-ring date (BP). The left hand part of the diagram is based on the bristlecone pine chronology and is applicable back to 4500BP. The right hand part is tentative, being based on the assumption that the error is 950yr at the beginning of the Holocene (~11000BP).

Reliable varve-series are now available for the past 15000yr, and the use of those at Schleinsee (Geyh et al., 1971; also see Merkt, in Schove and Fairbridge, 1983), had made it possible to estimate, in 1978 (Figure 2-5), that the deviation at 8000BP was ~800yr, at 9000BP (8000bp) ~ 900yr and at 10000BP (9000bp) possibly ~

1000yr. At the Lake of the Clouds, Stuiver (1970) had already suggested, from varve-counts that the radiocarbon deviation was 800yr and if we now add 200 for 'varves missed', we obtain the same result for USA and for Germany.

The varves at Schleinsee have not yet been measured and those at the Lake of the Clouds are too thin and complacent to allow us to obtain a satisfactory teleconnected overlap with glacial varves in Canada. Thus we need more tree-ring data and varve-thickness time-series to fill up the 'Missing Millennium.' However, I have assumed in Figure 2-4 (cf. Figure 2-5 and 2-6), that the deviation in the Late Glacial was 950yr throughout, so that 9000bp can be conveniently equated with 8000BC (i.e. 9950BP). The reality of the large-scale wiggles, suggested for the preceding millennia (Schove 1978), cannot yet be confirmed.

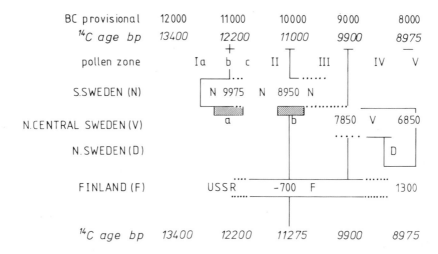

Figure 2-6: Glacial varve chronology in Sweden and Finland in radiocarbon chronology and in a tentative absolute chronology.
(Based on Schove, 1971, Table 5, where a tentative radiocarbon deviation of 950 years at 11250bp was assumed).
(a) Varves palaeomagnetically investigated by Noel and Tarling (1975).
(b) Varve counts confirmed by Stroemberg (1977).

The Late Glacial varve-chronologies (Figures 2-6, 2-7) expressed in a joint Canadian-Finnish time-scale (Schove, 1971- Table 5), extends over 4000yr and the radiocarbon positions have been confirmed by subsequent ^{14}C dates. The conventional Swedish dates before Christ are indicated by the abbreviation f. Kr, and their ^{14}C position is shown in Figure 2-6. The pollen-zones are those generally adopted in NW Europe.

2.4.3. Glacial Varve-Chronologies

These have been listed in Schove (1978, p.211) and are illustrated in Figure 2-7. They include:

Series A (Pre-12800bp). This series from southern New England has not yet been tied to the Scandinavian time-scale. The magnetic results of Johnson et al. (1948) have been summarized in Schove (1978, p.214), where the results of McNish and Johnson (1938), applied to the separate floating chronology of New Haven, are also given. In general, the varves belong to the relatively milder period between 17200 and 13000bp. The cold phase at varves numbered 2650/4800 corresponds to the Rosendale Advance for which Schove (1969) provisionally suggested a date of 14100bp but this date is now believed to be about 14800bp so that an alternative formula would be:

A + bp = 19400 ± 100

If we adopt the latter version (and the palaeomagnetic results), the declination must have been east of north at 15700bp (inclination near 40°), west of north at 15200bp (inclination minimum 40°), east of north at 14300bp and west of north from 13600bp to 13100bp (inclination rising to 55°). However, both dates and measurements will remain uncertain (see Figure 2-6, bottom row), until an overlap can be obtained with Series B.

Series B. This, series from northern New England (Figure 2-7), is teleconnected with Finland, Sweden and Canada, and belongs to the Boelling period (Zone 1b). Varve 6900 thus corresponds to a Mid-Boelling (Ib) date of 12350bp (Schove, 1969 and 1983, Paper 39); no magnetic flip has been found in these Boiling varves which showed decreasing declination about 10° east of north and an inclination of about 55° (Johnson et al., 1948).

Series N. The contemporary series from S. Sweden (Figure 2-6) has been investigated in part by Noel and Tarling (1975) and Noel (1975), and if we add 2250 to their f.Kr. dates to get approximations to bp, we have intensity maxima at intervals of about 24 years at 12401bp, 12386bp, 12363bp, 12324bp, 12301bp and 12279bp, a major maximum at about 12170bp and further low values at 12005bp. An apparent reversal at 12403/12377bp might be due to indirect effects of the rapid temperature rise on the sediments (cf. Schove, 1978, Figure 3).

Series F. Finnish varves, at present based on Sauramo's work from 1918AD to 1940AD (Schove, 1971, p.215), provide a long, reliable series (subject to any minor corrections that the latest cores may suggest).

Series V and D. North-central Swedish varves contain small chronological errors,

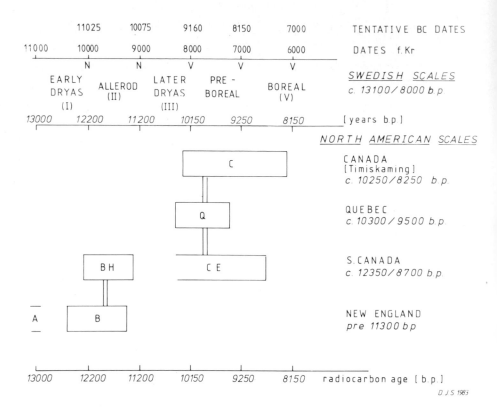

Figure 2-7:   Glacial varve chronologies indicated by f.kr. the
Swedish abbreviation meaning 'before Christ'.
Series A and B are from southern New England (floating chronology) and northern
New England respectively (see Schove, 1969 and 1983) and have been
studied palaeomagnetically. See section 2.4.3 for description of other
labelled varve series.

but in the far north the several (D) series measured from the Doeviken zero are good and have been teleconnected exactly with the later part of the excellent Canadian (C) series (Schove, 1971, pp. 229-230).

Series C and Q. These two series (Figure 2-7) overlap precisely, thus confirming the original Antevs numbering.   The Canadian and Baltic series together, constitute a chronology of about 4200 radiocarbon years or about 3900 varve-years, and the Canadian series may yet be teleconnected with varves in Germany or with tree-rings in the Middle East.

The dates (except for the three incorrectly ascribed to Dansgaard) in Table 5 of Schove (1971) still stand, but the ^{14}C deviations given there may yet prove to be slightly too large (see Figure 2-5). If the assumptions are acceptable, the 'Swedish' errors are 1140/1170 in Zone IV and ~1090 in Zone II (see Schove 1971, Figure 8).

2.4.4. Holocene Tree-Ring Chronologies

The Bristlecone pine chronology developed by Ferguson (Ferguson and Graybill in progress) is shown extending back to 7850bp in Figure 2-4. This series is weather-sensitive, wide rings at high altitudes requiring both moisture and summer warmth. A comparison of ring-widths since 1796AD with weather information (Schove and Berlage, 1965; Wright, 1977, Schove, 1983, Paper 35), shows that wide rings are of the 'Pacific type' – that is they correspond to what meteorologists term 'a negative Southern Oscillation', and are associated with above normal pressure in the Indian Ocean in the first half of the year. The Bristlecone series is filtered, and partly for this reason, wide rings are associated also with opposite conditions in the winter of the preceding year, (high pressure over the central Pacific). The relation with global barometric patterns makes it possible to cross-date this series with almost any other long time-series of proxy meteorological data. Varve-series such as this, may in future be correctly counted and palaeomagnetically measured, and can thus be dated in absolute years.

As an example of the possibilities, we instance the cross-dating of this Californian series with a floating chronology for juniper found in the tombs of the Phrygian Midas kings at Gordion, Asia Minor [Schove (1979), with corrected explanation in Schove (1983), Paper 40]. The tree-rings were measured by Bannister (1977) and later by Kuniholm (unpublished thesis). The date of the final ring number 1764 had been estimated from archaeology and radiocarbon as 725±25BC (astronomical year (AY) = – 724) which would have been near 7276 on the Bristlecone scale, leading to an expected lag of approximately 7276 – 1764 = 5512. The final tests only will be mentioned. Outstandingly wide rings at Gordion gave non-random results for USA maxima and minima at 5469 (Scores 5 and 27 respectively) and 5470 (Scores 30 and 2 respectively); if 5470 is correct, wide rings in both regions thus tend to agree. Nevertheless, narrow rings in the two regions did not agree, and indeed there was non-random minimum of such coincidences (Score only 3 against a random background of about 15) at 5470. 5470 was thus assumed to be the best answer with an a implied dating of the thesis chronology at 1562BC to 767BC. This dating fitted the corrected dating (see section 2.4.6) of the Lake Saki varves, which similarly reflected winter rainfall. A future publication with his chronology, (extended now to several thousand years) to be published by Kuniholm, will prove invaluable in varve dating (e.g. Dalmatia, etc.,) in Mediterranean countries.

The European tree-ring series (see Figure 2-4) extend beyond the Bristlecone series back to 7160BC, 9100BP or 8200bp (Bruns et al., 1983). However, at present the European series consists of several floating chronologies based on the

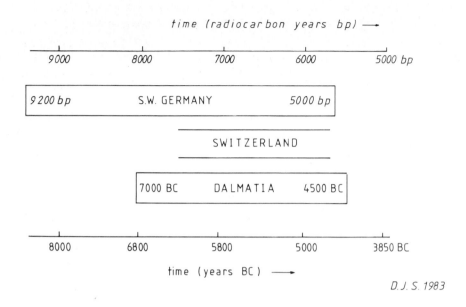

time (radiocarbon years bp) ⟶

| 9000 | 8000 | 7000 | 6000 | 5000 bp |

9 200 bp S.W. GERMANY *5000 bp*

SWITZERLAND

7000 BC DALMATIA 4500 BC

| 8000 | 6800 | 5800 | 5000 | 3850 BC |

time (years BC) ⟶

D. J. S. 1983

Figure 2–8:   Good varve series of Early and Middle Holocene not yet studied
palaeomagnetically and their approximate position in ^{14}C and
tree-ring chronologies.

Merkt and Beug for Germany and Dalmatia respectively - in Schove and
Fairbridge, 1983).

collections of Pilcher at the Queen's University of Belfast (N. Ireland) and on those of
Becker at the University of Hohenheim (W. Germany) at Stuttgart.

The way in which one of these floating chronologies was cross-dated with the
Bristlecone scale is explained here.   The published Bristlecone series is numbered
forwards in such a way that 4001BC (AY = − 4000) is termed + 4000.   The Irish
numbering was somewhat similar, so that the year 6000 (Pilcher et al. 1977) was
radiocarbon dated to a bp date that, when calibrated would lie within a century of
2000BC. Only two of the final tests will be mentioned.   Outstanding peaks and troughs
in the two series were weighted, and the coincidences were noted (one way or the
other) for different lags.   With lags of 20 and 21, the results were non-random.   At
20, the peaks in the American series often fitted the troughs in the Irish series (Score
ratio 170:127); at lag 21 the situation was reversed. A final test was then made by
noting the ring-numbers when steep rises or steep falls took place.   Non-random
results were found especially at 20 and also at 22 (ascribed to the greater auto-
correlation and the biennial cycle in the Irish series), with scores 303 and 295
compared with a random background of about 235.   The steep rises in Ireland were

most significant, and corresponded to an excess of steep falls in America. Lag 20 was thus accepted as better than 21, and year 2001BC (− 2000) or 6000BY (Bristlecone Year), must be Irish ring number 6020. Radiocarbon wiggles had already suggested (Lerman, cited by Pilcher et al. 1977), that Irish ring number 6500 lay near 1460BC and this makes it 1481BC (AY = − 1480), with the series extending from 4018 to 1070BC.

Intermittent floating chronologies are available for pine trees in Scandinavia, and in the Alps where some X-ray densitometer measurements are available from 8000bp. Tree-rings from pollen- zones IV and V (see Figure 2-4) are expected to cross-date with the Canadian varve series. As it will be difficult to find remains of trees in the colder pollen-zones I and III, this is as far as dendrochronology can be expected to go. Annual densitometric values are not yet fully published, but moving means are plotted for Renner's (1982) curve and the years are given only on the radiocarbon time scale. However, my comparison with the Irish curve suggests that the ring plotted by Renner at 3150BP corresponds to 1488±2BC or 3438BP, a very hot period in temperate latitudes. The displacement of the Swiss curve is thus assumed to be 288 true years throughout. Renner's results for this period could thus be used to add more details to the last part of the climatic curve (Figure 2-9). This climatic curve, and fuller lists of climatically significant radiocarbon centuries, is useful (e.g. in Figure 2-4 and in Schove, 1983) for dating marginal pollens in sediments that have so far been dated only in uncertain radiocarbon years. Measurements of pine-trees in the Scandinavian mountains are found above the tree-line, at dates when glacial varves were still being found in Canada, and provided no problem arises from missing rings, Karlen's work in Scandinavia (unpublished) should prove useful in extending tree-rings to the period before 8250bp.

Another series from USA that will be useful for cross-dating, is that reported as extending back to 3240BC at a site in the White Pine Range, east-central Nevada. This series is at the lower, rainfall-dependent, range of the Bristlecone pine (Ferguson and Graybill, in progress), and should reflect winter conditions.

2.4.5. Holocene Varve Chronologies

Holocene varve-chronologies are being prepared in North America, but no good series has yet been described in the published literature and there is still difficulty in finding precise time horizons. Pacific marine varves are useful, but some years are thought to be missing through drought; promising reconstructions of temperature and rainfall back to 1850AD were made by Soutar and Crill (1977), whereas a rough chronology (with a gap) published by Pisias (1978) in chart form, extended through most of the Holocene. Some long series have been included in unpublished North American theses (e.g. Kalil (Los Angeles) 1264AD-1570AD for a Santa Barbara core).

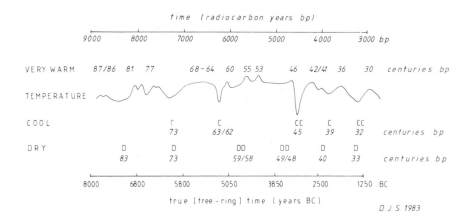

Figure 2-9:   The summer temperature curve in North America and Europe
(some details based on Alpine tree-ring densities after Renner, 1982).
Very warm, cool and dry radiocarbon centuries (60 = 6099bp – 6000bp)
based on Schove(1978). Dry centuries based on peat stratigraphy in the
North European plain. Calibrated dates are given at the bottom and
[14]C deviation of about 950yr assumed before 7200BC.
The Ulmus and Tsuga declines at 5000bp and 4700bp correspond to
substages 50/59 and 47/56 respectively.

The method of freezing cores is proving useful in both Canada and Finland for finding varve-series that extend to the present day, and volcanic horizons may soon be dated exactly by identification with acid-layers in Greenland ice-cores (cf. Clausen , in Schove and Fairbridge, 1983).

Some varve chronologies that may be useful for palaeomagnetic chronology are shown in Figures 2-4 and 2-8.   The German series from Schleinsee (Geyh et al., 1971), accepted as justifying the tabulated chronology of the Late Glacial (Schove, 1971), implied an assumption that the hard-water effect on ^{14}C dates (Schove, 1978, p. 212) was constant, and gave greater radiocarbon deviations than seemed acceptable at the time.   The varve-counts given by Stuiver (1970) for the Lake of the Clouds had already suggested a deviation of 800 at 10000BP, but even his figure, compared to the 1000 given in Figure 2 of Schove (1978 and 1983, Paper 39) had seemed, to many, impossibly large, until 1982.

The dried up lake of Faulensee in Switzerland contains thousands of varves, but there also appear to be unconformities (cf. Schove, 1978, 219).   The laminated sediments running from 7660 to 5700bp appear to be annual varves of the period 6560-4600BC which is 2300 years older than Welten supposed.

The only good Holocene varve-series that has been investigated magnetically is a

Swedish series for which the dates had been placed as 1000 f. kr. to AD '700' in a Swedish time-scale which is no longer accepted as accurate. Schove (1978, p. 221) made a provisional revision to 1400BC/300AD, based on estimated dates of Finnish non-varved sediments, as these had suggested a short westerly geomagnetic 'excursion' at about 300BC. However, we still await confirmation from the Finnish varved sediments now being investigated by Saarnisto, that the 300BC date is correct. Meanwhile, new investigations by Cato in Sweden (cf. Schove and Fairbridge, 1983) have provided a reliable Swedish varve chronology back to before 1400AD, and investigations by Saarnisto in Finland (in Schove and Fairbridge, 1983) promise to provide a varve chronology of 9000 years. In the Old World it is possible to use documentary evidence of weather and volcanic activity to provide fixed-points in Late Holocene varve-series. The varve-series from Lake Van, measured by Kempe and Degens (1979), thus uses a mediaeval volcanic eruption as a date check, and provides a weather-chronology that appears to be reliable back at least to 1500BC (see Kempe in Schove and Fairbridge, 1983). In Scandinavia and Finland, the tephrochronological horizons of Hekla ash have been found in unvarved sediments already, and are dated as follows (Hammer et al., 1980):

2700±70 BC   Hekla 4:
1150±50 BC   Hekla 3, etc.

Other volcanic horizons in Greenland are dated 536AD, 620AD, 797AD, etc. The date of Thera is likewise estimated from an acid layer as 1390±50BC.

2.4.6. Teleconnection of tree-ring and varves

Varves are not as easy to teleconnect as tree-rings. Holocene varve-counts are seldom accurate enough to use even the standard Hamburg and Belfast computer tests for local cross-dating, and this had to be done century by century in the case of the Lake Saki varves referred to below.

The Lake Saki varves in the Crimea, north of the Black Sea, could nevertheless be teleconnected approximately with the varves at Gordion in Asia Minor south of the Black Sea, the errors varying between 147 and 144 in the period 1500BC to 1000BC and falling to 135 by the 9th century BC (Schove, 1978, 1979, 1983 Papers 39 and 40).

2.4.7. Climatic chronology in radiocarbon centuries

Sediments which have been studied magnetically and palynologically often have radiocarbon dates attached; such dates can often be shown to be wrong if the fluctuations of climatically sensitive pollens disagree with global climatic fluctuations.

the latter being based on consistent radiocarbon evidence from different parts of the north temperate zone. Lists of outstanding radiocarbon centuries have therefore been given (Schove, 1967; 1978, pp 232–233 and 1983, Paper 39), and a standard curve is presented in Figure 2–9. This curve can be used for both North America and Europe, but all early phases are colder in eastern Canada and hotter in Europe (or Alaska) than the absolute height suggests. The reasons for the radiocarbon errors are not always clear, but lack of adequate coverage or carbonate error may explain some discrepancies. The so called four Little Ice Ages (Schove, 1967), are usually easy to locate in pollen diagrams, as follows:

Substade	45	4575bp	3375BC	Short but severe
	32	3225bp	1680/1520BC	Longer but less severe
	28/27	2800bp	~1050BC	Severe
	24/22	2300bp	6th/5th century BC	Prolonged

2.4.8. Provisional Absolute Chronology

The provisional absolute chronology adopted in the scales at the top of Figures 2–6 and 2–7 is based on glacial varve teleconnections (Schove, 1971 p. 230). The radiocarbon deviation was assumed to be 950yr at the beginning of the Holocene (Pollen–zone III/IV interface), as 300yr later. Although varve-counts at the Lake of the Clouds (Stuiver, 1970) had suggested a deviation at 800yr, the climatic chronology of the Swiss and German varves had suggested at least 100yr more (Schove, 1978 p. 220). We now know that the century centred at 6450BC (8400BP) is 7630bp, so that the radiocarbon deviation then was 770 to 775 years (Bristlecone dates from Ferguson, measured by Suess – see Bruns et al., 1983). We know also that the deviation increased during the preceding millennium to 900yr.

Glacial varves can usually be reliably counted in Canada and Finland, and as soon as dated tree–rings for several centuries prior to 8250bp are available, the cross-dated varve complex for North America and the Baltic (Figure 2–4) will be dated in absolute chronology back to ~12850bp. An extension to before 16750bp will be possible once a bridge is found to overlap the varves for northern (Series B) and southern (Series A) New England. Palaeomagnetic studies should prove very helpful in this respect.

2.5. LEAD-210 AND OTHER ISOTOPES

By Frank Oldfield,
University of Liverpool, UK

2.5.1. Introduction

This section outlines the range of radio-isotope based dating methods (other than ^{14}C) available for establishing chronologies of sedimentation in lakes. Particular attention is given to ^{210}Pb dating in view of its increasing importance in limnochronology. A more comprehensive account of the full range of possible techniques can be found in Krishnaswami and Lal (1978).

2.5.2. Lake sediment dating by ^{210}Pb

^{210}Pb which has a half-life of 22-26yr is a member of the ^{238}U decay series. ^{226}Ra in the lithosphere decays to yield the inert gas ^{226}Rn which then diffuses from rock and soil surfaces into the atmosphere. The radon gas which has a much shorter half-life of 3.83 days then decays through a series of short-lived isotopes, to ^{210}Pb. The ^{210}Pb reaching lake sediments thus comprises two components: (i) 'supported' ^{210}Pb yielded by the in-situ decay of ^{226}Ra deposited in the sediments as part of the particulate mineral input from the catchment and (ii) 'unsupported' or excess ^{210}Pb which reaches the sediments via the atmosphere.

The supported ^{210}Pb may, for the purpose of dating by ^{210}Pb, be regarded as in secular equilibrium with its parent isotope ^{226}Ra. The unsupported ^{210}Pb provides the dating parameter and in order to evaluate its reliability it is necessary to outline the routes by which it reaches the sediments. Figure 2-10 shows both the particulate input of ^{226}Ra (Route A) which yields the supported ^{210}Pb, and the main pathways by which unsupported ^{210}Pb reaches the sediments. Route B indicates deposition by direct atmospheric fallout which can take the form of both wet fallout or dry deposition. The relative importance of the two has not been conclusively established though most authorities believe that wet fallout is the more significant. Route C consists of deposition by indirect atmospheric fallout on to the catchment. A convenient but rather arbitrary distinction is here made between ^{210}Pb which is immediately incorporated into the drainage net (C_1) and ^{210}Pb which is delivered to the lake after deposition on to subsequently eroded soil particles (C_2). Whereas the former will reach the lake with little or no delay, the latter may have been detained in the catchment for a long and unknown period. Route D comprises deposition from radon decay in the water column. Radon in the water column will arise from upward diffusion across the mud-water interface, and from the decay of ^{226}Ra in the waters of the lake and inflowing

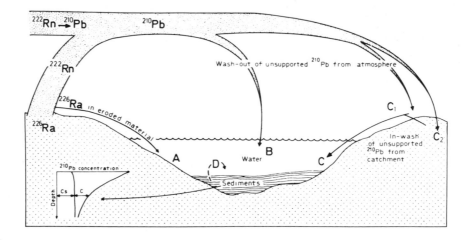

Figure 2-10:   The pathways by which ^{210}Pb enters lake sediments.
Pathway A: erosive input of ^{226}Ra, yields supported
^{210}Pb through in situ decay within the sediment column.
Pathways B, C and D yield unsupported (excess) ^{210}Pb (see text).
The inset shows a stylized plot of ^{210}Pb concentration
versus depth. C_2 is the supported concentration, C the
unsupported concentration

streams. Only part of the ^{226}Ra will decay to ^{210}Pb in the water, and this route is not thought to be a major contributor to the total flux of ^{210}Pb to the sediment.

The balance of evidence suggests that in most lakes the main source of unsupported ^{210}Pb reaches the sediments via Route B, after the ^{210}Pb falling on to the lake has become attached to sedimenting particles (Oldfield and Appleby in press). Once in the sediments, the ^{210}Pb is subject to continued radioactive decay, and its activity will tend to decline from the surface downwards.

In dating by ^{210}Pb, the supported ^{210}Pb activity is estimated by ^{226}Ra assay (assuming secular equilibrium), or by determination of the ^{210}Pb activity at depths below the lowest level yielding significant unsupported ^{210}Pb activity. Total ^{210}Pb is measured for a sequence of depths down a sampled profile by direct α assay, by β assay of the daughter isotope ^{210}Bi, or (most often) by γ assay of the grand-daughter isotope ^{210}Po. The unsupported ^{210}Pb is then determined for each assayed level by subtraction of the estimated supported activity. (Figure 2-11).

In deriving dates from the changes in unsupported ^{210}Pb concentrations in a core of sediments, the simplest case is one in which:

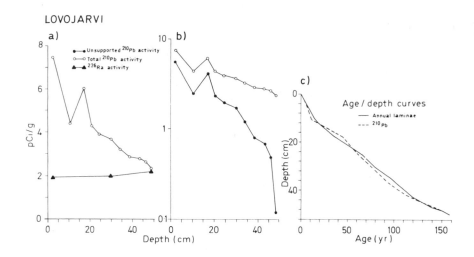

Figure 2-11:   LOVOJARVI, S.Finland. The derivation of dates from ^{210}Pb assuming
a constant rate of supply of unsupported ^{210}Pb to the sediment irrespective
of changes in dry mass sedimentation rate (CRS model): (a) plot of both
total ^{210}Pb and ^{226}Ra concentrations versus depth. Subtraction
of the radium supported ^{210}Pb activity as estimated from the ^{226}Ra
assay, assuming secular equilibrium allows calculation of the unsupported
^{210}Pb concentration at each depth as shown in (b). In (c) the CRS based
age/depth curve is compared directly with the true age determined by counting
annual laminae in the sediments. (from Appleby et al. 1979).

a. the net flux of unsupported ^{210}Pb to the sediment surface has
   remained constant for the whole period of measurable activity ($\sim$
   150yr).

b. the rate of dry mass sedimentation has been constant,

c. sediment mixing has been negligible and

d. the unsupported ^{210}Pb, once incorporated in the sediment, has not
   migrated or diffused vertically.

Under these conditions the unsupported ^{210}Pb concentration will vary exponentially
with the cumulative dry mass of sediment; therefore the plot of the logarithm of the
unsupported ^{210}Pb concentration against declining cumulative dry mass with depth will
be linear. Such simple linear profiles generally provide a reliable, non-controversial
basis for calculating mean sedimentation rates for the past 100-150yr, using a
constant flux : constant sedimentation rate (CF:CS) model. Much more controversial
are non-linear profiles, which give quite widely varying estimates of sedimentation
rates and age/depth relationships, according to the assumptions made and the dating
model used (Robbins 1978; Appleby and Oldfield in press; Oldfield and Appleby, in

press).

Non-linear profiles require some modification of the assumptions set out above. All available evidence is strongly in favour of the non-mobility of ^{210}Pb once deposited in the sediment, and most contention surrounds the validity of the assumptions of constant flux and constant sedimentation rate, and the effects of mixing.

Where unsupported ^{210}Pb concentration profiles are non-linear but nevertheless monotonic, several possible models are available. Pennington et al. (1976) have, implicitly, assumed that the initial concentration (and not the flux) of unsupported ^{210}Pb in the sediments is constant. This could be envisaged if Route C in Figure 2-10 were the main route by which the unsupported ^{210}Pb reached the sediment, or if the unsupported ^{210}Pb were scavanged from the water column by sedimenting autochthonous particles in proportion to their flux density at the mud-water interface. Under the assumption of constant initial concentration, (CIC model, sensu Appleby and Oldfield, 1978), the flux of unsupported ^{210}Pb to the sediment will be proportional to the rate of sedimentation.

Robbins et al. (1978) have approached non-linear profiles which are more or less flat in the upper levels, by assuming steady state mixing over a fixed depth, superimposed on the assumptions of constant unsupported ^{210}Pb flux and constant sedimentation. An alternative approach, retaining the assumption of a constant net flux of unsupported ^{210}Pb to the sediment is that first outlined by Goldberg (1963) and developed in recent years by Appleby and Oldfield (1978 and in press) as the CRS (constant rate of supply) model. In this model, dry mass sedimentation is allowed to vary and dates are calculated from the cumulative total residual activity of the unsupported ^{210}Pb in the sediment at each depth.

Where the profile of ^{210}Pb concentration versus accumulated dry weight is non-monotonic (i.e. 'kinked'), both the CF:CS and CIC models used alone will fail to provide a completed basis for calculating dates and sedimentation rates. The CRS model may be applied to such profiles but can only be used confidently where tests for internal consistency and compatibility with independently derived dates are carried out successfully. Oldfield and Appleby (in press) give a full account of the range of models available for treating ^{210}Pb profiles and of the empirical tests which can be used to explore their validity in any particular lake. Although the CRS model has been sustained in a wide range of contexts (Figures 2-11 and 2-12), problems may arise from processes such as sediment resuspension and focussing, slumping, and short and/or variable water residence times: thus caution is required in the evaluation of all non-linear profiles.

Used in conjunction with studies of the sediment record of recent geomagnetic secular variation, ^{210}Pb can help to provide a sound chronology for the last few centuries of sedimentation (cf. Oldfield et al. 1980). For chronological purposes the

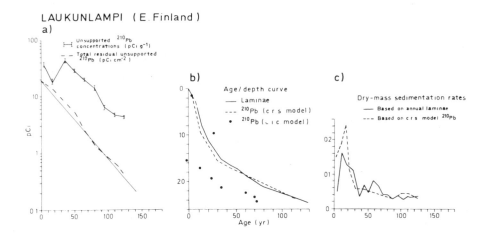

Figure 2–12:   LAUKUNLAMPI, E. Finland. Alternative ^{210}Pb models and the
estimation of dry mass sedimentation rates. Note how in (a) despite the
strongly inflected concentration vs. depth curve, the total cumulative
residual unsupported ^{210}Pb curve approximates to the log: linear
relationship expected if it is to be used as the dating parameter.
In (b) CRS based dates compare well with true age but no reliable chronology
is possible assuming a constant initial concentration. In (c) dry mass
sedimentation rates based on annual laminae and on CRS calculations
are compared (from Appleby et al. 1979).

two techniques can be regarded as independent and mutually validating especially in
view of the comprehensive analysis of post 1600AD secular variation world-wide
presented by Barraclough (1978) and by Thompson and Barraclough (1982).   The
availability of well established ^{210}Pb chronologies of sedimentation for lakes which
have a well resolved recent secular variation record may shed some light on the
question of time–lags between sediment deposition and the time of acquisition of a
stable NRM.   Areas with recent 'turning' points in both declination and inclination
would be especially suitable for this type of study. In addition, where ^{210}Pb provides
a reliable time–scale, but the validity of the recent NRM record is in doubt, the
radiometric chronology available for the last 150 years facilitates a direct comparison
of the secular variation record recoverable from the sediments with the known pattern
based on historical observations.

2.5.3. Other radioisotopes used in limnochronology

For extending the ^{210}Pb based record and providing a link between it and the
optimal time scale for resolution by ^{14}C, the most attractive radio–isotope is ^{32}Si

(half-life ~300yr). ^{32}Si is produced in the atmosphere through the spallation of argon by high-energy cosmic ray particles. Its incorporation into lake waters is thought to be by routes similar to those outlined for ^{210}Pb (Figure 2-10). Subsequent removal to the sediment is mainly through its incorporation in the siliceous tests of phytoplankton such as diatoms. Although the feasibility of selectively extracting the biogenic silica from lake sediments has been confirmed by Kharkar et al. (1963), no conclusive demonstration of the practical use of ^{32}Si in dating studies has yet been made.

On a much longer time-scale, ^{230}Th, a daughter isotope of ^{238}U, has provided chronologies of sedimentation for major 'pluvial' lakes laying outside the areas of former glaciation. Dates are derived from the ^{230}Th/^{238}U ratio and the assumptions and adjustments required to derive dating parameters are outlined in Krishnaswami and Lal (1978). Lacustrine carbonate and salt samples are both suitable for dating by this method and time-scales back to 300000 years ago have been established (Kauffman and Broecker, 1965; Goddard, 1970). As more long term records of geomagnetic secular variation are obtained from ancient and persistent non-glacial lakes, this technique will be of increasing importance for palaeomagnetists.

Finally mention is made of short-lived and artificial radioisotopes which may serve the palaeomagnetist both as an indication of sediment mixing processes and as confirmation of the retrieval or non-retrieval of the most recent sediments. The ^{238}Th/^{232}Th ratio can be applied to the last decade of sedimentation, since ^{238}Th (half-life 1.9yr) is present in sediment in excess of its grandparent ^{232}Th (Koide et al. 1973). Fall-out products resulting from nuclear testing (especially ^{137}Cs and, less commonly ^{90}Sr, ^{3}H, ^{55}Fe and ^{239,240}Pu) postdate 1953, and can be used in similar way.

2.6. THERMOLUMINESCENCE

By Ann Wintle,
University of Cambridge, UK

2.6.1. Introduction

The best type of technique to use for dating the magnetic record is one which can be applied to the material containing the magnetic minerals and which is influenced by the same event. The thermoluminescence (TL) dating method is such a technique. In the case of baked clay, the heating empties out the TL due to the clay's earlier geological history. In the case of sediments, exposure to sunlight prior to sedimentation has a similar, but not such an efficient, effect. The last ten years have seen several applications of TL dating in a geomagnetic context and results have often been obtained close to the limits of the technique. These will be discussed in detail later.

2.6.2. General principles of TL dating

Much effort was expended in the 1960s, particularly at the Research Laboratory for Archaeology in Oxford, in order to develop an absolute dating technique for pottery from archaeological sites. This led to two principal experimental techniques which form the basis of all current TL dating studies. These methods rely on the ability of certain crystalline minerals, in particular quartz and feldspar, to retain information about their past radiation history. The minerals contain a record of the amount of ionizing radiation (α, β and γ) to which they have been exposed since they were last heated. This radiation is due to the decay of naturally occurring radioactive elements in the pot itself and in the surrounding soil, the major sources being the uranium and thorium decay chains and ^{40}K. When the minerals are exposed to radiation, either in their natural environment or in the laboratory, free electrons and holes are produced. A small fraction of these become trapped at defects in the crystal and it is these trapped charges which give rise to the TL signal used in the age determination. When the minerals are heated, trapped electrons are released and are able to recombine within the crystals; the fraction that does so in a radiative process causes the TL. The TL signal is thus proportional to the number of trapped electrons prior to heating which is itself directly proportional to the amount of radiation exposure. This past radiation exposure, known as the equivalent dose (ED), is obtained by comparing the natural TL with that induced in the sample by laboratory irradiation as shown schematically in Figure 2-13. The age is then obtained by the use of a simple equation:

$$\text{age (years)} = ED/ADR$$

where ADR, the annual dose rate, is obtained from radioactivity analyses. For pottery the dose rate is usually constant because of the long lifetimes of the parent radionuclides. The cosmic ray component is usually about 5%. For further details the reader is referred to Aitken (1974) and Fleming (1979).

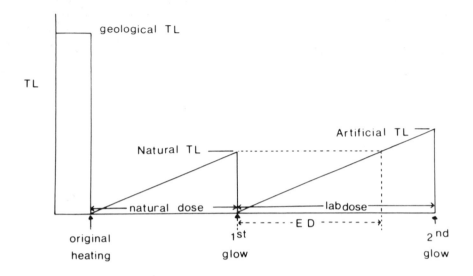

Figure 2-13:   Basis of TL dating of baked clay.

2.6.3. Fine grain method

Because pottery is inhomogenous, containing grains of different minerals of various sizes, perhaps up to 1mm diameter, and because the natural radiation contains α radiation which has a range of about 20μm, it was soon realized that TL dating would have to take account of the internal dosimetry of the pottery. Two techniques were developed, the fine grain method by Zimmerman (1971) and the quartz inclusion technique by Fleming (1970). In the fine grain method the outer 2mm of the pot is removed and the interior crushed. 1-8μm grains are then deposited out of an acetone suspension on 1cm diameter aluminium discs ready for measurement. Some of these discs are given laboratory β doses before heating to obtain the TL (see Figure 2-14) and this enables the construction of a sensitivity curve for unheated pottery (see Figure 2-15). The discs are heated to 500°C at about 10°C/s in a nitrogen atmosphere and the TL is observed with a photomultiplier tube. When discs that have been heated in the laboratory during the measurement of their natural TL (as shown schematically in Figure 2-13) are subsequently irradiated, a second response curve is obtained, known as the second glow growth curve. Sometimes this shows a sensitivity change. However, a second glow growth curve is always constructed so that possible non-linearity of the initial TL response to dose may be taken into account as indicated in Figure 2-15. A first glow growth curve is also obtained after exposure to α radiation which is known to be less effective at producing TL than β or γ radiation. In the pottery itself, most of the dose deposited in the grains is due to α decays in the uranium and thorium chains; however, because of their lower efficiency (about 10%) in producing TL, they are responsible for only about 30% of the TL in fine grains.

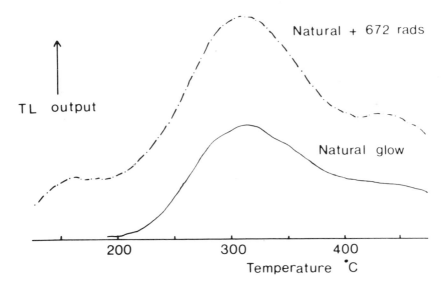

Figure 2-14: Typical glow curves for fine grain pottery showing increased TL due to an added laboratory dose (after Gunn and Murray, 1980).

The growth curves are constructed at 25°C intervals and the ED is thus obtained as a function of temperature. The ED is seen to increase with temperature up to 350°C and then gives a constant value, which is that used in the age equation. The lower values below 350°C are evidence for the loss of natural TL below this temperature due to the emptying of the shallower electron traps at ambient temperature. Loss of high temperature TL, known as 'anomalous fading', has been found in volcanic feldspars and occasionally in pottery. It is checked by storing a few laboratory irradiated samples for at least two weeks before measurement; if the samples show more than 10% loss of the laboratory induced signal, then the pottery sample is rejected from the dating programme.

To obtain an age it is now necessary to measure the annual dose rate. The α dose rate from the uranium and thorium chains is usually obtained by a thick source α counting technique. In this method the crushed pot is spread on a ZnS(Ag) screen

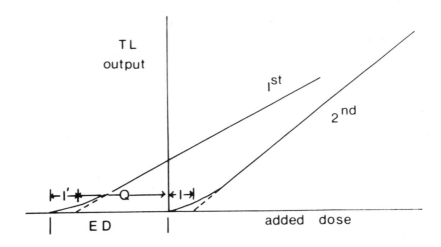

Figure 2-15: Idealized β growth curves showing first to second glow
sensitivity change and effect of nonlinearity. From laboratory measurements,
ED= Q + I where I is assumed equal I', (after Gunn and Murray, 1980).

which is placed on a photomultiplier tube; the scintillations produced in the ZnS(Ag)
by α particles coming from decays in the 20μm layer of pottery next to the screen are
counted for about 24 hours. This technique is described in more detail elsewhere
(Aitken, 1974; Fleming, 1979). The annual β dose rate from uranium and thorium can
also be calculated from the same data and then combined with the β dose rate from
the ^{40}K in the pot. The latter is obtained from the K_2O content measured by flame
photometry or atomic absorption analysis. The β dose rate can also be obtained by
dosimetry (Bailiff and Aitken, 1980). The γ dose rate, which comes from the
surrounding soil, can be obtained by α counting and K_2O analysis of the soil.
However, it may be better determined by encapsulating a dosimeter, such as natural
CaF_2 , in a copper tube and placing it as close as possible to the burial site of the
pot. It is then left for up to a year to record the average annual dose rate. It has the
advantage of taking into account reductions in dose rate due to the presence of water
in the soil (Mejdahl, 1970). If it is not possible to use a dosimeter, instant readings
may be obtained using a scintillation counter at the time of sample collection; this
would take into account any inhomogeneity in the radioactive environment.

As can be appreciated from the above description, a large number of
measurements have to be made in order that a TL date may be calculated.
Associated with each of these is a random and a systematic error. Appropriate

combination of the errors allows dates to be obtained with random and total error limits, the former being used when comparing dates obtained in the same laboratory on similar material. These measurement associated errors do not include errors due to any changes in the burial surroundings such as major fluctuations in the water content due to alteration of the water table.

2.6.4. Applications of the fine grain method

In spite of the potential of combined TL and magnetic studies on pottery, only one systematic study has so far been carried out. In 1975 ceramics were collected in Peru from 50 archaeological sites which were in the process of being excavated. Samples were taken for both TL and magnetic intensity measurements. Ages covered the last 3000yr, with most of the results falling in the period 0-1500AD (Gunn and Murray, 1980). Twenty TL dates were obtained altogether from 9 different sites. One site which had 7 sherds dated by TL showed that stratigraphy alone was a bad indicator of the age of a particular sherd and that an individual TL date for each archaeomagnetic sample was desirable. Concentrating on the sherds which gave good magnetic results and firm TL dates, Gunn and Murray (1980) presented a plot of geomagnetic intensity versus time for Peru which may be compared with those constructed in other parts of the world (see chapter3, section3).

Huxtable and Aitken (1977) tried fine grain dating of baked clay used as ovenstones from prehistoric aboriginal fireplaces at Lake Mungo, Australia. However they were not satisfied with the TL measurements which they considered to be affected by 'spurious' luminescence and which were not improved by further cleaning of the fine grains. Other TL studies were carried out on baked sediments from below the fireplaces. Although a fine grain date of 30400±3300yr was reported by Adams and Mortlock (1974), these samples were subsequently shown to exhibit 'anomalous fading' (W.T. Bell, personal communication, 1980).

Successful dating by the fine grain technique was obtained for clay baked by two lava flows in the French Massif Central (Huxtable et al, 1978). Oriented baked clay at Royat gave similar magnetic directions to the overlying magnetically reversed lava flow. Earlier TL studies of feldspars extracted from lava flows, including those from several flows in the Massif Central, were unsuccessful because of anomalous fading (Wintle, 1973). However, the baked clay samples showed negligible anomalous fading and a mean age of 25800±2200yr was determined by combining the results from seven samples. This date is not inconsistent with other TL, ^{14}C and K-Ar dates for other reversed flows in the region (Gillot et al, 1979).

2.6.5. Quartz inclusion method

Instead of using the 2–8μm grains as described above, TL dates have also been obtained on quartz grains of 90–125μm (Fleming, 1970). These quartz grains have several advantages. First, quartz has never been found to exhibit anomalous fading and second, the combination of larger grain size and acid pretreatment considerably reduces any chance of spurious luminescence. 90–125μm grains are extracted from the pot, or other baked material, by crushing and sieving. The grains are then put through a magnetic separator and the crystalline, nonmagnetic fraction is etched in concentrated HF to totally dissolve the feldspars and remove the surface of the quartz grains. Depending upon the etching time (usually 10–40 min) the surface of the grain which has been exposed to α radiation from the surrounding clay is almost totally removed. Hence, the grains may be treated as though they had only received the β dose from within the pot and the γ dose from the surrounding soil. This results in a disadvantage over the fine grain technique since the total dose rate is therefore more dependent on any changes in the environment, e.g. water content. The β and γ dose rates are obtained in the same way as for the fine grain samples described in section 2.6.3.

Once again the grains are irradiated in the laboratory and first and second glow growth curves constructed in order to obtain the ED for the age equation. The main difficulty encountered in the application of this method to older material, in particular that from Lake Mungo, is the non-linearity of the growth curves due to onset of saturation of the TL effect in quartz. This limits the use of the technique for burnt materials.

2.6.6. Applications of quartz inclusion method

This technique was also used by Huxtable and Aitken (1977) and Bell (personal communication, 1980) for the baked clay ovenstones and underlying sediments at Lake Mungo. Both studies reported non-linear TL growth of the quartz extracts. Huxtable and Aitken (1977) used an exponential fit to obtain seven dates which were combined to give a weighted mean age of 33500±4300yr. The dominant contribution to the total error was thought to be the lack of precision in the description of the exponential growth curve.

Having established non-linearity for total doses of greater than 9krad, Bell (personal communication, 1980) kept within the linear region using additional doses of less than 2krad to obtain the first glow growth curve, rather than doses of 4.8krad and 9.6krad used by Huxtable and Aitken (1977). Greater precision was possible by this approach, and dates were obtained for baked sediment from beneath four fireplaces. The mean age was 33500±2100yr, in excellent agreement with the results on the ovenstones.

2.6.7. TL dating of sediments

Several geomagnetic 'events' have occurred during the Brunhes epoch and some of these have been recorded in aeolian, marine and lacustrine sediments. However, they rarely all occur in the same section so that it is difficult to be sure of the identity of any particular geomagnetic 'event'. TL dating has been applied to sediments in the USSR for about 15 years but rather naively and without thought for the dosimetry. In the west, TL dating of sediments was developed independently by Wintle and Huntley (1979) and was based on the knowledge obtained from ten years of studies on pottery. A review of the current state of TL dating of sediments has been published recently (Wintle and Huntley, 1982).

The main difference between dating pottery and dating sediments is the fact that the latter were not heated in the past. Instead the original geological TL signal was bleached by exposure to sunlight during weathering, transport and depositional processes. Typical glow curves obtained after bleaching the natural TL for different lengths of time are shown in Figure 2-16. It can be seen that the bleaching initially occurs quite quickly and then slows down. After very long bleaching times there is still a finite TL signal, and this has been found in recent glacial and fluviatile deposits (H. Proszynska, personal communication, 1981). This means that for young sediments (~50000yr) this residual signal must be estimated and then subtracted from the natural TL. Another effect of the sediment not having been heated in the past, is that when it is heated in the laboratory for the first time in the course of a glow curve determination the TL sensitivity is changed and second glow measurements are therefore meaningless.

Advantages of working on sediments include the fact that large amounts of material are available and that they are often very uniform, particularly when compared with human occupation sites. Mineralogically the silt fractions of recent sediments are very similar, containing predominantly quartz and feldspar, and this leads to the similarity of the TL properties of the sediments studied, which include aeolian (Wintle, 1981), marine (Wintle and Huntley, 1980), lacustrine and fluviatile. Their natural radioactivity is also similar except that in the case of marine deposits there is an excess of ^{230}Th precipitated on to the ocean floor. This led to the development of a time dependent age equation for marine sediments. A similar equation to allow for the in-wash of ^{226}Ra may be necessary for lacustrine samples, but they have yet to be studied in detail, since there is the added complication of the uncertainty in the past water content.

Dating has been carried out using a variety of grain sizes depending upon the type of sediment being studied. The major TL mineral has been shown to be feldspar rather

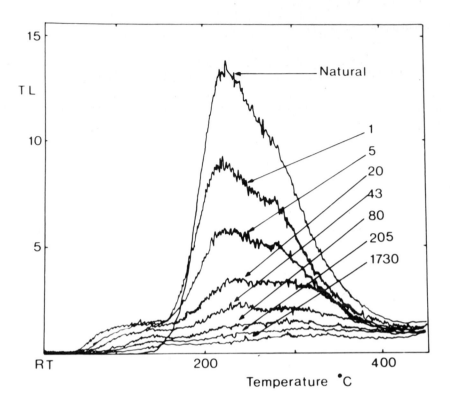

Figure 2-16:   Glow curves of a deep-sea sediment sample which has been exposed
to a sun-lamp for various lengths of time, given in minutes
(after Wintle and Huntley, 1980).

than quartz; however, most samples have shown no anomalous fading and it is
thought that only feldspars of volcanic origin show this annoying property. Growth
curves at a particular temperature show non-linearity of TL response above about
15krad but this does not conform with exponential growth towards saturation and is
more likely to be connected with the non first order kinetic behaviour of some
feldspars. This non-linearity means that the ED cannot be obtained (except for
samples less than 20300 years old) by methods using linear extrapolation of the first
glow growth curve, as is done for pottery. Instead, the samples are bleached using
an artificial light source and doses added until one is found which matches the natural
TL. This is taken to be the ED to be used in the age equation.

2.6.8. Applications to sediments

Until now, the only applications of TL dating to magnetic events in sediments have been in the USSR, and are considered in many cases only as preliminary results. The Chegan geomagnetic event, which is thought to be contemporary with Biwa II in Japan (Champion et al, 1981), was dated in varved clays at the type site of Chegan in the Gornyy Altai at 266000±30000yr (Shelkopas et al, 1973) and also at Likhvin at 280000±30000yr (Faustov et al, 1974). An earlier event, the Ureki, was dated at 330000±60000yr by Shelkoplyas (Zubakov and Kochegura, 1976) although this is thought to be too young (Kochegura and Zubakov, 1978).

Tentative dates have also been produced by Shelkoplyas for loess in Tadzhikistan (Davis et al, 1980). A magnetic reversal, thought to be the Blake event, was recorded in loess dated at 110000±13000yr. This loess was above the fifth palaeosol at one of the sites, which had nine soils above the Brunhes–Matuyama boundary. At Luochan in Northern China where there are twelve soils above the Brunhes–Matuyama boundary, the loess above the fifth palaeosol also showed a magnetic reversal (An and Wei, 1980) Preliminary TL dating by Li et al (1977) gave an age 178000±24000yr for the top of the ancient soil and 212000±16000yr for the loess just under the soil and hence this event was correlated with Biwa I in Japan rather than with the Blake event.

Western applications have been confined to the period since the last interglacial, with emphasis on dating of samples of known age to check the validity of the technique. However a recent study on a Polish loess, thought to contain a record of the Blake event (Tucholka, 1977) has given a preliminary date of 120000 ±18000yr and 112000 ±17000yr by Proszynska (1983). More work on samples of known age needs to be carried out before TL dating of sediments, particularly those older than 100000yr, can be carried out routinely.

REFERENCES

Adams, G. and Mortlock, A.J., Thermoluminescent dating of baked sand from fire hearths at Lake Mungo, New South Wales. *Archaeology and Physical Anthropology in Oceania*, 1974, *9*, 236-237.

Aitken, M.J.   *Physics and Archaeology*.   Clarendon Press, Oxford, 1974.   (2nd ed.).

An, Z.-S. and Wei, L.-Y. The fifth palaeosol in the Lishi loess and its palaeoclimatic significance. *Acta Pedologica Sinica*, 1980, *17*, 1-10. in Chinese.

Appleby, P.G. and Oldfield, F. The calculation of lead-210 dates assuming a constant rate of supply of unsupported ^{210}Pb to the sediment. *Catena*, 1978, *5*, 1-8.

Appleby, P.G. and Oldfield, F. The assessment of ^{210}Pb data for use in limnochronology. In *Proceedings of the IIIrd International Symposium on Palaeolimnology, Finland 1981.* , in press.

Appleby, P.G., Oldfield, F., Thompson, R., Huttuinen, P. and Tolonen, K. ^{210}Pb dating of annually laminated lake sediment from Finland. *Nature*, 1979, *280*, 53-55.

Barraclough, D.R. Spherical harmonic analyses of the geomagnetic field for eight epochs between 1600 and 1910. *Geophys. J. R. astr. Soc.*, 1978, *36*, 497-513.

Berner, W., Oeschger, H. and Stauffer, B. Information on the CO_2 cycle from ice core studies. *Radiocarbon*, 1980, *22*, 227-235. 10th Int. Radiocarbon Conf. - Bern.

Birks, H.J.B. Holocene (Flandrian) Chronostratigraphy of the British Isles: a review. *Striae*, 1982, *16*, 99-105.

Birks, H.J.B. and Birks, H.H.   *Quaternary Palaeoecology*.   Edward Arnold, 1980.   298 pp.

Bruns, M., Rhein, M., Linick, T.W. and Suess, H.E. The Atmospheric Carbon 14 level in the 7th Millenium BC. In *Proceedings of Conference: Carbon 14 in Archaeology.* , 1983.

Bucha, V. Influence of the Earth's magnetic field on radiocarbon dating. In Olsson, I.U. (Ed.), *Nobel Symposium*.   Vol. 12:   *Radiocarbon, Variations and Absolute Chronology*. John Wiley and Sons, New York, 1970.

Champion, D.E., Dalrymple, G.B. and Kuntz, M.A. Radiometric and paleomagnetic evidence for the Emperor reversed polarity event at 0.46±0.05my in basalt lava flows from the eastern Snake River Plain, Idaho. *GRL*, 1981, *8*, 1055-1058.

Davis, M.B. Pleistocen biogeography of temperate deciduous forests. *Geoscience and Man*, 1976, *13*, 13-26.

Davis, M.B. Outbreaks of forest patheogens in Quaternary history. *Proceedings IV International Palynological Conference, Lucknow (1976-77)*, 1981, *3*, 216-228.

Davis, R.S., Ranov, V.A. and Dodonov, A.E. Early man in Soviet Central Asia. *Scientific American*, 1980, *243*, 92-102.

Faegri, K. and Iversen, J.   *Textbook of Pollen Analysis*.   Blackwell, 1975.   3rd edition, 295 pp.

Faustov, S.S., Ilichev, V.A. and Bolshakov, V.A. Paleomagnetic and thermoluminescence analyses of the Likhvin section. *Doklady Akademii Nauk SSSR*, 1974, *214*, 1160-1162. (in Russian, English translation, 214, 123-125).

Fleming, S.J. Thermoluminescent dating: refinement of the quartz inclusion method. *Archaeometry*, 1970, *12*, 133-145.

Fleming, S.J.   *Thermoluminescence Techniques in Archaeology*.   Qxford University Press, 1979.   233 pp.

Flenley, J.R.   *Equatorial Rain Forest: A Geological History*.   Butterworth, 1979.   176 pp.

Geyh, M.A., Merkt, J. and Mueller, H. Sedimentological, pollen-analytical and isotopic studies of annually laminated sediments in the central part of the Schleiensee, Germany. *Arch. Hydrobiol.*, 1971, *69*, 366-399.

Gillot, P.Y., Labeyrie, J., Laj, C., Valladas, G., Guerin, G., Poupeau, G., and Delibrias, G. Age of the Laschamp paleomagnetic excursion revisited. *Earth Planet. Sci. Lett.*, 1979, *42*, 444-450.

Goddard, J. $^{230}Th/^{234}U$ dating of saline deposits from Searles Lake, California. Master's thesis, Queens College, New York, 1970.

Goldberg, E.D. Geochronology with ^{210}Pb. In *Radioactive Dating*. International Atomic Energy Agency, Vienna, 1963.

Gordon, A.D. and Birks, H.J.B. Numerical methods in Quaternary Palaeoecology. *New Phytologist*, 1972, *71*, 961–979.

Gove, H.E. (ed). First Conference on Radiocarbon Dating with Accelerators, *Unpublished Proceedings*, Rochester, 1978.

Griffiths, D.H. The remanent magnetism of varved clays from Sweden. *Mon. Not. R. astr. Soc. Geophys. Suppl.*, 1955, *7*, 103–114.

Gunn, N.M. and Murray, A.S. Geomagnetic field magnitude variations in Peru derived from archaeological ceramics dated by thermoluminescence. *Geophys. J. R. astr. Soc.*, 1980, *62*, 345–366.

Hammer, C.H., Clausen, H.B. and Dansgaard, W. Greenland ice sheet evidence of post-glacial volcanism and its climatic impact. *Nature*, 1980, *288*, 230–235.

Hedges, R.E.M. Radiocarbon dating with an accelerator, review and preview. *Archaeometry*, 1981, *23*, 3–18.

Heide, K.M. *Late Quaternary Vegetational History of North-central Wisconsin, USA: Estimating Forest Composition from Pollen Data*. Doctoral dissertation, Brown University, 1981. 213 pp.

Huntley, B. and Birks, H.J.B. *An Atlas of Past and Present Pollen Maps for Europe: 0-13000 years ago*. Cambridge University Press, 1983.

Huxtable, J. and Aitken, M.J. Thermoluminescent dating of Lake Mungo geomagnetic polarity excursion. *Nature*, 1977, *265*, 40–41.

Huxtable, J., Aitken, M.J. and Bonhommet, N. Thermoluminescence dating of sediment baked by lava flows of the Chaine des Puys. *Nature*, 1978, *275*, 207–209.

International Study Group. An inter-laboratory comparison of radiocarbon measurements in tree rings, *Nature*, 1982, *298*, 619–623.

Jacobson, G.L. Jr and Bradshaw, R.H.W. The selection of sites for paleovegetational studies. *Quaternary Research*, 1981, *16*, 80–96.

Johnson, G.L., Murphy, T. and Torreson, O.W. Pre-history of the Earth's magnetic field. *Terr. Magn. Atmos. Elec.*, 1948, *53*, 349–372.

Kauffman, A. and Broecker, W.S. Comparison of ^{230}Th and ^{14}C ages for carbonate materials from Lake Lahontan and Bonneville. *J. geophys. Res*, 1965, *70*, 4039–4054.

Kempe, S. and Degens, E.T. Varves in the Black Sea and in Lake Van (Turkey). In Schluchter, Ch. (Ed.), *Moraines and Varves*. Balkema, Rotterdam, 1979.

Kharkar, D.P., Lal, D. and Somayajulu, B.L.K. Investigations in marine environments using radioisotopes produced by cosmic rays. In *Radioactive Dating*. International Atomic Energy Agency, Vienna, 1963.

Klein, J., Lerman, J.C., Damon, P.E. and Ralph, E.K. Calibration of radiocarbon dates: Tables based on the consensus data of the workshop on calibrating the radiocarbon time scale. *Radiocarbon*, 1982, *24*, 103–105.

Kochegura, V.V. and Zubakov, V.A. Palaeomagnetic time scale of the Ponto–Caspian Plio–Pleistocene deposits. *Palaeogeogr. Palaeoclimatol. Palaeoecol.*, 1978, *23*, 151–160.

Koide, M., Bruland, K.W. and Goldberg, E.D. Th-228/Th-232 and Pb-210 geochronologies in marine and lake sediments. *Geochim. Cosmochim. Acta*, 1973, *37*, 1171–1187.

Krishnaswami, S. and Lal, D. Radionuclide Limnochronology. In Lerman, A. (Ed.), *Lakes: Chemistry, Geology, Physics.*, 1978.

Kuniholm, P.I. *Dendrochronology at Gordion and on the Anatolian Plateau*. Doctoral dissertation, University of Pennsylvania, Philadelphia, Pa., 1977.

Kutschera, W. (Ed.). *Symposium on Accelerator Mass Spectrometry*. Proceedings, Argonne National Laboratory, 1981.

Lal, D. and Peters, B. . In *Handbuch der Physik.*, 1972.

Li, J-L., Pei, J-X., Wang, Z-Z., and Lu, Y-C. A preliminary study of both the thermoluminescence of the quartz powder in loess and the age determination of the loess layers. *Kexue Tongbau*, 1977, *22*, 498–502. (in Chinese).

Liritzis, Y. and Thomas, R. Palaeointensity and thermoluminescence measurements of Cretan kilns from 1300 to 2000 BC. *Nature*, 1980, *283*, 54–55.

McNish, A.G. and Johnson, E.A. Magnetization of unmetamorphosed varves and marine sediments. *Terr. Magn. Atmos. Electr.*, 1938, *43*, 401–407.

Mejdahl, V., Measurement of environmental radiation at archaeological excavation sites. *Archaeometry*, 1970, *12*, 147–159.

Mook, W.G. Carbon-14 in hydrogeological studies. In Fritz, P. and Fontes, J.Ch. (Eds.), *Handbook of Environmental Isotope Geochemistry*. Elsevier, Amsterdam, 1980.

Moore, P.D. and Webb, J.A.   *Illustrated Guide to Pollen Analysis*. Hodder, 1978. 192 pp.

Noel, M. The palaeomagnetism of varved clays from Blekinge, southern Sweden. *Geol. Foren. Stockh. Fohr.*, 1975, *97*, 357–367.

Noel, M. and Tarling, D.H. The Laschamp geomagnetic 'event'. *Nature Phys. Sci.*, 1975, *253*, 705–706.

Oeschger, H. The contribution of ice core studies to the understanding of the environmental system. *Radiocarbon*, 1983. in press, 11th International Radiocarbon Conference – Seattle 1982.

Oldfield, F. and Appleby, P.G. Empirical testing of ^{210}Pb dating models for lake sediments. In H.J.B. Birks (Ed.), *Festschrift volume for Dr. W. Pennington*. Leicester U.P., in press.

Oldfield, F., Appleby, P.G. and Thompson, R. Palaeoecological studies of lakes in the Highlands of Papua New Guinea: 1. Chronology of sedimentation. *J. Ecol.*, 1980, *68*, 457–477.

Otlet, R.L., Huxtable, G., Evans, G.V., Humphreys, G.D., Short, T.D. and Conchie, S.J. Development and applications of the Harwell small counter facility for the measurement of ^{14}C in very small samples. *Radiocarbon*, 1983. in press, 11th International Radiocarbon Conference – Seattle 1982.

Pearson, G.W. High precision Radiocarbon Dating by Liquid Scintillation Counting applied to Radiocarbon Time-Scale Calibration. *Radiocarbon*, 1980, *22*, 337–345.

Pennington, W., Cambray, R.S., Eakins, J.D. and Harkness, D.D. Radionuclide dating of the Recent sediments of Blelham Tarn. *Freshwater Biol.*, 1976, *6*, 317–331.

Pilcher, J.R., Hillam, J., Baillie, M.G.L. and Pearson, C.W. A long sub-fossil oak tree-ring chronology from the North of Ireland. *New Phytol.*, 1977, *79*, 713–729.

Pisias, N.G., Palaeoceanography of the Santa Barbara Basin during the last 8000 years. *Quat. Res.*, 1978, *10*, 366–384.

Proszynska, H. TL dating of some subaerial sediments from Poland. *PACT*, 1983, Vol. *9*. in press.

Rackham, O. *Ancient Woodland its History, Vegetation and Uses in England*. Edward Arnold, London, 1980. 402 pp.

Renner, F. Beitrage zur Gletscher – Geschichte des Gotthardgebietes und dendroklimatologische analysen zur fossile Holzern. Geographisches Institut der Universitat, Zurich. *Phys. Geog.*, 1982, Vol. *8*.

Robbins, J.A., Krezoski, J.R. and Mosley, S.C. Radioactivity in sediments of the Great Lakes. *Earth Planet. Sci. Lett.*, 1978, *36*, 325–333.

Schaeffer, O.A., Davis, R., Stoeneer, R.W. and Heymann, D. . *Proc. Int. Conf. on Cosmic Rays.*, 1963, *3*, 480–489.

Schove, D.J. and Berlage, H.P., Pressure Anomalies in the Indian Ocean Area, 1796–1960. *Pure Appl. Geoph.*, 1965, *61*, 219–231.

Schove, D.J. World climate, 8000 – 0 BC. In *Proceedings Int. Symposium on World Climate*. Roy. Met. Soc., 1967.

Schove, D.J., Varve telecpnnection across the Baltic. *Geogr. Ann.,,* 1971, *53A*, 214–234.

Schove, D.J., Tree-rings and Varve Scales combined c. 13500 BC to AD 1977. *Palaeogeogr. Palaeoclimatol. Palaeoecol.*, 1978, *25*, 209–233.

Schove, D.J. Teleconnection of varves in N. America and Europe. In C. Schluechter (Ed.), *Varves and Moraines*. A.A. Balkema, Rotterdam, 1979.

Schove, D.J. (ed.).   *Sunspot Cycles*. Hutchinson and Ross, 1983. in press.

Schove, D.J. A varve teleconnection project. In M. Ters (Ed.), *Etudes sur le Quaternaire dans le Monde*.

, 1871b.

Schove, D.J. and Fairbridge, R.W.(Eds.).   *Ice cores, Varves and Tree-rings*.   A.A.Balkema, Rotterdam, 1983.

Shelkoplyas,V.N., Il'ichev,V.A. and Svitoch,A.A, Thermoluminescence dating of recent sediments of the Ob Plateau and Gornyy Altai. *Doklady Akademii Nauk SSSR*, 1973, *212*, 935–937.

Shotton, F.W. An example of hard-water error in radiocarbon dating of vegetable matter, *Nature*, 1972, *240*, 460–461.

Siegenthaler, U., Heimann, M. and Oeschger, H. [14]C Variations caused by changes in the global carbon cycle. *Radiocarbon*, 1980, *22*, 177–191.

Smith,A.G. and Pilcher,J.R. Radiocarbon dates and vegetational history of the British Isles. *New Phytologist*, 1973, *72*, 903–914.

Soutar, A. and Crill, P.A., Sedimentation and climatic patterns in the Santa Barbara Basin during the 19th and 20th centuries. *Geol. Soc. Am. Bull.*,, 1977, *98*, 1161–1172.

Stuiver, M, Tree ring, varve and carbon-14 chronologies. *Nature*, 1970, *240*, 454–455.

Stuiver, M. A high-precision calibration of the AD radiocarbon time-scale. *Radiocarbon*, 1982, *24*, 1–26.

Thompson, R. and Barraclough, D. Geomagnetic secular variation based on spherical harmonic and cross validation of historical and archaeomagentic data. *J. Geomag. Geoelectr.*, 1982, *34*, 245–263.

Tucholka,P. Magnetic polarity events in Polish loess profiles. *Biuletyn Instytutu Geologicznego*, 1977, *305*, 177–123.

Tucholka, P. Short period secular variations (SPSV) of the geomagnetic field recorded in highly scattered palaeomagnetic records of holocene lake sediments from north Poland. *Earth Planet. Sci. Lett.*, 1980, *48*, 379–384.

Turner, J. The tilia decline: an anthropogenic interpretation. *New Phytologist*, 1962, *61*, 328–341.

Turner,C. and West,R.G. The subdivision and zonation of interglacial periods. *Eiszeitalter und Gegenwart.*, 1968, *19*, 93–101.

Walker,M.J.C. and Lowe,J.J. Postglacial environmental history of Rannoch Moor, Scotland III. Early-and mid-Flandrian pollen stratigraphic data from sites on western Rannoch Moor and near Fort Willian. *Journal of Biogeography*, 1981, *8*, 475–491.

Watson,R.A. and Wright,H.E. Jnr. The end of the Pleistocene: a general critique of chronostratigraphic classification. *Boreas*, 1980, *9*, 153–163.

Webb, T. III. Temporal resolution in Holocene pollen data. In *Proceedings*.   Vol. 2:   . Third North American Paleontological Convention, 1982.

West,R.G. Palaeobotany and Pleistocene stratigraphy in Britain. *New Phytologist*, 1981, *87*, 127–137.

White, N.R.   Isotopic fractionation in accelerator based radiocarbon dating. *Unpublished Proceedings of Symposium on Accelerator Mass Spectrometry*. Argonne, 1981.

Wintle,A.G. Anomalous fading of thermoluminescence in mineral samples. *Nature*, 1973, *245*, 143–144.

Wintle, A.G. Thermoluminescence dating of late Devensian loesses in southern England. *Nature*, 1981, *289*, 479–480.

Wintle,A.G. and Huntley,D.J. Thermoluminescence dating of a deep-sea sediment core. *Nature*, 1979, *279*, 710–712.

Wintle,A.G. and Huntley,D.J. Thermoluminescence dating of ocean sediments. *Canadian Journal of Earth Sciences*, 1980, *17*, 348–360.

Wintle, A.G. and Huntley, D.J. Thermoluminescence dating of sediments. *Quaternary Science Reviews*, 1982, *1*, 31–53.

Wright, P.B.   *The Southern Oscillation – Patterns and Mechanisms of the Teleconnections and the Persistence*. Hawaii Inst. of Geophysics (Publication HIG 77-13), University of Hawaii, 1977.

Zimmerman,D.W. Thermoluminescent dating using fine grains from pottery. *Archaeometry*, 1971, *13*, 29–52.

Zubakov, V.A. and Kochegura, V.V. On the reverse magnetization of the Palaeo-Euxine layers of the Urek section. *Gerzenovskie chtenya*, 1976, *XXVIII*, 31–35. in Russian.

3.
Archaeomagnetism of Baked Clays

3.1. INTRODUCTION

By Daniel Wolfman,
Arkansas Archeological Survey, USA

The archaeomagnetic study of baked clays began towards the end of the 19th century with the work of Gheradi (1862), Folgheraiter (1896, 1897a, b, 1899) and Mercanton (1902, 1910, 1918a, b). The foundations of the modern subject were laid by Emile Thellier and his students in the period between 1930 and 1960. They described the basic magnetic properties of baked clay (Thellier, 1951; Roquet, 1954), they developed sampling techniques (Thellier, 1936) and laboratory apparatus for archaeomagnetic determinations (Thellier, 1933, 1938; Pozzi and Thellier, 1963) and they also developed alternating field (Thellier and Rimbert, 1954, 1955; Rimbert, 1959) and thermal (Thellier and Thellier, 1959) demagnetization techniques for the removal of secondary components of remanence. Thellier (1938) also suggested the possibility of a correlation between remanence and cooling rate. There followed, in France, many years of fruitful interaction between Thellier, the experimentalist and Néel, the theoretician, which led to the development of the first comprehensive theories of the acquisition of viscous and thermoremanent magnetization by baked clays and volcanic rocks (Néel, 1949, 1955).

One of the main objectives of archaeomagnetic research has been the construction of curves depicting secular variations in both direction and intensity. Following the construction of such curves, questions about the nature of the geomagnetic field and its secular variation can be addressed. Common research questions include the relative importance of the non-dipole and dipole components, westward drift, rate of dipole wobble, possible displacement of the dipole with respect to centre of the earth, and periodicity of the secular variation record. In many situations, when some indication of approximate age is available, archaeomagnetic direction results can be used to date baked features uncovered at archaeological sites with high precision. An indication of approximate age is necessary because geomagnetic direction is not a unique function of time.

On the whole, the accumulation of archaeomagnetic data has proceeded rather slowly. There are three main reasons for this: first, the difficulty of collecting adequate sets of oriented samples for investigation of directional changes; second, the time-consuming laboratory procedures for intensity determinations, and third, the relatively rare occurrence of suitable archaeological sequences with adequate chronologies. While lake sediments offer the attraction of long continuous records from a diversity of geographical locations (see Chapter 4), there are nevertheless several distinct advantages in using baked clays rather than lake sediments for secular variation studies. These include the following: (a) baked clays can yield absolute determinations of the ancient field intensity, (b) the NRM is generally stronger and more stable than that of lake sediments, (c) baked clays acquired their NRM almost instantaneously whereas the NRM of sediments is acquired over an extended period of time (see Chapter 1), (d) archaeological sites are often fairly well dated, either directly by radiocarbon (and occasionally by thermoluminescence) or by stylistic cross-dating. In contrast lake sediments are notoriously difficult to date (see Chapter 2).

This chapter presents the results of archaeomagnetic investigations carried out in many parts of the world. The underlying objectives of the editors have been (a) to include updated presentations of results which were originally published in journals which are not readily accessible and (b) to invite well established research workers in the field to make a contribution describing some of their most recent work.

3.2. BASIC TECHNIQUES FOR ARCHAEOINTENSITY DETERMINATION

By M.J.Aitken,
Oxford University, UK.

3.2.1. Thermal remagnetization

The thermoremanent magnetization (TRM) acquired by a sample of baked clay is proportional to the strength of the magnetic field in which it cooled down, at any rate for the fields up to $100\mu T$ (1 Oe) that are encountered on the Earth's surface. Consequently if a sample is thermally remagnetized in a laboratory field F^L, the ancient field is given by:

$$F^A = (M^A/M^L) \; F^L \tag{1}$$

where M^A is the moment that the sample acquired by cooling in the ancient field and M^L that acquired on laboratory remagnetization.

The fundamental difficulty in using this equation as it stands is that during the laboratory heating, mineralogical changes may have taken place. The effective sample that carried M^L may then have TRM properties different from the original sample that carried M^A, so that the equation is not valid because of a change in 'TRM capacity' – the TRM acquired in a given field. Such changes may be an actual chemical oxidation or reduction (for example, between magnetite and haematite) or merely an alteration in the condition of the magnetic grains, such as grain growth. There are other effects also that upset the validity of the equation 1. For instance in addition to the TRM acquired on cooling in antiquity, M^A may have a component of chemical magnetization (CRM) as well as a second component of TRM if the sample has been heated in antiquity. Because M^A is liable to be a composite of different types of magnetization it is customary to refer to it as NRM (natural remanent magnetization). Although for unbaked clay the major component of the NRM is TRM, the term TRM is often used in a restricted sense signifying thermal remagnetization in the laboratory.

All these effects, but most commonly mineralogical change on laboratory heating, mean that intensity determinations are likely to be unreliable unless tackled with a time–consuming series of measurements per sample, as well as with patience and a willingness to reject samples showing indications of the effects just mentioned. Most intensity techniques used for baked clays are variants of that developed by Thellier and Thellier (1959). More recently another method, incorporating alternating field demagnetization, has been developed by Shaw.

3.2.2. The Thellier technique and its variants

As was established by Thellier (1938) the magnetic grains that acquire a remanent magnetization on cooling through a certain temperature lose it at the same temperature on reheating – the blocking temperature (T_B). Within a ceramic sample there is usually such a variety of grains that the distribution of blocking temperatures is continuous from the Curie point downwards. Thus equation 1 can be used to determine values for F^A from a succession of laboratory reheatings to increasingly higher temperatures until the Curie point is reached. A given reheating temperature yields the value for F^A as recorded by the grains having blocking temperatures in the range up to the reheating temperature. Such stepwise remagnetization is the essential feature of the Thellier technique (including variants) and though developed more than two decades ago is still the method of accepted reliability. This reliability stems from the fact that the method allows detection of the reheating temperature at which mineralogical change interferes with the validity of the equation 1, data from higher temperatures then being rejected.

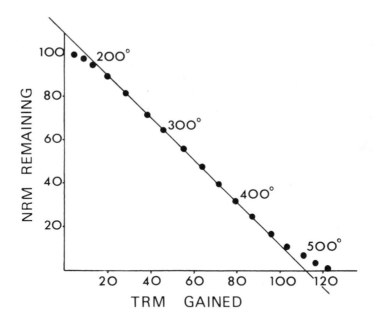

Figure 3-1:   Arai plot of measurements made in the Thellier thermal demagnetization technique. The numbers indicate the temperatures(°C) used in the heating steps.

The ratio between ancient field (F^A) and the laboratory oven field $(F)^L$ is derived from the slope of the best fitting line drawn through the points

It is usual to make an Arai plot of the results as in Figure 3-1: NRM remaining after a given temperature step versus TRM gained by cooling in laboratory field from the temperature of the step. Denoting $NRM(T_1)$ and $NRM(T_2)$ for two successive steps at temperatures T_1 and T_2, likewise for TRM, we have the differential form of equation 1, i. e.

$$F^A/F^L = NRM(lost)/NRM(gained) = [NRM(T_1) - NRM(T_2)]/[TRM(T_2) - TRM(T_1)] \quad (2)$$

Except in the method of Kono (see below), there are two heatings and coolings at each temperature step in the usual version of the Thellier method now employed – the zero field variant of Coe (1967). The first of these is in zero field, and the second in F^L; if the moment measured (always at room temperature) after each of these coolings is denoted by $M^o(T)$ and $M(T)$ respectively then:

$$NRM(T) = M^o(T) \quad (3)$$

and

$$TRM(T) = M^L(T) - M^o(T) \quad (4)$$

It is assumed that M^L and M^o are in the same direction, otherwise vector subtraction is used. In the original Thellier method, zero field was not employed but between each pair of heatings at a given temperature the orientation of the sample was reversed, the NRM and TRM being obtained by appropriate addition and subtraction of the moments measured after each cooling.

In the Kono variant of the Thellier technique (Kono and Ueno 1977), there is only one heating and cooling at each temperature step. The sample is positioned so that the TRM acquired is at 90° to the NRM; in this way the NRM and TRM can be evaluated by vector subtraction without the need for a second heating. However, while the saving in time is substantial, this method is at the mercy of intrinsic anisotropy (see below), and in pottery this is usually strong enough to cause serious upset to accurate results. In extreme cases the result can be in error by a factor of 2.

In a version of the Thellier technique developed for archaeomagnetic use of a SQUID cryogenic magnetometer (Walton 1977), the field used for remagnetization, F , is set close to the expected value of F^L on the basis of earlier samples. This has a number of important advantages, the first of which is that any error due to the magnetization not being strictly proportional to the applied field is made negligible. A second is that the Arai plot can be modified so as to allow enlargement of scale; the NRM is plotted against M as in Figure 3-2. In the version now used at the Oxford Research Laboratory for Archaeology (Aitken et al. 1981), the order of the two heatings at each temperature step is reversed, i. e. the in-field heating is done before the zero field heating. This has two immediate advantages: first, it avoids the possibility of error due to zero field memory effect (Fox, 1979); second, it avoids the type of error due to the presence of multidomain grains discussed by Levi (1977).

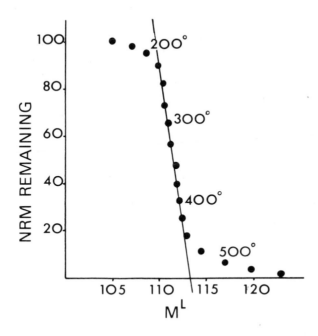

Figure 3-2: Modified Arai plot for the data of Figure MAITF1
allowing expanded horizontal scale.

The latter shows the value pf M^L, the observed magnetization
(TRM gained plus NRM remaining) after each temperature step.

If $F^L=F^A$ the best fit line will be vertical; in practice the ratio F^A/F^L
is derived from the slope of the line,
F^L having been chosen to be close to the expected F^A

Whatever variant is used, the criterion for an intensity determination to be judged reliable is that the Arai plot shows a linear portion extending over a substantial fraction of the demagnetization range. In the low temperature region, up to 150 or 200°C, there is liable to be curvature due to viscous components. At the high temperature end there is likely to be deviation from linearity due to mineralogical alteration during the laboratory heating. In Figures 3-1 and 3-2, which represent the same data, there has been an enhancement of TRM capacity; in other samples the curvature might be in the opposite direction, due to a diminution of TRM capacity. In a substantial proportion of samples (more than 50% for some types) the onset of curvature is so early that there is no real linear portion; a reliable estimate for the ancient intensity is then not possible and the temptation to make the best of a bad job must be firmly resisted.

So far there has not been any comprehensive investigation into the mineralogical changes responsible for the non-linearity. It does not seem to be a straightforward matter of the laboratory heating taking place in an atmosphere (usually air) that is different from the atmosphere in the ancient kiln. It seems likely that a more common cause of non-linearity is due to hydration of magnetic oxides during burial, with dehydration occurring during the laboratory heating (some relevant discussions have been given by Barbetti et al. 1977 and by Sigalas et al. 1978). Non-linearity can also result from there being different degrees of cooling-rate effect (see below) for grains blocking in different temperature regions; likewise in respect of intrinsic anisotropy. With these two effects a reliable result can be obtained nevertheless, in principle for all cases, in practice only in some, by making careful correction (in respect of anisotropy see Aitken et al, 1981, for instance).

The criterion of linearity in the Arai plot is of dominant importance, but it is not necessarily sufficient to ensure reliability. For instance, if a chemical magnetization is carried by the sample and it has the same spectrum of blocking temperatures as the TRM, then although this will not upset the linearity it will usually give rise to an erroneous answer for F^A. The tendency of a sample to acquire chemical magnetization is often revealed by irregularity of direction as the demagnetization of the NRM proceeds and constant direction is a useful additional criterion for reliability. Another check sometimes employed, and indeed it was incorporated in the original Thellier procedure, is some form of monitoring of TRM capacity. For instance, after all regular steps of the remagnetization have been completed a further step is added in which the TRM from, say, 300°C is measured again and compared with the TRM from 300°C that was measured during the regular steps. This suffers from the drawback that there are few samples for which mineral alteration does not occur by some temperature and this type of monitor only indicates whether or not it has occurred during the whole sequence of steps. An alternative 'running monitor' has been used in the work described in Section 3.8 but this suffers from the defect that it monitors only the grains with blocking temperatures in the lower part of the demagnetization range. Measurement of alternating field susceptibility is another form of monitor, but of course the grains concerned are even more remote from the grains that are carrying the intensity information.

The new technique being developed by Walton (see Walton 1981 and Section 3.3) is an attempt to avoid the need for laboratory heating much above 200°C (thereby avoiding interference from mineralogical alteration) and at the same time take account of cooling-rate dependence of TRM. Essentially it involves heating the sample to a fixed temperature and holding it there in field, F, while observing any slight changes in magnetization that occur. An understanding of its basis can be obtained from the mathematical description given in Section 3.3 but as an introduction the reader may find it useful to envisage it by ignoring, in the first place at any rate, the existence of any dependence of TRM on cooling rate. If a sample is rapidly heated to a fixed temperature, T, and held in zero field then after the abrupt drop in magnetization due to unblocking of grains having blocking temperatures lower than T, and the fall in intrinsic magnetization of the grains remaining blocked, there will be a small, slow decrease due to viscous demagnetization, i.e. the unblocking of grains having relaxation times (at T) that are comparable with the duration of the experiment. On the other hand if instead of zero field the sample is held at T in a field stronger than the ancient field there will be an abrupt rise followed by a small slow increase.

If $F^L = F^A$ then there will be an abrupt fall due to the drop of intrinsic magnetization, but no subsequent change. Hence by finding the value of F for which this condition pertains, the ancient field can be determined. In practice this simple explanation is complicated by cooling rate effects and the reader is referred to the details given in Section 3.3. For the method to be feasible there are severe demands on the sensitivity and stability of the detector used and a high grade SQUID magnetometer is essential. Furthermore facilities are required for measuring while the sample is in the field and while it is hot.

3.2.3. The Shaw technique: thermal remagnetization with ARM monitoring

In this technique (Shaw, 1974) alternating field (AF) demagnetization is employed instead of thermal demagnetization, while retaining thermal remagnetization (but in a single step, from above the Curie point). By this means the ratio of NRM to TRM can be obtained for different parts of the coercivity spectrum. In addition, after AF demagnetization of the NRM, and before any heating takes place, the sample is given an anhysteritic remanent magnetization (ARM) – by subjecting it to a strong AF with a weak DC field applied. The coercivity spectrum of this ARM is obtained by AF demagnetization. The TRM is then emplaced and AF demagnetized so as to be able to compare its coercivity spectrum with that of the NRM. Finally a second ARM is emplaced and AF demagnetized. Comparison of the coercivity spectrum of the second ARM with the first ARM is used to indicate which portions of the spectrum have been affected by mineral alteration during emplacement of the TRM. These portions are excluded from the comparison of the NRM and TRM spectra from which the ratio of NRM to TRM is obtained for evaluation of the ancient intensity.

The steps are as follows:

A. Measure NRM and repeat this measurement after AF demagnetizations using successively increasing maximum fields (typically up to 150mT).

B. Emplace ARM1 using the same maximum AF and a DC field of the order of 0.1mT.

C. Measure ARM1 and repeat measurement after each step of AF demagnetization (as in A).

D. Emplace TRM by heating to above the Curie point (say 700°C) and cooling in known weak field – usually chosen close to expected value of ancient field.

E. Measure TRM and repeat measurement after each step of AF demagnetization.

F. Emplace ARM2 using same conditions as in B.

G. Measure ARM2 etc as in C.

Comparison of G with C indicates which parts of the coercivity spectrum have been altered by the heating. These parts are then discarded from A and E before using the ratio of NRM to TRM to evaluate the ancient intensity. As proposed by Shaw the origin is taken as a data point in the NRM/TRM plot and with the samples measured by him it was the higher coercivity part of the spectrum that showed no alteration effects. In the variant developed by Kono (1978) the incremental ratio of ARM1/ARM2 is used not to discard portions of the NRM/TRM plot but to correct them for the effects of alteration, i.e. it is assumed that such effects are quantitatively the same in both ARM and TRM. This approach has been exhaustively investigated by Thomas (1981) with encouraging indications for its validity.

In some samples a substantial part of the magnetization is carried by grains having coercivities beyond the range of the highest AF available. In such cases the decision whether or not to use the origin of the NRM/TRM plot as a data point can have a significant effect on the value obtained for the ancient intensity. One reason why it might be wrong to use the origin is that in the ancient firing the temperature did not reach the highest Curie point of all the minerals present, leaving some high coercivity grains unmagnetized in the NRM whereas they are included in the TRM. Another reason is that the grains beyond the range of the AF are unchecked by the ARM monitoring: use of the origin assumes those grains to be invulnerable to alteration and this may be erroneous.

Another difficulty with the Shaw technique is that, unlike the Thellier method, it is not possible to separate out the component of magnetization acquired from a partial heating that took place subsequent to the original firing. However, because it can separate 'good' minerals from 'bad' ones, it can be powerful in getting results from samples for which the Thellier method fails. Even for samples which pass the Thellier reliability tests it is desirable to use also the Shaw technique on some of them as confirmation that no unforeseen interferences have crept in. A high degree of confidence can be placed in results for which the two techniques are in agreement. Such confidence is desirable before results are used for geophysical interpretation.

3.2.4. Dependence of TRM on cooling rate

For a group of similar grains, the slower the cooling the lower the temperature at, which effective blocking occurs. This may be understood qualitatively by considering that if grains which during cooling blocked at, say 100°C, are held for a long time at room temperature in a new field, their magnetization will gradually change to that corresponding to the new field. This is the familiar viscosity effect that was well established by Thellier in the early days.

For single domain grains of given characteristics cooling in a fixed field, the lower the temperature at which blocking occurs the stronger the TRM acquired. This is because the fractional alignment increases as the temperature decreases; for a weak field the fractional alignment is approximately proportional to (J_s/T) where J_s is the intrinsic magnetization of the grains at the temperature concerned and T, is the absolute temperature. Thus the alignment increases with decreasing T not only because of less thermal agitation, but also because of the magnetic moment of the individual grains becoming stronger. Theoretical discussions of the effect have been given by Dodson and McClelland-Brown (1980) and by Walton (1980). Experimental observation has been reported by Fox and Aitken (1980).

It appears that the effect is not a strong one. As reported in Section 3.8, for some fifty pottery samples of various types there is an average difference of about 3% between cooling over two days, such as may have happened for ancient pottery, and cooling over four hours, as is typical for many intensity determinations. It is only with rapid cooling, such as is liable to be employed for the small samples used with SQUID magnetometers, that the effect becomes significant; there is a difference of about 11% between cooling over five minutes and cooling over two days.

Theoretically, grains that block near their Curie point are likely to show a smaller diminution of blocking temperature for slow cooling than for grains blocking far from the Curie point. On the other hand for a given change of blocking temperature the change in fractional alignment will be greatest nearest to the Curie point due to the rapid variation there of J_s. Thus, even if the TRM of a sample is carried wholly by grains of a given mineral, variation of cooling rate effect between different blocking temperature ranges may give rise to non-linearity in the Arai plot of Thellier determinations. Such non-linearity can be corrected for by making a subsequent investigation of the cooling-rate behaviour of the sample. The fact that linear plots can be obtained without doing this is indicative that the effect is a rather slight one in general. Another related cause of non-linearity could be irregularity in the ancient cooling rate - due for instance to opening up the kiln prematurely - but the effect of this too is likely to be slight. An advantage of the new Walton technique (Section 3.3) is that correction for cooling rate effects is intrinsic to the procedure.

For multi-domain grains it has been suggested by Stacey (1963) that the cooling rate effect is likely to be in the opposite direction, i.e. a stronger TRM is produced by rapid cooling. This is on the basis that the domain configuration, that is determined by the demagnetizing field within the grain, should almost nullify the external field and the higher the temperature the greater the effective alignment represented by this configuration. As with single domain grains the effective blocking temperature decreases for slower cooling.

3.2.5. Intrinsic anisotropy

This is a dominant effect with pottery, but less so for bricks and tiles. If it is not taken into account the intensity evaluated for pottery can be in error by up to a factor of 2 in extreme cases, but for bricks and tiles only by about 10%. It appears to arise through preferential alignment of the magnetic grains during fabrication, with further enhancement perhaps occurring during firing. Turning on the potter's wheel or pressing into a mould have both been shown to produce such alignment. A similar effect is seen in X-ray diffraction studies, presumably arising from alignment of the clay platelets (Pentecost and Wright 1964).

Whatever the mechanism, it appears that with pottery, and to a lesser degree with bricks and tiles, the TRM emplaced by a given field depends on the direction of the field with respect to an 'easy plane' of magnetization (Rogers et al. 1979). For pottery this easy plane is usually, but not always, parallel to the surface of the pottery (i.e. it is a different effect from the shape anisotropy that occurs with strongly magnetized samples through internal demagnetizing fields). The TRM emplaced by a given field parallel to the easy plane is higher than that emplaced by the same field when perpendicular to the easy plane. The ratio, a, between the two can be as high as 2 for some types of pottery (e.g. Samian ware) though a ratio of 1.5 is more typical. Within the easy plane there may also be variations, though these are usually much weaker.

To avoid error due to this effect the TRM needs to be emplaced in the same direction as the NRM. If it is not, then correction needs to be made (such as described in Aitken et al, 1981). The precision with which the two magnetizations need to be aligned depends of course, not only on the anisotropy ratio, a, but also on the orientation of the NRM with respect to the easy plane. From experience at the Oxford Laboratory with more than a thousand pottery samples, it seems that for alignment to within 10° it is unusual to get a correction factor different from unity by more than 5%, and that for alignment to within 25° it is unlikely to be different by more than 10%. A complication in getting good alignment is that as a result of the anisotropy the direction of the TRM is liable to be significantly different to that of the field that produced it. For instance for a sample having a = 1.5, the two directions can differ by as much as 11° for some orientations with respect to the easy plane.

3.2.6. Internal demagnetizing fields (shape anisotropy)

For a thin slab of pottery magnetized perpendicular to its plane there is an internal demagnetizing field of magnitude 1.5σ mT where σ is the specific magnetization in Am^2kg^{-1} . Thus for samples having a specific magnetization of $10^{-3}Am^2kg^{-1}$, or more, there is risk that the actual field seen by the sample will be lower than the external field by upwards of 5%. This is an upper limit to the effect for two reasons: first that the magnetization may not be perpendicular to the plane, and secondly that at elevated temperature, due to the decrease of J_s, the specific magnetization is less. Hence grains blocking at high temperature will be less affected than those blocking at a lower temperature. This will tend to give a curved Arai plot and, as long as the criterion of an extensive linear portion is stringently enforced, a sample for which the effect is substantial will be rejected.

Experimental observation of the effects of internal demagnetizing fields in respect of ARM have been reported by Schmidbauer and Veitch (1980).

Distorting fields may also be produced by adjacent pottery or iron tools or by the kiln structure. Thus, although the sample itself may be too weakly magnetized for its own demagnetizing field to be significant, the possibility of such external distortion cannot be ruled out. This external distortion can be both positive and negative. The material producing these external distorting fields does not necessarily cool in step with the sample itself, so there may be a linear Arai plot despite a significant error in the ancient intensity that is evaluated. It is unlikely that all samples of a specific date range will be affected, or affected to the same degree. Such effects may be the cause of otherwise unexplained scatter in such a group. Hence the importance of multiple sampling so as to obtain warning that all is not well.

3.2.7. Concluding remarks

As we have just seen, an erroneous result cannot be absolutely ruled out even with stringent application of reliability criteria. There is also the possibility of a sample being wrongly dated, a residual from an earlier period, or having been fully refired at a later time. Hence internal consistency and coherence in a set of results is a further important criterion of reliability; it is prudent to resist the temptation to draw exciting geophysical conclusions from sparse data, particularly in respect of determinations reported without explicit statement of the reliability criteria employed.

3.3. THE RELIABILITY OF ANCIENT INTENSITIES OBTAINED BY THERMAL DEMAGNETIZATION

By Derek Walton,
McMaster University, Canada

3.3.1. The problem

The intensity of the Earth's magnetic field is easy to measure directly but is notoriously difficult to obtain by indirect methods. Most intensities available at the present time, and those reported in this contribution, have been obtained using thermal techniques derived from the pioneering work of Thellier and Thellier (1959). Since mineral alteration on heating would obviously invalidate the results, the Thelliers devised a stringent test to ensure that no alteration had occurred in any sample from which they obtained an intensity. Few samples can pass this test, and in an attempt to reduce the rejection rate, a number of techniques have since been devised which attempt to isolate a sub-set of grains which has not altered, and obtained the intensity from them. To do this, one must obviously identify those grains which have altered and ignore them. In general, this identification can be done in at least two ways: by measuring their susceptibility, or by measuring their coercivity. The second method forms the basis of the ARM method introduced by Shaw (1974) and its variants.

Most other techniques do not measure the susceptibility directly, but arguing that the ancient intensity must have remained constant, look for changes as the demagnetization temperature is raised. If a Thellier plot is constructed of NRM removed, against PTRM introduced, a non-linearity is taken to signal the onset of alteration. Then, it is hoped that the straight line through the points up to the temperature at which the slope changed should yield the ancient intensity.

The low temperature points often reveal non-linearities, and are ignored, a "best" straight line through the remainder being used to determine the ancient intensity. An example is shown in Figure 3-3. This procedure can only be justified if the low temperature points are affected either by viscosity, by a temperature dependent cooling rate correction, or some other reversible effect.

Unfortunately, it will be shown that alteration can occur, and a good straight line will still be obtained in a Thellier plot leading to intensities which are completely wrong. Therefore, the fact that a straight line is obtained is, by itself, no guarantee that mineral alteration has not occurred. Thus, unless the susceptibility has been shown to have remained unchanged, results obtained using thermal demagnetization techniques alone, cannot be trusted to yield results which are even approximately reliable.

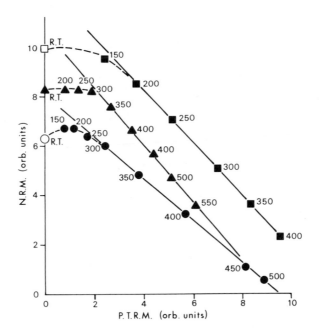

Figure 3-3:   A plot of the original moment remaining after
heating to the temperature shown against moment acquired
on cooling in the laboratory field, showing low temperature
non-linearities possible due to 'viscous demagnetization'
in samples of Attic Pottery from three different sources.
The time spent at each temperature was ~10s.
Viscous demagnetization is due to random re-orientation
of the original sherd in the Earth's magnetic field.

In this contribution, the detection of mineral alteration will be discussed in detail
and a technique suggested to partially correct for its effect.   The effects of viscosity
and cooling rate on the apparent ancient intensity obtained using thermal methods will
also be reviewed.   Procedures will be suggested for identifying and correcting for
them.

3.3.2. Mineral alteration

Mineral alteration can be detected by measuring the susceptibility.   Thus, if the
change in magnetization ΔM, produced by a change in field ΔH, is measured, for our
purposes mineral alteration has NOT taken place if the susceptibility $k=\Delta M/\Delta H$ remains
unchanged.

It is often assumed that if the Thellier plot of NRM against PTRM yields a straight line, the slope of this line leads to a reliable value for the ancient intensity. That is, k has not changed. The effects of a changing cooling rate correction and viscosity do not invalidate this test because they can be expected to be confined to a limited temperature range, and the non-linearity introduced should be obvious (see Figure 3-3). Unfortunately, this may not necessarily be true of mineral alteration. In general, it introduces changes in the susceptibility that increase with increasing temperature. When these are of the order of the experimental error, thay lead to an error in the slope of the NRM/PTRM plot used to obtain the ancient intensity. It is difficult to estimate a maximum value for this error, but if the points used in obtaining the intensity only cover a small loss of the NRM, it can be large. In fact, if the detection of alteration depends on observing a non-linearity, it may escape detection altogether, with the altered points lying on a perfectly adequate straight line. In other words, $\Delta k / \Delta T$ which can be expected to increase with temperature, will lead to a linear change in M but will be the wrong M. A particularly extreme artificial situation is illustrated for a hypothetical sample in Figure 3-4 which were obtained by starting with data obtained from a sample from an Attic sherd dated between 575 and 540 BC (the solid circles). Then the effect of mineral alteration on these points was calculated as follows: assuming (i) that the sample was maintained at each temperature for the same length of time and (ii) that the change in moment due to alteration at each temperature was thermally activated,

$$\Delta M = Ae^{-T'/T} \tag{5}$$

where A is the maximum change in susceptibility if all grains which are capable of altering do so and T' is equal to the activation energy divided by Boltzmann's constant.

In obtaining Figure 3-4, it has been assumed that the result of alteration is to produce a non-magnetic mineral, and that A is equal to the initial moment. Thus, if alteration is complete, the moment of the sample disappears. (It should again be emphasized that this model is entirely hypothetical.)

The total change in moment at each temperature, then, will be the accumulated change at the preceding temperatures:

$$\Delta M_{total} = -\sum Ae^{T'/T}{}_i \tag{6}$$

The solid circles represent the original data. The crosses represent results obtained using equation (6) for T' + 1773°K, and the open circles for $T' = 1273$°K. If it is assumed that the initial data yielded a correct value for the ancient field (which in itself is not a defensible assumption), then if $T' = 1733$°K, the field deduced would be twice the correct value. But if $T' = 1273$°K, it would be almost six times the correct value.

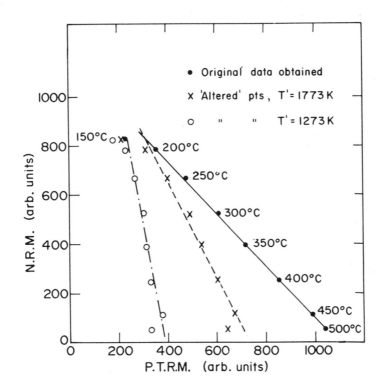

Figure 3-4: A plot of NRM against PTRM for a hypothetical sample which
has undergone alteration. It is assumed that alteration results in
a lower moment, and that it is thermally activated. T' is the
activation energy divided by Boltzmann's constant.
For the sake of simplicity the NRM has not been demagnetized.
This corresponds to a situation in which only those grains which
have become unblocked have altered, which in turn corresponds to
assuming that the smaller grains alter first. On the other hand, it
is also possible to demagnetize the NRM. If this is done, the "altered"
points move down slightly, for T' = 1773°K. For T'= 1273°K the
changes are too radical, and the sample is effectively completely
demagnetized at too low a temperature

Since the 'altered' points lie on quite reasonable straight lines, it must be
concluded that the linearity of a Thellier plot is not a sufficient condition for the
reliability of the ancient intensity. This reliability can only be established by
measuring the susceptibility. In fact, this is precisely what the complete Thellier
method does. However, it would be desirable to monitor k as the sample is
demagnetized. In this way data obtained prior to alteration could be preserved.

Unfortunately, if k is measured, say at room temperature, it will not necessarily
include the effect of grains which have altered but are blocked at room temperature.

Thus, it is important to measure k at high temperature.

If k is measured at the highest temperature which the sample has reached during the measurement process, it will obviously reveal nothing, since the susceptibility may already include that of grains which have altered. On the other hand, if the sample is returned to the preceding temperature, then alteration will be signalled by a change in k at that temperature.

This suggests a way of partially correcting for alteration: assume that the highest temperature reached at this point in the measuring process is T. On returning to the preceding temperature, the susceptibility at $T - \Delta T$ (where ΔT is the temperature interval between successive measurements) is found to have changed from k to $k + \Delta k$. If this process is continued, a plot of $\Delta k / \Delta T$ can be made. Integration will then yield k at any temperature, and the measured values can be corrected.

3.3.3. Cooling Rate

It is possible for the effective cooling rate to depend on the temperature difference. This is simply due to the finite time required for the sample to begin to cool after its surroundings have cooled. This can result in non-linearities if the cooling rate correction is substantial. The solution is to employ a long cooling time in order to minimize the correction, or to actually estimate the ancient cooling rate and correct at each temperature. A method of estimating the ancient cooling rate (Walton, 1982) is now described.

The analysis uses results derived in a previous paper (Walton, 1980) which will be referred to as W. As derived in W, the moment produced on cooling is:

$$M = H_A \, _{v1}\int^{v2} (J_B J / 3kT_B) \, v^2 N(v) \, dv \qquad (7)$$

where H_A is the ancient field, J_B the saturation magnetization at the blocking temperature T_B, v is the grain volume, $N(v)$ the grain size distribution and v_2 and v_1 the volumes of grains blocked at the higher temperature and lower temperature respectively.

As shown in W, the quantities J_B and T_B depend on the rate of cooling. Let a variable α be defined as:

$$\alpha = d(K/T)/dT \qquad (8)$$

where K is the anisotropy constant. The moment is then proportional to $[\ln(cK/v\alpha)]^n$ where c is a frequency factor.

It is convenient to define a 'cooling time constant' τ:

$$\tau = k/v\alpha \tag{9}$$

In general, of course, τ is temperature dependent. In particular if the sample cools according to Newton's Law of cooling, τ decreases as the temperature increases. The coefficient n results from approximating the temperature dependence of J as:

$$J = J_0(1-T/T_c)^n \tag{10}$$

For magnetite $n \sim 0.5$ while for haematite $n \sim 0.25$.

The thermal techniques for obtaining the ancient field pioneered by Thellier, involve comparing a moment introduced on cooling in the laboratory field with that removed on heating to successively higher temperatures. A knowledge of the ancient cooling rate is necessary for an accurate determination of the ancient field intensity. This knowledge need only be approximate because the function $[\ln(c\tau)]^n$ is only weakly dependent on τ.

3.3.4. A new technique for obtaining H_A and τ

Assume that the sample is heated rapidly from a temperature T_0 to a higher temperature T. When the sample reaches T part of the magnetic moment will reside in grains whose blocking temperatures are both above and below T. Those grains whose blocking temperatures are lowest will first come to equilibrium with the applied magnetic field. But the approach to equilibrium of grains whose blocking temperatures are less than T will result in a loss of moment. The result is a decrease in moment followed by an increase when the distributions of those grains with blocking temperatures higher than T begin to reach equilibrium.

The rate of change of the moment with $\ln(ct)$ is given in W as

$$\partial M/\partial[\ln(ct)] = (\pi\sqrt{2}/e)(n_0-n_{eq})J(kT/K)^2[\ln(ct)][N(kT/K)\ln(ct)] \tag{11}$$

n_0 is determined by the moment distribution produced on cooling and is

$$n_0 = (J_B v H_A)/(3kT_B) \tag{12}$$

and

$$n_{eq} = (JvH_L)/(3kT) \tag{13}$$

where H_L is the laboratory field.

As the sample is held at the temperature T, initially smaller grains will come to equilibrium followed by grains of increasingly larger volume. Initially, then, the distribution of the grains with low blocking temperature come to equilibrium followed by grains which blocked at higher temperatures. Since T_B increases and J_B decreases as v increases, n_0 will decrease with time. On the other hand, n_{eq} increases with time as v increases.

Now $(\partial M/\partial \ln(ct))$ will be zero when $n_{eq} = n_0$, i.e. when

$$H_A/H_I = JT_B/J_BT \tag{14}$$

$$K_Bv/3kT_B = \ln(c\tau) \text{ and } Kv/3kT = \ln(ct) \tag{15}$$

where t is the time at which $(\partial M)/(\partial \ln(ct)) = 0$ and K_B is the anisotropy constant evaluated at the blocking temperature.

so if

$$K = aJ^p \tag{16}$$

$$J/J_B = [(T \ln(ct))/(T_B\ln(ct))]^{1/p} \tag{17}$$

Using equation 10:

$$J_B/J = (1-T_B/T_c)/(1-T/T_c)^{\,n} = 1 + (T-T_B)/(T_c-T)]^{\,n} \tag{18}$$

Bearing in mind that T_B is close to T this equation may be expanded. After some algebra equation 14 becomes

$$H_A\ln(c\tau_A) = H_L\ln(ct)\left[1+([(p-1)n]/[(pn-1)+T/T_c])\,(\ln(t/\tau_A)/\ln c(\tau_A))\right] \tag{19}$$

where τ_A is the ancient cooling time constant.

An example of this behaviour is shown in Figure 3-5, which displays the moment as a function of time for a sample of Attic pottery. The data were obtained with a highly sensitive magnetometer capable of obtaining moments, while the sample is held at an elevated temperature, with an accuracy of better than 0.1%.

Since there are two values for H and two values of t, an approximate value for c may be deduced, if the second term in the brackets in equation 19 is neglected. c is approximately $7 \times 10^7 s^{-1}$.

It is, of course, impossible to obtain the ancient field unless τ_A is known, although it need not be known precisely because it enters equation 19 only as $\ln(c\tau_a)$. This requires another relation which can be obtained from the dependence of the moment on the cooling rate since the moment gain on cooling is proportional to $[\ln(c\tau_L)]^n$. For example: on further heating to 250°C the sample referred to above revealed a decrease in moment when cooled to 190°C to that observed after it had been cooled from 195°C. On decreasing the cooling rate by an order of magnitude (increasing τ_L by a factor of 10) half the lost moment was recovered. Thus a two order of magnitude increase in τ_L would be required to completely restore the moment lost between

195°C and 250°C. As shown in W the moment introduced on cooling is proportional to $H_L[\ln(c\tau_L)]^n$. Hence,

$$H_A[\ln(c\tau_A)]^n = H_L[\ln(c\tau_L')]^n \qquad (20)$$

where τ_L' is that cooling time constant that would be required to completely restore the moment lost by heating to 250°C.

Using equation 19, and taking n = 0.5 which is appropriate for magnetite, τ_A ~ 100s. This value appears to be rather low. Other samples of Attic pottery yielded values about an order of magnitude higher. Typically τ_A appears to be on the order of 1000s for this pottery.

H_A for this sample is 46μT. The modified Thellier method used for a previous determination yielded H_A = 63μT for a sample from the same sherd. The difference is rather more than can be accounted for by differences in cooling rate. It is likely that alteration affected the old result.

3.3.5. Summary

The technique proposed here consists of holding the sample at an elevated temperature and monitoring the change in its moment with time. If the laboratory field is within about 10% of the ancient field the moment will show a characteristic decrease followed by an increase. The time at which the inflexion occurs yields the product $H_A\ln(c\tau_A)$. The values for $\ln(c\tau_A)$ and H_A can be separated by obtaining the laboratory cooling rate for which no change in moment occurs on cooling to a fixed temperature in the chosen laboratory field.

3.3.6. Viscosity

A typical low temperature non-linearity which may be due to viscosity is shown in Figure 3-3. It is postulated that the sample has undergone a 'Viscous Demagnetization'.

The viscosity can be expected to disappear after a time at a temperature T given in W, by

$$([TJ^2(T_R)]/[T_RJ^2(T)])^{2+r} = (\ln(ct_A)/\ln(ct_T))^{3+r} \qquad (21)$$

where c is a rate constant ~$10^8 s^{-1}$, and t_A is the age of the sample, J is the saturation magnetization, T_R is room temperature ~300°K, and r is an exponent dependent on the grain size distribution. The details of the derivation of this expression are given in W.

Some idea of the values of t_T can be obtained if it is assumed that the carriers of the magnetization are monodomain grains of magnetite. The dependence of J on temperature is approximated by

$$J = J_o \; 1 - (T/T_c)^{0.5} \qquad\qquad\qquad (22)$$

Then, if $c = 10^8 s^{-1}$, and $t_a \sim 3000$ yr, and if t_T is 2 mins, the temperature T required is 270°C. If t_T is 20min, it is still 240°C. Some indication that viscous moments persist to such high temperatures was obtained by Wilson and Smith (1968).

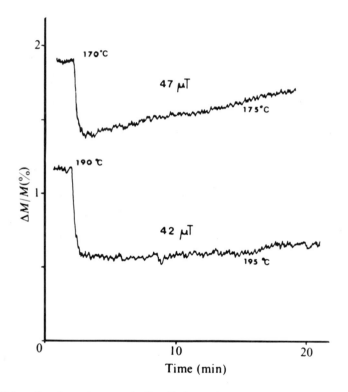

Figure 3–5:   The change in moment, M, with time, of a sample of Attic pottery on
heating by 5°C from two temperatures at two different fields.
The zero level of both curves has been shifted by an arbitrary amount.

Consider a much shorter $t_A \sim 3yr$. The viscous moment produced will survive to a temperature ~190°C for a heating time of 2 mins. Now it is conceivable that many archaeomagnetic samples are reoriented at intervals on the order of 3 to 30 years. This would have the effect of randomizing the orientations of all those grains whose relaxation times at room temperature are less than this. These grains, in effect, have been demagnetized. A glance at Figure 3-3 reveals that, if this were the case, it would account for the virtually horizontal line up to 200-250°C.

The conventional way of detecting a viscous moment is to look for a change in the direction of the magnetic vector. Unfortunately, this will not reveal the effect of frequent re-orientation of the material from which the sample was taken. This 'viscous demagnetization' would of course result in the magnetic vector being in the same direction as the original. Of course an ancient viscous component will remain which will, in general, have a different direction. The magnitude of this component will be equal to the moment removed between the temperature at which the effect of viscous demagnetization disappears and that at which the last remaining viscous component disappears. If this is small, the change in direction may be difficult to detect.

Another way to detect the disappearance of viscous moments is to monitor the change in moment with time. The moment is proportional to $\ln(t)$ [actually it is proportional to $\ln(ct)$, but c is so large that the power law is difficult to observe] so a plot of M against $\ln(t)$ will reveal a change in slope when the viscous moment disappears.

3.3.7. Discussion

To summarize, non-linearities in the low temperature region can occur for at least three reasons: cooling-rate effects, viscosity, and alteration. The correction for a faster cooling rate in the laboratory becomes appreciable with small samples which cool in a time on the order of a few minutes.

3.4. MINERALOGICAL PROBLEMS

By Ruth Thomas,
University of Edinburgh, UK.

3.4.1. Introduction

In their earliest forms, archaeointensity methods were designed to reject results from samples which did not behave 'ideally'. Ideal behaviour during archaeointensity determination occurs if the NRM has remained unchanged since its original acquisition and the laboratory TRM is acquired by the same mineral grains and in the same way as the NRM. Although archaeointensity data may be rejected if, for example, the sample has been poorly fired or refired after the original firing, it is non-ideal behaviour of magnetic minerals that is the greatest single cause of rejection. As the acquisition of any set of archaeointensity measurements is a lengthy process it is important that the rejection rate of the data be as low as possible. This rejection rate can be decreased, either by pre-selection of samples and avoiding samples likely to show non-ideal mineralogical behaviour or by devising a method which makes use of previously rejected data. Obviously the second option is much the better , as it opens the way to the use of many more archaeological samples in archaeointensity determinations.

It is important that, if archaeointensity data are obtained from results which do not conform to the rigid guidelines of methods accepting only ideal results, there should be some way of measuring the degree of their deviation from the ideal and that some objective criteria of their reliability be stated. In fact, a method and the interpretation of the data should be such that any person using them would obtain the same result and assign that result the same level of reliability.

3.4.2. Non-ideal mineralogical behaviour

3.4.2.1. Chemical alteration before TRM acquisition

Weathering, which includes all physical and chemical changes in the sample between the time of the original firing and the sampling, is important in archaeointensity work because if it alters the magnetic minerals , it invalidates the basic assumption that the originally measured remanence represents an unchanged NRM.

Weathering products of magnetic minerals which have relevance to archaeointensity determination are goethite ($\alpha FeOOH$) and lepidocrocite ($\gamma FeOOH$). Both these oxyhydroxides are antiferromagnetic and are not therefore important remanence carriers, although goethite can be weakly ferromagnetic. They can be introduced to the sample from external sources or can be formed by alteration of oxides produced during the original firing. A chemical remanence (CRM) may be produced if the resulting oxyhydroxides dehydrate and the oxide grows through the superparmagnetic single domain critical grain size and is blocked in the direction of the ambient field at that time.

When a sample containing oxyhydroxides is refired in the laboratory the oxyhydroxides undergo the following reaction :

$$\alpha \text{ FeOOH} \xrightarrow{\text{300-350°C}} \alpha \text{ Fe}_2\text{O}_3$$
(goethite)   (haematite)

$$\gamma \text{ FeOOH} \xrightarrow{\text{300-325°C}} \gamma \text{ Fe}_2\text{O}_3 \quad + \quad \alpha \text{ Fe}_2\text{O}_3$$
(lepidocrocite)   (maghemite)   (haematite)

3.4.2.2 Stress and annealing

One possible form of non-ideal mineralogical behaviour which has not been studied in great detail is the effect of stress on the magnetic properties of minerals. Domain-wall movements are influenced by magnetostatic and stress effects.

Stress can be caused by crystal defects and dislocations. It could be present despite the original firing of the pottery if the annealing during that firing was not completed, if the minerals were in a state of partial chemical alteration or if weathering had occurred. It is possible to produce structural defects if a critical cooling rate is exceeded. Lowrie and Fuller (1969) have calculated the critical cooling rate for magnetite to be 30°C/s. This rate would only be exceeded if the pottery were quenched, which is unlikely. Stress can also be produced by drilling, cutting or removal of a sherd from a whole pot. Stress induced by physical or chemical means could be annealed out during the laboratory firing. Lowrie and Fuller (1969) have demonstrated the dependence of IRM, TRM, coercive force and AF demagnetization in magnetite on defect structure. They correlate changes in coercive force and other magnetic parameters with the type of annealing they believe to be occurring at certain temperatures. The IRM and coercive force decrease on annealing at about 400°-500°C, increase at 600°C and decrease at 700°C and 745°C. Lowrie and Fuller (1969) suggest that the annealing occurring at 400-500°C is caused by cation reordering and that the increase in magnetic parameters at 600°C is caused by defects coalescing. Defects are subsequently annealed at 700°C and 745°C causing a reduction in coercive force. R.W. Smith (1967) shows the existence of similar effects in haematite.

3.4.3. Improvement of acceptance rate

It is obviously advisable not to use samples which appear to have been weathered and sub-samples should be taken from the interior of sherds to minimize the chances of weathering. Pre-selection of only red, fine-grained material on the grounds that it only contains haematite as the main remanence carrier and is therefore unlikely to alter, does not increase the acceptance rate (e.g. Thomas 1981). Sometimes it is possible to assess the stability of a specimen to refiring by using opaque microscopy.

Typical uses of magnetic parameters as indicators of stability are the three criteria which P.J.Smith (1967) suggested should be fulfilled for a specimen to give a reliable archaeointensity result using the Thelliers' method: (i) the heating and cooling saturation magnetization versus temperature curves for the specimen should be similar in shape; (ii) the Curie temperature should change by no more than 10°C during heating; (iii) the value of the saturation magnetization (M_s) should change by no more than 15%. Ade-Hall et al (1968) and Lawley (1970) suggested that there should be a single Curie temperature above 500°C.

Use of such criteria reduces the number of samples on which it is possible to carry out archaeointensity determinations. These criteria are effective when applied to samples with the non-ideal behaviour types listed above, although viscous samples would not necessarily be rejected.

If the range of samples to be used is to be increased then it is necessary to be able to determine what process is producing the non-ideal behaviour and to have a set of objective criteria to determine how to produce a valid result from the data. The same kind of non-ideal behaviour will produce different effects depending on whether a thermal or AF demagnetization method is used. There has been a gradual trend in archaeointensity methods over the past years to make use of more non-ideal samples and to introduce reliability criteria, for example the Thelliers' method and the AF demagnetization method of Shaw (1974) as modified by Kono (1978) and by Thomas (1981).

3.4.4. Description of methods

3.4.4.1. The Thelliers' Method

In its original form this method was virtually fool-proof as it rejected all data which did not conform to rigid guidelines. However, many samples had to be rejected and as the method was time consuming, it gradually became customary to accept less ideal data. Attempts were made to recognize types of non-ideal behaviour and to predict which portion of the PTRM-PNRM plot could provide a valid result. If an attempt is made to carry out an archaeointensity determination using the Thelliers' method without use of additional measurements as suggested by Smith (1967) then the only way of assessing the reliability of the results is with the PTRM–PNRM plot itself. This is in contrast to the ARM correction method: at no point in the Thelliers' method is it possible to unequivocally separate effects such as VRM or low temperature refiring from mineralogical alteration during refiring and any attempt at interpretation of non–ideal mineralogical behaviour from the PTRM–PNRM plots becomes very difficult.

It is possible that non–ideal mineralogical behaviour could cause only a portion of the PTRM–PNRM plot to deviate from the ideal straight line and that a valid result could be obtained from the remainder of the results. VRM can be removed by lower temperature demagnetization but alteration of magnetic minerals becomes more likely as the temperature increases and reliable results would generally fall in the region unaffected by either VRM or alteration.

For weathered material, it is difficult to predict what type of deviations would result on a PTRM–PNRM plot from the Thelliers' method or from which portion of the graph a correct result may be obtained. Indeed, two separate groups of authors, (Barbetti et al., 1977; and Sigalas et al., 1978), with similar weathering models disagree on which portions of the graph would yield the correct result. The process of detecting non–ideal behaviour is complicated by systematic curvatures of the PTRM–PNRM plot which can arise from interacting single domain grains (Dunlop and West, 1969) or multi–domain grains (Coe, 1967).

No use can be made of samples which have altered either chemically or physically unless this alteration only begins to occur at high temperatures, and there is no VRM or low temperature TRM contaminating the NRM. The process of selecting which portion of the graph provides the correct result is thus highly subjective.

3.4.4.2. AF demagnetization methods

A viable archaeointensity method using AF demagnetization of NRM and TRM must incorporate a test to ascertain whether magnetic grains have been changed by the laboratory heating, because, unlike the Thelliers' method there is no point at which the method compares both the low temperature TRM and NRM before the sample is fired through the Curie temperature. The AF demagnetization of ARM may be used as a monitor of the coercivity spectrum of the sample before and after laboratory firing (ARM1 and ARM2) and a region over which the coercivity spectrum was unaltered by the TRM acquisition process may be located (Shaw, 1974; also see section 3.2.3 of this chapter). Two graphs are thus produced, ARM2–ARM1 and TRM–NRM.

VRM, annealing, chemical alteration and weathering, if any of them resulted in alteration during refiring, would be detected by this method and the result would be rejected unless a portion of the coercivity spectrum remained unchanged in both magnitude and shape. Weathering of the sample before collection remains a problem.

The method removes an element of subjectivity from interpretation of results but makes no attempt to use samples where alteration is so drastic that the whole coercivity spectrum is affected. Thus it avoids samples where mineralogical problems occur rather than making use of less than ideal results. This method has been further developed, e.g. by Kono (1978) and Thomas (1981), to make use of samples, which alter during laboratory refiring. The ARM demagnetization results are used to produce a correction factor which corrects for differences in NRM and TRM caused by refiring in the laboratory. Where the ARM2-ARM1 relationship is linear the samples have obviously altered because the ability to acquire ARM has changed, the size but not the shape of the coercivity spectrum has changed. It is assumed that a similar change in the rate of acquisition of low field TRM occurred and the following correction is applied:

$$B_{anc} = B_{lab} \times (NRM/TRM) \times (ARM2/ARM1) \qquad (23)$$

where B_{anc} is the ancient field and B_{lab} is the laboratory field. This correction has been shown to be valid by Kono (1978) and by Thomas (1981).

Kono divided his results into several classes according to the relationship shown by the ARM2-ARM1 and the TRM-NRM graphs. The criteria are as follows:

1. the ARM2/ARM1 plot is linear with gradient = 1.0

2. the ARM2-ARM1 plot is linear but gradient ≠ 1.0

3. the ARM2-ARM1 plot is non-linear,

and

a. the TRM-NRM plot is linear and the best fitting line passes through the origin,

b. the TRM-NRM plot is linear and the best fitting line does not pass through the origin,

c. the TRM-NRM plot is non-linear.

Class (1a) results were the only ones accepted by Shaw (1974). Kono used samples for which the palaeointensity values were known and showed that classes (1b), (2a) and (2b) also gave valid results. Kono (1978) made no attempt to assess the reliabilities of these groups of results.

Having accepted that this method can cope with alteration occurring during laboratory firing, the next stage in increasing its usefulness is to make use of samples in which the linearities of the ARM2-ARM1 and TRM-NRM plots are obscured by other forms of non-ideal behaviour. In the method used by Thomas (1981) it is only necessary for portions of the TRM-NRM and ARM1-ARM2 graphs to be linear, the linear segments being compared between the same demagnetization limits. The ARM plots monitor only the alteration occurring in the laboratory and differences in shape between ARM and TRM plots at low and high demagnetization fields give clues to the type of non-ideal behaviour which has occurred. For example differences at low fields may indicate the presence of VRM or low temperature refiring and at high temperatures could indicate that the sample was poorly fired originally. For a particular type of non-ideal behaviour, the deviation may vary in magnitude from sample to sample because the blocking temperature spectrum of each unblocking field interval for a sample may vary. However the general shape of the deviation will be similar. Thomas (1981) divided the results into groups according to theoretical predictions of the shape of the ARM and TRM graphs and the stability of the remanence vector on demagnetization. Each group of results was assigned a reliability rating on the basis of theoretical and experimental work.

The ARM correction method was specifically designed to cope with alteration, either physical or chemical, which caused changes in the remanence acquiring capacity of the sample. VRM is readily detectable because of changes in direction of the NRM vector and because it causes deviations on the NRM-TRM curve but not on the ARM2-ARM1 plot. If, during the course of the weathering, magnetic grains are removed or altered in size and no CRM is produced then the ARM correction method will produce the correct archaeointensity result as it will correct for alteration caused by the dehydration of weathering products. No use can be made of samples where a CRM is produced or where the coercivity of the sample is altered due to the production of stress during weathering as these occurrences invalidate the basic assumption that the NRM is a TRM. No archaeointensity method can cope with this problem. Samples with a CRM will be rejected if the CRM has a dissimilar AF demagnetization curve to the TRM.

3.4.5. Summary

Mineralogical problems have limited the range of samples from which it has been possible to obtain archaeointensities. Inevitably, attempts which were often subjective were made to obtain results from non-ideal material. The Thelliers' method is less useful when dealing with non-ideal behaviour than the AF demagnetization methods. The only data available from a Thelliers' method determination is the PTRM-PNRM plot and the direction of the NRM and TRM vectors on demagnetization and as the same deviation can apparently be produced by a number of mechanisms, any decision as to which mechanism is producing the behaviour and which portion of the PTRM-PNRM plot will yield the correct result is difficult. Without the routine measurement of other data during the Thelliers' method, it seems unlikely that this method will in future be applicable to many more sample types. The ARM correction method, because it provides the additional check of the coercivity before and after laboratory refiring, provides more information in interpretation of what kind of non-ideal behaviour is producing deviations. Once the behaviour is identified it becomes a simple matter of using theoretical prediction to ascertain how to obtain the correct result.

3. 5. EXISTING ARCHAEOINTENSITY DATA FOR GREECE

By Derek Walton,
McMaster University, Hamilton, Canada

Archaeointensities obtained some years ago on well-dated material from the Athenian Agore (Walton, 1979) are reproduced in Figure 3-6. In obtaining this data a SQUID magnetometer, small samples and the modified Thellier method similar to the technique presently employed by Aitken (see section 3.3 this volume) were used. The only test for alteration was based on linearity of a plot of moment lost in heating to a given temperature against an expanded plot of moment remaining after the sample was cooled from that temperature in a known field. On this basis, roughly 40% of the points correspond to the "α" criterion, while the remainder were in the "$\beta+$" category. Since the linearity of such plots is not an adequate test for alteration, it is possible that some of the results have been affected by alteration. Furthermore, data obtained below 250°C were sometimes included, and at those temperatures remanent viscous moments could still have an effect.

On the other hand, the reproducibility of the results might encourage more faith in the data (note that, although the necessity for reduction in size on reproduction makes it somewhat difficult to see, the points on the steeply rising and falling portions of the curve in Figure 3-6 are in sequence and not scattered.

However, it should be recalled that Athenian potters used similar techniques in a given period and clay from the same beds. Thus, all the pottery from a given period probably has similar chemical and physical characteristics. Therefore, it is possible for alteration to affect a group of sherds in a similar fashion. Comparison of the Greek results with those from surrounding areas indicates that this may just possibly be the case: while the reproducibility of the Greek data is ±4% the discrepancy between these results and others is about 20% to 30%. While it is possible for differences in experimental technique to account for some of this, it does not seem likely that it can account for all of it.

There is no alternative except to repeat the measurements, this time with a technique which can detect alteration. A method which depends on monitoring the susceptibility appears to work, and it is hoped that the Greek data can be corrected in the near future.

Thus, the Greek results are reproduced here in this volume solely for the sake of

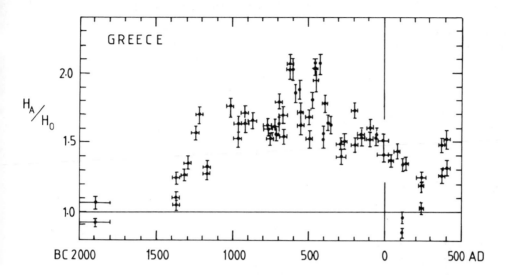

Figure 3-6: The change in the ratio of the ancient intensity of the geomagnetic field H_A, to the present day value H_0, in Athens.

Each dot represents a single measurement made on a core taken from a different sherd.

completeness, since it is impossible to have confidence in any results which rely on non-linearities in NRM-PTRM plots to reveal mineral alteration.

3.6. ARCHAEOMAGNETIC DATA FROM BULGARIA AND SOUTH EASTERN YUGOSLAVIA

By Mary Kovacheva,
Geophysical Institute, Sofia, Bulgaria

3.6.1. Introduction

   Bulgaria, situated at the crossroads between the Western World and the Orient has a rich historic and prehistoric past. Extensive archaeomagnetic investigations have been carried out there during the last fifteen years.   Most of the sites illustrated in Figure 3-7 contain more than one dated point. Results from south-eastern Yugoslavia are also included (Kovacheva and Veljovitch, 1977).

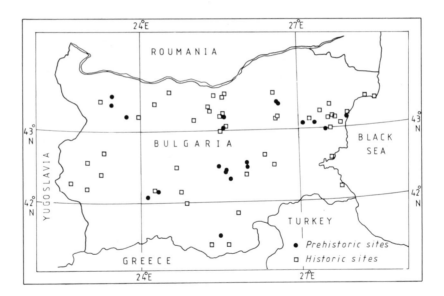

Figure 3-7:   Geographical distribution of dated
Bulgarian archaeological sites.

Samples from ancient ovens, furnaces or burnt layers were used for obtaining the inclination, declination and intensity of the ancient geomagnetic field. For the historic period bricks were also used.   Samples were orientated with a magnetic compass on a horizontal plate set in Plaster of Paris. Three cubical specimens of side 24mm or less were taken from each sample and measured with an astatic magnetometer with sensitivity 2nT per scale division. The Thellier and Thellier (1959) method was combined with the graphical method of Arai (Nagata, 1961). Non-linear behaviour of the NRM–TRM curve was avoided by excluding the upper points of heat treatment (Kovacheva, 1977, 1980) assuming that prehistoric samples were not properly baked. Low temperature points were avoided to remove viscous components. As a general rule, at least six successive points on the NRM–TRM line were required.

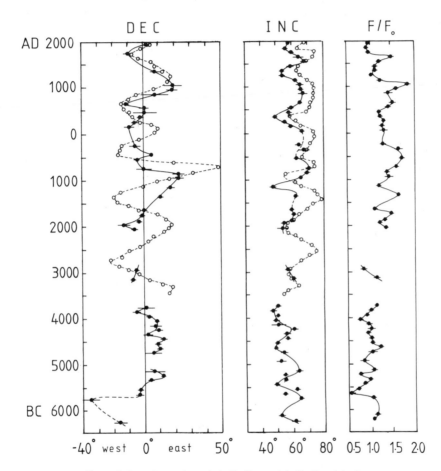

Figure 3–8:   Comparison of declination and inclination data for Bulgaria (black dots and solid lines) with those for Ukraine (open circles and dashed lines). Relative intensity curve (F/F$_o$) for Bulgaria also shown on right.

As archaeomagnetic investigations require a continuous integration of new and old results, all the individual determinations have been combined to form weighted means by the relation :

$$\sigma_{av} = (\Sigma p_i (x_i - x)^2)^{0.5} / \Sigma p_i (n-1) \qquad (24)$$

where n is the number of included sites in a given century, p the number of samples examined from every site:

$$x = \sum p_i x_i / \sum p_i \qquad (25)$$

where x_i is the average result for every site.

3.6.2. Results

Declination, inclination and intensity curves from Bulgaria are shown in Figure 3–8. They show quite an irregular change with time although they represent smoothed values of the measured parameters. Even after ignoring the smallest variations, the remainder show a very complicated motion. The amplitudes of variations in declination are bigger than those of inclination. The declination curve is based on fewer samples than the inclination curve in which the results from bricks are included. The extreme value of declination obtained by us in Bulgaria for the 11th century AD is observed in France in about 12–13th century AD (Thellier 1981) and for England in about the 13th century AD (Aitken 1978). The apparent periodicities in directional results are of several hundred years and appear longer for declination than for inclination.

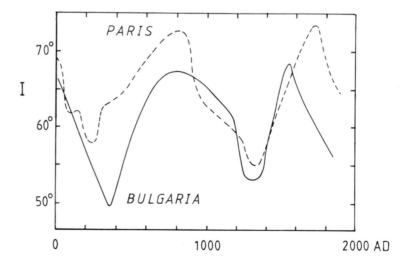

Figure 3–9:   Comparison of inclination curve for Bulgaria (solid line)
with that for Paris (dashed line).

The intensity curve clearly shows an overall increasing trend from 6000BC to 1000AD, since when the intensity has decreased. It seems that intensity variations with periods of several hundreds of years are superimposed on this long term variation which has a periodicity of the order of thousands of years. It must be emphasized that the real existence of the shorter periods depends critically upon the accuracy of dating.

Age determination constitutes one of the principle difficulties in all archaeomagnetic investigations. As far as the historic period is concerned the situation is not so difficult but going back in time the difficulties of dating increase because the ages of prehistoric sites are mostly based on radiocarbon dates and several alterations have had to be made to the archaeological ages (Kovacheva and Veljovitch, 1977). Some sites still do not have any radiocarbon dates: their age can only be estimated using the relative chronology of the prehistoric cultures which is well established in Bulgaria from archaeological evidence. Radiocarbon dates used in this work are calibrated (Clark, 1975) so the early Neolithic period extends back to before 6000BC.

3.6.3. Comparison with other European results

In Figure 3-8 the Bulgarian declination and inclination curves are compared with those obtained by Rusakov and Zagny for the Ukraine (Zagny, 1981). The Ukrainian inclination curve shows steeper values than the Bulgarian curve as would be expected for a dipole field as the Ukraine is situated to the north of Bulgaria. Similarly the inclinations for Paris (Thellier, 1981) are also steeper than those for Bulgaria as shown in Figure 3-9.

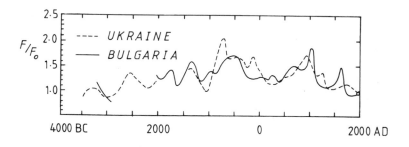

Figure 3-10:   Comparison of relative intensity curves for Bulgaria (solid line) and Ukraine (dashed line).

The increasing intensity trend that is shown in the Bulgarian curve appears also in the results from Ukraine (Figure 3-10).   For the time interval 400AD – 400BC, the results for Athens agree with the Bulgarian ones (Figure 3-11) but the two peaks in Walton's curve at about 500BC and 700BC are not found in the Bulgarian data (nor in Aitken's results for NW Europe).   The general trend (Figure 3-12) suggests that the

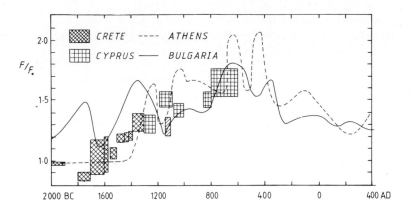

Figure 3-11:   Comparison of relative intensity curves for Bulgaria (solid line)
and Athens (dashed line). The Athens curve is that of Walton (1979)
Data for Crete and Cyprus (Aitken, personal communication) are
also shown.

geomagnetic dipole moment was considerably greater than the present-day value back
to 2800BC (broken line on the graph). However the curve reflects fluctuations of both
strength and direction of the geomagnetic field. It is clear that the form of the
geomagnetic field variations is rather complicated and differs from Bucha's (1970)
reduced dipole moment (RDM) curves (Smith, 1967).

The results presented in this section are expressed in terms of the so called 'long
chronology' in prehistory corresponding to calibrated radiocarbon dates.

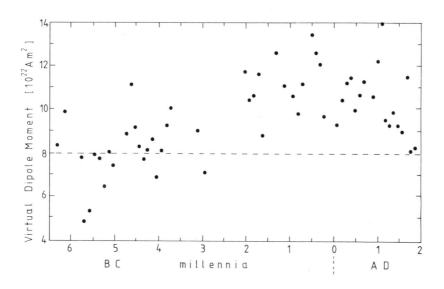

Figure 3-12:   Virtual dipole moments (VDMs) for Bulgaria from 6500BC to the present.

3.7. ARCHAEOMAGNITUDE RESULTS FROM EGYPT

By Ken Games,
University of Liverpool, UK.

3.7.1. Introduction

One of the most critical factors in archaeomagnetic studies is the accuracy of the dating of the samples used. In Egypt we can rely on archaeological dates, since the chronology that has been worked out is very accurate, certainly in the relative and often in the absolute sense. Also there is an abundance of well-dated adobe bricks which we used to test a new technique devised by Games (1977) and we were able to compare the results obtained by applying the more convenient Shaw (1974) method to many well-dated sherds over the same period. It happens that more well-dated adobe bricks are available than well-dated pottery samples. In all cases, only adobe bricks or sherds which could be accurately dated were used. An explanation of the basis of the dating is given in the Appendix of Games(1980) where most of the results are presented.

3. 7. 2. Archaeomagnitude Method for Adobe Bricks

The sun-dried adobe bricks were analysed using the method described in Chapter 1, section 1.5. To determine the magnitude of the ancient geomagnetic field, B_A, a 2cm core was taken from the adobe bricks. The NRM of the core was measured, and it was then stepwise AF demagnetized and remeasured to a maximum AF of up to 150mT. The sample was then given an ARM called ARM(1). The ARM is then AF demagnetized in the same way as the NRM. The dry sample was then crushed, and a known amount of water was added. The correct amount of water for a particular brick has to be found experimentally, so that the mud has the right consistency when mixed, and so that when it dries out, no cracking occurs. The thoroughly mixed sample was put into a specially devised plastic piston arrangement, from which it could be ejected or 'thrown' into a cylindrical container by depressing the piston forcibly. This was the best way we could find experimentally of simulating the original brickmaking process. It was tested on the modern brick from Lima, and four samples gave an ancient field value of $29 \pm 2 \mu T$ for this brick, compared with the known field value of $28 \mu T$ as recorded at the nearby Huancayo observatory.

The re-throwing process was carried out in a set of Rubens coils with the uniform field inside set at $50 \mu T$. The sample was left to dry for 4 days in the Rubens coils. The SRM was then AF demagnetized in the same way as the NRM. Finally a second ARM – ARM(2) was given under the same conditions as ARM(1), and this was also AF demagnetized and remeasured in the same way as the NRM.

Figure 3-13 shows an example of a graph of NRM vs SRM for sample E10.1 using the demagnetizing field as a parameter. Since the low coercive force region is susceptble to secondary magnetizations such as VRM, the data in this region are rejected on the basis of the non-linearity of the plot and the stability of the NRM direction. Figure 3-14 shows a plot of the accepted data for E10.1. From this graph, the ancient geomagnetic field magnitude, B_A, is calculated from $B_{anc}/B_L = \Delta NRM/\Delta SRM$, where $B_L = 50 \mu T$.

As a check that the magnetic properties of the sample have not changed, ARM(1) was plotted against ARM(2) as shown in Figure 3-15. This should give a straight line of gradient = 1.00. Any variation from unity is a measure of the degree of alteration undergone by the sample during resetting of the SRM. If the gradient differed from unity by more than 5% the sample was rejected.

In most cases, two determinations of B_A were made using two cores from each brick, and the error on the mean value of B_A quoted is half the difference between the two determinations. The average value of this error is ± 4.5 %.

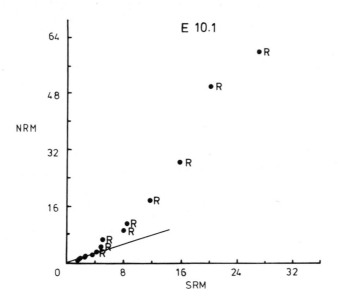

Figure 3-13: A plot of NRM/SRM using the demagnetizing field as parameter for sample E10.1. The points labelled R were rejected for reasons explained in text.

Units of intensity are $\mu Am^2 kg^{-1}$.

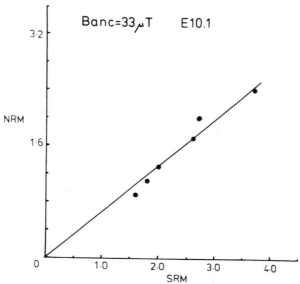

Figure 3-14: A plot of the accepted NRM/SRM values for E10.1. Units of intensity are $\mu Am^2 kg^{-1}$.

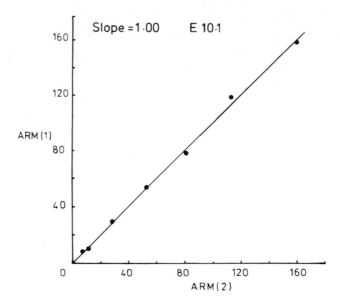

Figure 3-15:   A plot of ARM(1)/ARM(2) for the accepted data for E10.1. The slope
of the line is shown at the top of the figure. The sample is rejected
if the gradient differs from unity by more than ± 5%

Units of intensity are $\mu Am^2 kg^{-1}$.

3.7.3. Archaeomagnitude method for sherds

The method adopted is the same as previously applied to Peruvian sherds and is
described in section 3.14.2. Two basic improvements were made so as to avoid
errors due to cooling rate and anisotropy. In all cases, the TRM was installed in the
same direction as the NRM to within about 10°. This minimizes the error due to fabric
anisotropy as outlined by Rogers et al. (1979). Also, the samples were cooled from
700°C to room temperature in about 7 hours in order to reproduce as nearly as
possible the original cooling rate of the NRM. As with the adobe brick samples two
determinations of B_A were made using two specimens from each sherd wherever
possible.

3.7.4. Results

The archaeomagnitudes determined from adobe bricks and sherds are shown in
Figure 3-16. One of the most interesting features of these results is the fairly close
agreement between the values obtained from sun-dried adobe bricks and the values

obtained from the fired material.   Despite some differences, this fairly good
agreement between the two completely different source materials gives us more
confidence in the derived magnitudes of the geomagnetic field. There are a few points
which disagree with the rest of data, and we cannot with certainity explain these
discrepancies. They may be simple errors, but it is also possible that in some cases
the adobe bricks in a structure were actually re-used bricks from an earlier building.

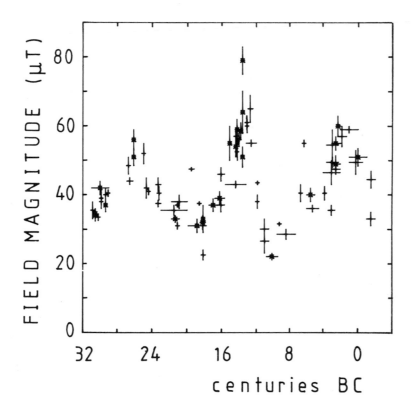

Figure 3-16:   A plot of geomagnetic field magnitude (mT) against time for Egypt.
Values derived from adobe bricks are shown by simple crosses, and those from
fired material by asteriks. The errors on the dates are those supplied by
archaelogists. Errors on field magnitude values are half the difference
between the two values determined for each sample

The graph in Figure 3-16 suggests that the magnetic field in Egypt fluctuated quite
rapidly over the 3000yr interval studied. During the period 3000BC to 0BC there are at
least three maxima in the field at approximately 2600BC, 1350BC and 200BC, with
possibly two others less well defined around 2000BC and 700BC. Near the large

maximum at 1350BC, the rate at which the field was changing reached 140nT/yr and this persisted for about 300yr either side of the maximum value. For the rest of the time, the rate of change was considerably smaller, being no greater than about 40nT/yr. As pointed out in section 3.7.3, these results are consistent with changes in the archaeomagnetic field strength found elsewhere, but it is interesting to look at the records from observatories to see how these results compare with present day changes in the magnetic field. Over the last 20 or 30 yr there has been a very large rate of decrease in the strength of the magnetic field centred on Paramaribo observatory ($5°49'N, 304°47'E$). The locality in question has shown a rate of change of the field of about 100nT/yr over an area of some 3000km^2, and this rate has persisted for at least 30yr. So it would appear that at this locality through historic time, large rates of change in the geomagnetic field which are comparable to those suggested by archaeomagnetic data have occurred though of course the direct observations do not cover as long a time period.

3.7.5. Conclusions

Values of the magnitude of the archaeomagnetic field obtained from sun-dried adobe bricks and from fired material are in fairly good agreement with each other. This fact, together with the accuracy of the dating of the Egyptian material, suggests that the fluctuations observed in the magnitude of the archaeomagnetic field in Egypt are real. There are still some gaps in the curve (Figure 3-16) where few or no data exist or where more data are needed to define the curve more precisely. Until this is done it is too early to try to 'date' archaeological samples using this curve, but in the near future application of the curve may be of some assistance to archaeologists.

3.8. SUMMARY OF PREHISTORIC ARCHAEOINTENSITY DATA FROM GREECE AND EASTERN EUROPE

By Ruth Thomas,
University of Edinburgh, UK.

3.8.1. Introduction

Sufficient archaeointensity data have been accumulated over the past few years that it is now possible to consider the behaviour of the Earth's magnetic field in greater detail. When looking at variations in intensity on a small time scale, or comparing archaeointensity curves from different areas, it is important that any differences in experimental technique and assessment of reliabilities of results between authors be considered. It is however not sufficient merely to have reliable archaeointensity data from a site: it is vital that they be accurately dated. Without accurate dating, archaeointensity data are useless. In the past, many archaeointensity data were archaeologically dated using a time scale of diffusionist chronology which was subsequently found to be in error. In addition to the data mentioned in the foregoing sections, data from Greece (Thomas, 1981) are included in this summary as they are well dated and cover the time span in question.

3.8.2. Additional data

Archaeointensities of samples of Neolithic pottery from Sitagroi (41°N, 24°E), and Seskla and Dimini (39.5°N, 23°E) in Greece, whose ages have been well established, were determined using a correction method based on ARM as described by Thomas (1981). The samples from Sitagroi were especially significant as they formed a chronological succession of samples from successive layers of occupation from 6000BC to 2000BC. They have been dated by the radiocarbon technique. The ages of the samples were thus well known and the intensity data were comparable in detail to directional data obtained from lake sediments. However, interpretation of archaeointensity data is complicated by the fact that the continuity of the occupation of the site at Sitagroi is not proven, so although the duration of each phase of occupation is known, it is not certain how long each level of occupation lasted. The archaeological relationships between the samples from Seskla and Dimini are unknown, and they have therefore been dated using thermoluminescence (see section 2.4). The results from Seskla and Dimini are incompatible with certain interpretations of the occupation of the site at Sitagroi. On this basis the present author believes that occupation of the site at Sitagroi was continuous with the possible exception of a small period between 3900BC to 3400BC. The method described by Thomas (1981), incorporates stringent reliability criteria and many of the results from the oldest samples were rejected in the construction of an archaeointensity curve for the period from 2000BC to 5000BC.

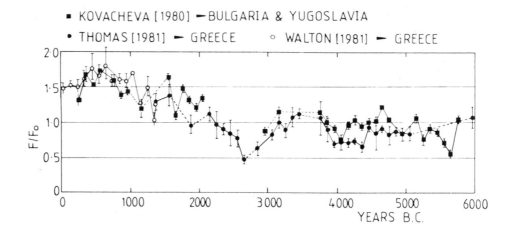

Figure 3–17:   Comparison of archaeointensities from Greece, Bulgaria and Yugoslavia

In order to compare these data with those of Kovacheva (section 3.6), they are presented as century means in Figure 3-17. Results from Crete (Thomas 1981) are also included in this figure but Burlatskaya's results (see section 3.10) are not included because they were derived from regions located some distance from the area of eastern Europe.   Burlatskaya's results do however show the same general trend as the other results in that the archaeointensity increases over the period 3000BC to 2000BC. Examination of Figure 3-17 shows that there is good agreement in the general trends of all results plotted. In fact all published data from this area, with the exception of those of Games (1980) from Egypt, agree with these trends. There is good overall agreement between the results from Bulgaria and Yugoslavia (section 3.6) and Thomas's (1981) results from Greece, but poorer agreement is observed when the Greek results are compared with those from further afield.

3.8.3. Implications for archaeological dating

It is possible to compare the two detailed archaeointensity curves of Kovacheva (section 3.6) and Thomas (1981) from different parts of the same geographical region which span several millennia. Such a comparison provides a measure of the reliability of an archaeointensity 'type-curve' for dating at a nearby site. It is important to resolve the maximum distance over which dated results from one site can be used to date samples from another site. If a single sample of known archaeointensity from Sitagroi were to be dated using the Bulgarian/Yugoslavian archaeomagnetic intensity curve from 3800BC to 500BC, it is likely that an erroneous age would be obtained even if its age were initially known to within ±200yr. Furthermore a result from just one sample could lead to a totally erroneous date if its measured archaeointensity were in error for experimental reasons. Even a series of archaeointensity values forming a chronological sequence (e.g. from Sitagroi) would not be very accurately dated if Kovacheva's curve for Bulgaria were used as a type curve. It seems therefore that even for sites as close together as the Bulgarian and Yugoslavian sites and Sitagroi, the results are not sufficiently similar to permit the use of a single type-curve for the region, at least when the results are averaged over 100yr intervals.

3.8.4. Virtual Axial Dipole Moment

VADMs averaged over 1000yr intervals are now compared with those calculated by Barton et al. (1979) from the Burlatskaya and Nachasova (1977) catalogue and with data from Games (1980) and Kovacheva (1980). A thousand year interval was chosen because it is long enough to ensure that a reasonably large number of results are included within each data set and that large dating errors should be immaterial. Cox (1968) suggested that averaging over 500yr intervals should smooth out scatter due to 'dipole wobble'. The VADMs were calculated using the relation

$$VADM = FR^3 (4 - 3 \cos^2 L)^{-0.5}$$

where F is the archaeointensity, R is the Earth's radius and L is the latitude of the site. The 1000yr averaged VADMs for the second to sixth millennia are shown in Figure 3-18. The standard deviations are shown in Table 3-1 but omitted from Figure 3-18 for the sake of clarity. It was not possible to calculate standard deviations from the Kovacheva (section 3.6) century mean data points. The results from Aitken et al (section 3.9) cover the third and fourth millennia and show the same trend as the other data (see Figure 3-19).

When averaged by millennia, the worldwide geomagnetic intensity curve from the Burlatskaya and Nachasova (1977) catalogue still exhibits an approximately sinusoidal form with a minimum in the fourth millennium BC (Table 3-1, Figure 3-18). However, the other data do not show this minimum: in fact those from Kovacheva show a slight increase in the VADM for the fourth millennium.

The standard deviation of the Kovacheva data could not be calculated, but all the other results overlap within the calculated standard deviations. With the exception of the result for the fifth millennium BC, all Kovacheva's results fall within the limits of the standard deviation of the results for a given millennium. There are some differences between the Sitagroi data and the Kovacheva data, but they show close

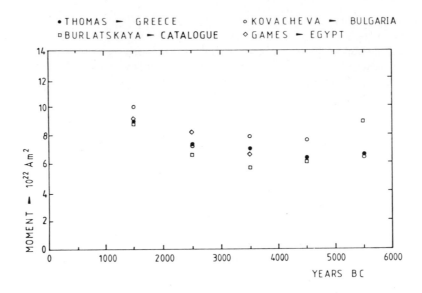

Figure 3-18:   VADMs averaged over 1000 year intervals for the second to the
sixth millennia

agreement for the third and sixth millennia BC and it is possible that the difference is
not significant.

Interestingly enough, when all the results are combined, the overall mean VADM for
the third to fourth millennia is about $7.4 \times 10^{22} Am^2$.   Prior to 1977, published data
for Europe dated earlier than 3000BC were confined to Bucha's (1967) data from
Czechoslovakia, and the evidence that the field varied sinusoidally with a period of
8000yr to 9000yr relied heavily on these data. This sinusoidal variation with a low
during the fourth millennium is not confirmed by either the results from Kovacheva or
Thomas (1981). Barton et al (1979) noted that there was a lack of coherence
between European and East Asian data prior to 3000BC in the Burlatskaya and
Nachasova (1977) catalogue. However data from Kovacheva (section 3.6) and
Thomas (1981) do not agree to any great extent with the East Asian data of Kitazawa
(1970). Bucha's (1967) Czechoslovakian data gave lower archaeointensities than all
the other data. The possibility of obtaining erroneously low results is greater with older
samples because the technology of pottery manufacture was less well developed and
pottery was often not well-fired. Also older pottery has been subjected to weathering
for a longer period. It is thus very easy to obtain erronously low results for older
pottery and without adequate screening of results the impression may be gained that
the field was much lower during Neolithic times. In fact much of the older Sitagroi
data were rejected because the material was poorly fired or weathered.

Table 3-1:   Comparison of VADMs averaged over 1000 yr intervals

Median interval	Data Source	VADM		
(yr BC)		mean	SD(%)	N
1500	A	8.88	26.2	13
	B	9.02	18.8	5
	C	9.17	32.8	17
	D	10.0	–	21
	A–D combined	9.4	–	21
2500	A	6.61	14.5	14
	B	7.31	33.3	12
	C	8.21	13.0	18
	D	7.21	–	18
	A–D combined	7.39	–	18
3500	A	5.62	44.0	13
	B	7.04	16.3	17
	C	6.69	5.0	6
	D	7.83	–	76
	A–D combined	7.39	–	76
4500	A	6.04	19.2	6
	B	6.33	14.3	17
	D	7.54	–	147
	A, B, D combined	7.36	–	147
5500	A	8.93	21.6	9
	B	6.52	11.6	2
	D	6.36	–	119
	A, B, D combined	6.54	–	130

A – Burlatskaya and Nachasova (1977);
B – Thomas (1981);
C – Games (1980);
D – Kovacheva and Veljovitch (1977).

3.8.5. Rate of change of the geomagnetic field

A knowledge of maximum rates of change of geomagnetic field is obviously of great importance to theoreticians in modelling the field and it is therefore pertinent to establish whether the rapid rates of change recently suggested on the basis of archaeointensity data are real. Shaw (1979) maintains that the field in central France changed from 55μT in 50BC to 100μT in 150AD, a rate of change of about 0.23μT/yr. Walton (1979) found that the field in Athens changed 0.13μT/yr during the first millennium BC. These rates of change are much faster than the maximum historically recorded rate of change of 0.05μT/yr averaged over the last 400yr for Mongolia (Yukutake and Tachinaka, 1969). To be considered reliable, evidence of rapid changes in intensity must be based on results from a sequence of individually dated samples from the same site. Kovacheva, Aitken and Burlatskaya all present data which form an archaeointensity curve, but which are from a variety of locations. Therefore it is not certain that estimated changes in magnetic field intensity are not due to the difference in location of the sites. The Sitagroi collection of data is ideal for investigating rates of change of the field. The maximum rate of change of the geomagnetic field at Sitagroi is 0.05μT/yr corresponding to an intensity change of from 47μT to 22μT between 3100BC and 2600BC. This rate is compatible with our knowledge of the time variation of the historic field.

3.9. PALAEOINTENSITY STUDIES ON ARCHAEOLOGICAL MATERIAL FROM THE NEAR EAST

By M.J. Aitken, P.A. Alcock, G.D. Bussel and C.J. Shaw
Oxford University, UK.

3.9.1. Introduction

In this contribution comparison is made of intensity results, within the time range 5000BP – 2000BP, obtained from four localities of the Near East. These localities span 23° of longitude:

Crete, specifically Knossos,	(35°N	25°E)
Cyprus, specifically Kition,	(35°N	33°E)
Egypt, specifically Thebes,	(25°N	32°E)
Mesopotamia, specifically Niniveh,	(35.5°N	43°E)
Babylon,	(23.5°N	44°E)
and Elam,	(31.5°N	48°E)

For Crete and Cyprus the samples were fragments of pottery and for the other localities the samples were taken from funerary cones (Egypt) and royal bricks (Mesopotamia) made of baked clay. As outlined below the dating of the samples, ultimately based on recorded observations of astronomical events, was of good reliability. The objectives in our project, of which these results represent an initial stage, are (i) to obtain results, closely-spaced in time, from each locality, with about half-a-dozen determinations for each reign or archaeologically distinguishable phase (usually of about half a century in duration, sometimes more and sometimes less); and (ii) to assess the coherence of the data, and the scatter, to give an external indication of overall reliability. With sparse data there is always uncertainty as to whether an abnormal result is truly a manifestation of what was going on in the Earth's core or whether its origin is more superficial, i.e. experimental error or a mistake in attribution.

3.9.2. Dating

The funerary cones from Egypt and the royal bricks from Mesopotamia carried inscriptions containing the names of individuals (e.g. high priests) which could be attributed to the reign of a king. The sequence and lengths of reigns are known from surviving lists and absolute dating is provided by recordings of datable astronomical events. For a discussion of the accuracy of the Egyptian dating the reader is referred to the appendix by H. Smith in Games (1980), and in respect of Mesopotamian dating to Sachs (1970).

The pottery sherds from Crete and Cyprus were dated on the basis of stratigraphy and/or style. The chronological framework for this dating is based on archaeological links with Egypt and hence it is on the same absolute scale as for the funerary cones except that in the second millennium there are uncertainties in the linkages of up to a century.

3.9.3. Technique

The determinations were made by the Thellier technique as modified for use in a SQUID cryogenic magnetometer. The procedure employed has been described elsewhere (Aitken et al. 1981) and only an outline will now be given. Sampling was by means of a diamond impregnated corer in a small battery driven drill which produced cylinders approximately 3mm diameter by 3mm high. For measurement these are mounted, using fireclay, on a 3mm diameter quartz rod, with the direction of the remanent moment (NRM) within 25° of the vertical.

The sample handling arrangements of the magnetometer are such that the mounted core can be moved, in about 10s, from within a miniature oven to the centre of the SQUID detecting coils (consisting of vertical and horizontal Helmholtz pairs – as described by Walton, 1977), the oven being located vertically below the detecting coils. The field in the oven can be adjusted within the range 0 to 0.15mT (1.5 Oe) as desired. In the Thellier remagnetization sequence the oven field is chosen, on the basis of the guessed value of the ancient intensity, so as to achieve as small a change as possible in the vertical component of the sample's remanent magnetization as the temperature of the steps of the sequence is increased. In the present work nine intervals between 150° and 550°C were used. At each temperature, T, there are three heatings each followed by measurement at room temperature:

1. M^I_V, after heating to T and cooling, both in field F,

2. $M^0_V(200°)$, after heating to 200°C and cooling, both in zero field,

3. M^0_V, after heating to T and cooling, both in zero field.

A plot is made of M^{l}_{v} versus M^{0}_{v} , and the linear portion of it used to determine the apparent ancient field F^{a}. If there is no change in M^{l}_{v} as the steps proceed, then $F^{a}\cos\rho = F^{l}$ where ρ is the angle to the vertical of the magnetization acquired in antiquity. Otherwise the gradient of the linear portion is used to calculate the value of F^{l} for which there would have been no change in M^{l}_{v}. The true value of the ancient field is obtained after correction for anisotropy (Aitken et al, loc cit) and for the difference in cooling rate (Fox and Aitken, 1980) between the ancient cooling, which probably took place over one or two days, and the rapid cooling during the Thellier determination, which takes place over about 5 min. Experimental evaluation of the effect for some fifty samples, by cooling each in the known laboratory field over two days and then carrying out a Thellier determination, indicated a ratio between apparent field and true field of 1.11, with a sample-to-sample scatter having a standard deviation of 0.02. This is the ratio used for correction. It may be noted that for laboratory cooling over 4 hours the ratio is 1.08 ± 0.03. Hence the correction is not very sensitive to the actual cooling time in antiquity, which is of course unknown. It should also be noted that in experimental techniques that do not employ a cryogenic magnetometer the cooling time is usually about 4 hours and hence for the associated results the cooling rate correction is barely significant.

The difference $[M^{l}_{v} - M^{0}_{v}(200°)]$ is used as a monitor of alteration. It represents the TRM capacity of grains having blocking temperatures in the range 200°C to 20°C. Compared to a single such measurement at the conclusion of the whole sequence this running monitor gives additional indication of the temperature at which alteration commences, additional that is to the evidence provided by the extent of the linear portion of the M^{0}/M^{l} plot.

3.9.4. Assessment of reliability

To be accepted as having good reliability, Class A (or B+), a determination must satisfy the following criteria:

1. The linear portion of the M^{0}/M^{l} plot must extend over at least 200°C and at least 60% of the demagnetization range (150°C and 50% for Class B+). The criterion of linearity is that the 68% confidence limits on the slope, as obtained from a least squares fit, must be such as to introduce an uncertainty of not more than 1% in the evaluated field.

2. The TRM capacity as given by $M^{l}_{v} - M^{0}_{v}(200)$ must not change overall by more than 10% (20% for Class B+) between the 200°C point and upper step of the linear portion.

3. Over the linear portion there must be no significant change in the angle of the ancient magnetization remaining after each temperature step.

4. There must not be more than one component of remanent

magnetization from 250° C upwards (relaxed for Class B+).

5. Unless there is a second component, or a viscous component, the 150°C point must not lie more than 10% away from the extension of the line through the linear portion (20% for Class B+).

6. The 550°C point must not lie more than 10% away from the extension of the line through the linear portion (40% for Class B+).

Of the samples measured, the percentages which fulfilled the above criteria were:

Region	A	B+
Crete	45%	23%
Cyprus	44%	29%
Egypt	20%	36%
Mesopotamia	41%	27%

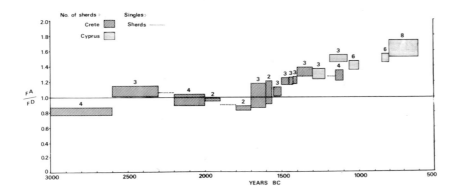

Figure 3-19: Archaeointensities for Crete and Cyprus: results of reliability A and B[+].
The vertical height of each box represents ± 1 standard deviation derived for the scatter of the individual results about the mean.
The horizontal extent indicates the span of the archaeological phase concerned. The numerals indicate the number of sherds in each box.

F^a/F^d is the ratio of the ancient field to the field at the site

due to an axial geocentric dipole of strength $8 \times 10^{22} Am^2$.

3.9.5. Presentation of results

In the figures the ratio F^a/F^d is shown where F^a is the ancient field (after correction for anisotropy and cooling-rate effect) and F^d is the calculated field at the site due to a geocentric axial dipole of strength $8 \times 10^{22} Am^2$. This corresponds to the VADM presentation of Barton et al. (1979) with the advantages as discussed by those authors.

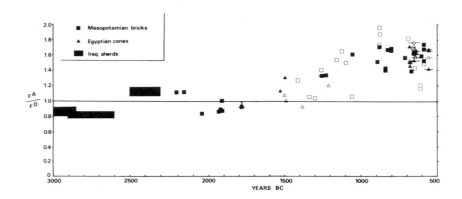

Figure 3-20:   Archaeointensities for Egypt and Mesopotamia.

Reliability Class A: solid symbols. Class B[+]: open symbols. Each point
represents 1 result except where otherwise indicated by numeral.
The date span is not more than 50 years except where otherwise indicated.

F^a/F^d is the ratio of the ancient field to the field at the site
due to an axial geocentric dipole of strength 8×10^{22} Am^2.

3.9.6. Discussion

The most complete record is in respect of Crete (Figure 3-19). The most striking
feature of this is the period 1500BC - 0BC during which the ratio F^a/F^d remained
above unity, reaching a value of around 1.5 in the middle of the period. It is of
interest to note that at 1000BC there is a maximum in the inclination record provided
by sediment cores from Lake Trikhonis in Greece (see Creer et al 1981, for
instance), and to some extent supported by the archaeomagnetic record for Bulgaria
(Kovacheva 1980). This suggests either a tilt of the central dipole towards Greece or,
if the increase in intensity is a localized effect, that the increase is predominantly in
the vertical component.

The results from the other localities (Figure 3-20) show agreement with the results
for Crete though it is not possible at the present stage to confirm or deny indications
of westward drift; at the present rate of 0.2 deg/yr there should be a 100yr lag of
Crete behind Mesopotamia. (Although to the eye there seem to be indications of this,
there are not yet enough data to make a statistical assessment). To our minds the

important result is that, contrary to the usual situation, contemporaneous series from different localities show values which, in general terms at any rate, are concordant. This gives a sound base from which to make comparison with other parts of the world, as well as giving confidence in the reliability of the techniques employed.

Support from the Science Research Council and from the Natural Environment Research Council is gratefully acknowledged.

3.10. ARCHAEOMAGNETIC INVESTIGATIONS IN THE USSR

By S. P. Burlatskaya,
Institute of Physics of the Earth, Moscow, USSR.

3.10.1. Introduction

Archaeomagnetic investigations in USSR were started in 1958 at the Geophysical Institute of the Academy of Science by Professor G. N. Petrova. From a compilation of geomagnetic field variations recorded at a number of observatories through the last 150–180yr (Weinberg, 1929), Tbilisi and Tashkent were identified as regions where variations of inclination had the greatest amplitudes. Both of these regions provide abundant sources of archaeological material such as bricks from many old buildings and baked clay objects. Ancient field directions were measured by the Thelliers' method.

3.10.2. Results from the Caucasus

The first Tbilisi collection consisted of samples from the 17th to 20th centuries AD (Burlatskaya 1965). Later on, sampling was extended to different regions of Georgia (Nodia et al., 1970 a,b), a small collection from Azerbaijan (Mirzachanov, 1970) and some samples from Armenia were included. The time-interval was extended back to the time of Christ and some samples were as old as 2800BC.

In Figure 3-21, 100 year means of geomagnetic field inclination and intensity in Georgia are shown. Mean dispersions for 84-95% of inclination and intensity values were 3.2° and 1.7µT respectively. Each value was assigned a 'weight' which was estimated from the reliability of the measurements and the number of samples and objects. The errors of these curves are about 1° for inclination (Curve a) and 0.5 µT for the intensity (Curve b). The variation of declination in Georgia from 300AD to

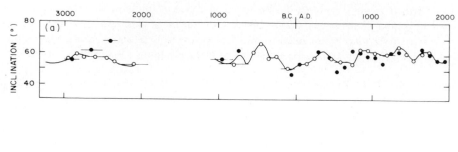

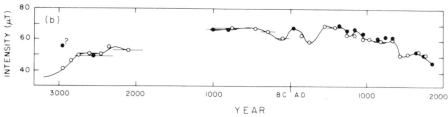

Figure 3-21: 100 year means of (a) inclination and (b) intensity
for the last 5500yr in Georgia, points after Burlatskaya (1965),
circles after Nodia et al (1970 a,b). Horizontal lines
show the limits of dating.

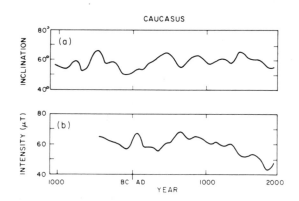

Figure 3-22: Change of the geomagnetic field for the last 2500yr in the Caucasus:
(a) inclination, (b) intensity for averaged data.

1800AD was studied by Nodia et al. (1970b), though the samples were only roughly
dated.

Data for inclination and intensity in the Caucasus for the last 2000yr are summarized in

Figure 3-22a. The inclination variations were calculated from the data of Burlatskaya (1965), Nodia et al. (1970a) and Mirzachanov (1970) and are accurate to about 2.7° . The intensity variations shown in Figure 3-22b are based on the results of Burlatskaya (1965), Nodia et al. (1970 a), Burlatskaya and Voronov (1974), Burlatskaya et al. (1976) and Burlatskaya et al. (1981). The weighted mean deviation is 2.3μT. The stability of the results was controlled by comparison of different sets of data.

It should be noted that each set of measurements is the result of the treatment of some hundreds of samples taken from tens of objects. In the papers referred to above, the quality of each collection and the reliability of the results are fully discussed. Data from different regions of Georgia show similar variations, hence the summary curves of Figure 3-22 are considered to represent the overall variations of the geomagnetic field in the Caucasus.

3.10.3. The Ukraine, the Crimea, Moldavia, Poland and Central Asia and Mongolia

The changes of intensity and inclination for the Ukraine, the Crimea and Moldavia (Burlatskaya et al.,1969a; Nechayeva,1970) and for the Ukraine and Moldavia (Rusakov and Zagny, 1970) for the last 2000yr are shown in Figures 3-23 a and b. Inclination variations for southern Poland for the last 1000yr made by Polish and Soviet scientists are shown in Figure 3-23c (Burlatskaya et al.,1969b).

The most recent inclination maximum for these two large regions shows a shift of phase from which the rate of westward drift has been estimated at about 0.12 deg/yr. For all investigated regions the time of the inclination maximum was plotted against the site longitude. For sites between 0° and 45°E longitude, westward drift at an average rate about 0.2 deg/yr was deduced. Data for regions situated to the east of 45°E could not be used in this test.

Investigations were made on a collection of samples from Samarkand and Buchara (64-67°E) in central Asia. Every specimen was divided into two parts one of which was measured in Moscow and the other in Leningrad. The difference in angular measurement was only about 1.8° and a systematic divergence of about 5% in intensity was attributed to the different experimental methods and data treatment used at the two laboratories. These results are shown in Figures 3-23 d and e, reproduced from Burlatskaya et al. (1970) and Burlatskaya (1972). However these results did not contribute to the estimation of drift rates.

To provide better coverage of eastern USSR, a study of Mongolian samples (103-110°E) was undertaken (Ajuushayev et al.,1976). Intensity and inclination variations are shown in Figures 3-23f and g respectively. The last inclination maximum occurs at about 1000AD.

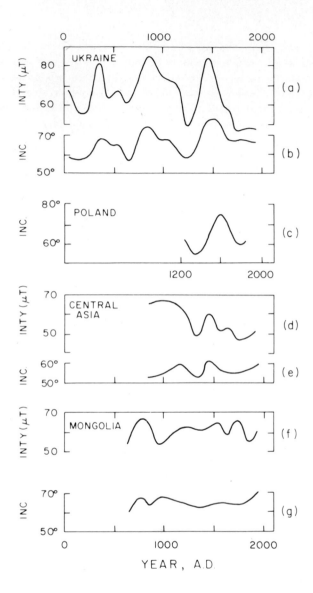

Figure 3-23:   Intensity and inclination for the last 2000yr in:
(a, b) the Ukraine (Nechayeva, 1970);
(c) Poland (Burlatskaya et al., 1969b);
(d, e) Central Asia (Burlatskaya et al., 1970);
(f, g) Mongolia (Ajuushayev et al., 1976)

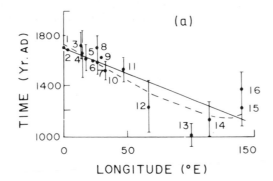

Figure 3-24:   Westward drift for: 1) London, 2) Paris, 3) Rome, 4) Sicily, 5) Venice,
6) Poland, 7) Lvov, 8) Kaunas, 9) Vilnus, 10) the Ukraine, 11) Caucasus,
12) Central Asia, 13) China, 14,15) Japan (after Burlatskaya et al., 1968)

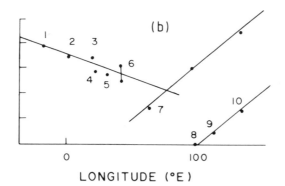

Figure 3-25:   Westward drift rates for: 1) Iceland, 2) England,
3) Central Europe, 4) Bulgaria, 5) Ukraine, 6) the Caucasus,
7) Central Asia, 8) Mongolia, 9) China, 10) Japan
(after Petrova and Burlatskaya, 1979)

3. 10. 4. The European part of the USSR and Siberia

The investigation of a rare collection of well-dated (±5 yr) specimens from the
Moscow region for 1500AD to 1900AD allowed us to reveal formally a 115yr variation
(Nachasova, 1972). The investigation of the geomagnetic field at Vologda for the last
190yr and at Gorky for the last 80yr showed the same time-characteristics (Burakov
and Nachasova, 1976). Analogous fluctuations were observed for Hiva from 1549AD
until 1906AD (Burakov and Nachasova, 1978; Nachasova, 1972).

For Kazan, Kaluga, Vladimir, Suzdal in central Russia the inclination change has
the same form but smaller amplitudes than for Moscow (Tarchov, 1963, 1970). A
wide variety of investigations was made for the European part of the USSR and Siberia

by Tarchov (1963, 1965, 1967) and by Tarchov and Ivanov (1965) providing results which were useful for comparison of data from different regions. Some inclination data shown in Figure 3-23f were taken from these papers.

3.10.5. The westward drift

Figure 3-24 shows the date and site longitude of the last inclination maximum obtained from archaeomagnetic data for sites ranging in longitude from ~10°W to ~150°E. Westward drift at an average rate of 0.23±0.06 deg/yr is suggested by results from those sites to the west of longitude 100°E but the points for China and Japan do not fit this simple pattern and suggest that the velocity of drift might change in sign as well as magnitude with time and longitude.

An analysis of newer measurements allowed a comparison of the inclination variations for: Ireland, England, Central Europe, Bulgaria, the Ukraine, the Caucasus, Central Asia, China and Japan (Burlatskaya and Nachasova, 1978). A plot of the time of the inclination maximum versus longitude (Figure 3-25) suggests westward drift at 0.33 deg/yr for the longitude range 20°W to 60°E and eastward drift at 0.12 deg/yr for the longitude range 100-140°E. A comparison of this picture with the configurations of isoclines of the East Asia anomaly shows that the drift velocity changes when the centre of the focus of the anomaly is crossed. This suggests that the sources of these drift variations are connected with the geomagnetic non-dipole anomalies, and that, in this case, the source is situated in the longitude interval about 80°-100°E (Petrova and Burlatskaya, 1979).

3.10.6. Summary of global archaeomagnetic data

The prime question for us was whether the ancient field had a dipolar symmetry like the historic field. The successive accumulation of data (Burlatskaya, 1965; Burlatskaya et al., 1968) showed that to a first approximation the distribution of the geomagnetic field over the Earth's surface corresponded to the dipole model for the last 1000yr – 2000yr, any deviations being commensurable with the non-dipole part of the present geomagnetic field.

Another question concerned the secular variation spectrum. The collection of archaeomagnetic results both from the USSR and other countries allowed us to compose a catalogue of archaeomagnetic data (Burlatskaya and Nachasova 1977) including results from 67 papers covering about 40 different regions from 3000BC to the present. Spectral analysis of archaeomagnetic curves for D, I and F derived from data averaged over eleven large regions, showed that the strongest variations have periods of 300-400yr and 600yr but variations with 900yr period also exist, as was found earlier (Burlatskaya and Nachasova 1978). Variations with longer periods were identified in curves of world-wide geomagnetic variations: the summarized intensity data (Burlatskaya, 1970) showed the existence of a period of about 8000yr. Spectral analysis of the three elements of the geomagnetic field (Burlatskaya, 1978), using curves composed of world-wide data revealed periods of 1200yr for direction and 1800yr for intensity.

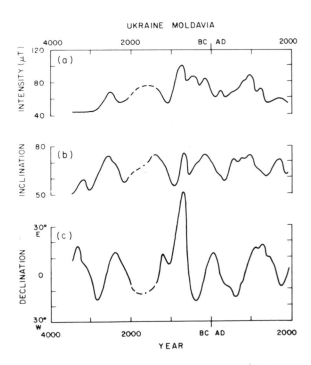

Figure 3-26:   Variations of: (a) intensity, (b) inclination and (c) declination
in the Ukraine — Moldavia for the last 5500yr (Zagny 1981)

Interesting results from a very rare collection of ancient kilns and ovens, baked walls and floors from Ukraine-Moldavia spanning the last 5500yr (Zagny 1981) are shown in Figure 3-26. The variations of all three elements have a component of 1200yr and there are also some changes with a characteristic time of 600yr which took place each 1200yr. These data as well as the results for Bulgaria-Yugoslavia (Kovacheva 1980) were used for summarizing up and averaging of world-wide data.

The variations of intensity, inclination and declination were calculated using the most complete set of averaged world data for the last 10000yr (Burlatskaya 1981) for which it was shown that the geomagnetic field had a dipolar structure. The non-

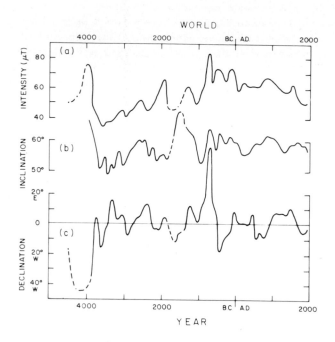

Figure 3-27:   Variations of: (a) intensity, (b) inclination and (c) declination
for trustworthy world archaeomagnetic data reduced to a common site
longitude of 42°N (Burlatskaya, 1981)

dipole part of geomagnetic field throughout this time was about 10–15% and had
regional character. We can choose two main regions: Europe and Japan-America.
The variations of the non-dipole part of geomagnetic field in these regions have
opposite phases.   In Figure 3-27 the secular variations of three elements of
geomagnetic field since 6500BP based upon the most trustworthy investigations are
shown. The intensity and inclination values have been recalculated for latitude 42°N.
Variations of the three elements of geomagnetic field had similar features, which were
noted   in   previous   papers   (Burlatskaya   et   al.,1969a;   Nechayeva,1970;
Nachasova, 1972; Zagny, 1981).

3.10.7. Virtual dipole moment and VGP paths

Figure 3-28 shows the quasi-sinusoidal oscillation of geomagnetic field intensities
averaged over 500yr intervals at 250yr steps (Burlatskaya, 1970) for the last 9000yr.
500yr means (triangles) and 1000yr means (circles) of the virtual axial dipole moment

(Barton et al., 1979) are in good agreement with this curve. Century means of the virtual dipole moment for the last 6000yr are shown in Figure 3-28. For comparison the above mentioned sinusoidal curve at 500yr means is shown, (dashed line).

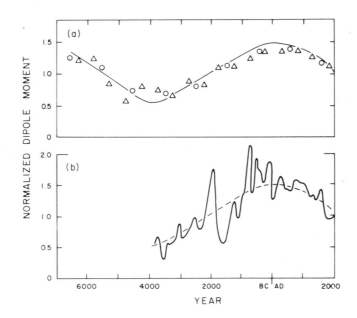

Figure 3-28:   Variations of geomagnetic moment:
(a) 500 year means of virtual dipole moment at steps of 250 year) for the last
10000 years – quasi-sinusoidal line (Burlatskaya, 1970), also 500 year means
(triangles) and 1000 year means (circles) of virtual axial dipole moments
(Barton et al., 1979);
(b) 100 year means of virtual dipole moments (and 500 year means – dotted line) based
upon the world archaeomagnetic data for the last 6000 yr (Burlatskaya, 1981).
Moments are all normalized with respect to the present day value.

Periodogram analysis on world–wide archaeointensity data revealed variations with a period of 3200-3700 yr (Burlatskaya, 1978).

The path of the global virtual geomagnetic pole (VGP) from 6400BC to the present based on world archaeomagnetic data is shown in Figure 3-29(Burlatskaya 1981). The dashed lines show uncertain parts of the path. Variations at about 1200yr and loops with characteristic times of 400-600yr may be seen. Only one loop in the 9th century BC descends to a latitude lower than 70°N. The longitudinal displacement of the virtual geomagnetic poles reveals a systematic eastward drift superimposed by variations of about 3000-4000yr, and shorter. The latitudinal dislocation displays a constant northerly component and 300-600yr fluctuations on its later background. The combination of westward and northward drift produces a spiral anticlockwise motion of virtual geomagnetic pole around the geographical axis with period of about 1200yr.

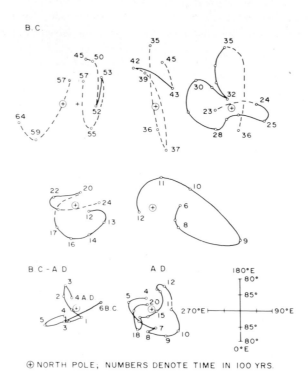

Figure 3-29:   The displacement of the virtual geomagnetic pole based on world-wide
                archaeomagnetic data for the last 8500 years (Burlatskaya 1981).
                Numbers on the curves denote ages in units of 100yr.

3.10.8. Analytical representation of the geomagnetic field

    About ten spherical harmonic analyses for the interval 1500 AD to the present are
available. A comparison of archaeomagnetic results with the analytical representation
of the field for the last 350 years was made by Braginsky and Burlatskaya (1972). The
intensities and inclinations were calculated from the set of Gauss coefficients taken
from Braginsky (1969), Fritsche (1889) and Leaton et al. (1965). The
archaeomagnetic results were taken from the papers by Burlatskaya et al. (1969a),
Nechaeva (1970), Nachasova (1972), Tarchov (1963, 1965) and Tarchov and Ivanov
(1965).

A comparison of archaeomagnetic intensity and inclination values for Bulgaria with the global analytical model for the last 500yr was made by Kovacheva et al. (1977). Average mean-square differences between calculated and experimental smoothed values in were estimated at $3-5\mu T$ for intensity and $2.4°$ to $3.2°$ for inclination.

The modelled interval was later expanded to 2000yr, and gave results which compared favourably with world-wide archaeomagnetic data (Burlatskaya and Braginsky, 1978). Benkova et al. (1977) compared archaeomagnetic results from Bulgaria with four analytical models of the field (Braginsky 1972, 1974; Kolomiitseva and Pushkov, 1974, 1975) and found mean square differences of 6.5 to $10.5\mu T$ in intensity and $4.6°$ to $8.1°$ in inclination. Expanding the time interval to 2000yr thus leads to a two-fold increase in divergence between the analytical models and the archaeomagnetic data.

Detailed comparison of two models of the main geomagnetic field (Braginsky, 1974; Kolomiitseva and Pushkov, 1976) with archaeomagnetic data for the last 2000yr (Benkova et al., 1979) showed that both models reflected the global distribution of intensity and inclination rather well with the above mentioned limits. They both described the general character of the trend, but they smoothed out variations with periods of hundreds of years. According to the model of Braginsky (1974) the zone of motion of the magnetic centre was located in the Pacific during the whole time-interval of 2000yr whereas Kolomiitseva and Pushkov's (1976) model required the magnetic centre to drift westwards. Benkova et al., (1979) note that differences between the model and measured values of all geomagnetic elements are greatest for India, Mongolia and China. This is in accordance with the existence of stationary but oscillating sources of the non-dipole field located in the outer part of the Earth's core (Creer, 1977; Petrova and Burlatskaya, 1979).

3.10.9. Conclusion

This summary of archaeomagnetic data shows that the ancient geomagnetic field was essentially dipolar with a spectrum of variations showing periods of 8000yr, 1800yr, 1200yr, 600yr and 400-300yr. Thanks are expressed to Prof. G.N. Petrova for discussions, advice and support of archaeomagnetic investigations.

3.11. RESULTS FROM CHINA

Q.Y. Wei, T.C. Li, G.Y. Chao, W.S. Chang, S.P. Wang and S.F.Wei,
Institute of Geophysics, Academia Sinica, Beijing.

3.11.1. Introduction

China, being one of the earliest civilized countries, contains many archaeological sites. These are to be found especially on the banks of Yellow and Yangtze Rivers. Many types of material suitable for archaeomagnetic research are available such as baked earth from pottery or brick kilns, ovens of residential structures, bricks from graves and city walls, pottery, tiles etc. The more important results of archaeomagnetic research in China to date are reported in this section.

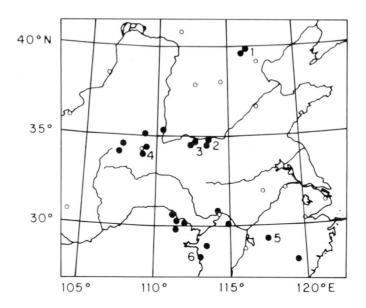

Figure 3-30:   Map of sampling localities:
1-Beijing 2-Zhengzhou 3-Luoyang
4-Xian   5-Jingdezhen 6-Changsha.

The samples used for the determination of both direction and total intensity of the ancient geomagnetic field were collected from more than 20 regions which may be identified in the sketch map of Figure 3-30. Their ages cover the last 7000yr (ca.

Table 3-2:   Description of sampling sites

Site No	Age			Λ (°,')	Φ (°,')	Sample type	N
01	1027BC	–	771BC	39 35	115 59	Baked earth	3
02	221BC	–	207BC	39 54	116 23	Tile	4
03	206BC	–	220AD	39 54	116 23	Brick	
04	24AD	–	220AD	39 54	116 23	Brick	10
05	222AD	–	280AD	39 54	116 23	Brick	14
06	265AD	–	420AD	39 54	116 23	Brick	17
07	618AD	–	905AD	39 54	116 23	Brick	7
08	916AD	–	1125AD	39 54	116 23	Brick	39
09	960AD	–	1276AD	39 54	116 23	Brick	12
10	1115AD	–	1234AD	39 54	116 23	Brick	4
11	1271AD	–	1368AD	39 54	116 23	Brick	14
12	1271AD	–	1368AD	40 29	116 45	Baked earth	6
13	1522AD	–	1566AD	39 54	116 23	Brick	24
14	1644AD	–	1911AD	39 54	116 23	Brick	8
15	1960.0	epoch		39 54	116 23	–	–
16	4495BC	±185		34 22	109 12	Baked earth	37
17	2000BC	–	1000BC	34 30	107 49	Baked earth	8
18	1180BC	±90		34 30	107 49	Baked earth	12
19	1027BC	–	771BC	34 10	108 57	Baked earth	18
20	221BC	–	207BC	34 20	109 17	Brick	24
21	206BC	–	220AD	35 28	110 26	Brick	5
22	420AD	–	589AD	34 16	108 57	Brick	8
23	618AD	–	907AD	34 16	108 57	Brick	6
24	960AD	–	1127AD	35 28	110 26	Brick	11
25	960AD	–	1127AD	34 17	107 36	Brick	9
26	1115AD	–	1234AD	35 05	109 05	Kiln brick	9
27	1271AD	–	1368AD	35 28	110 26	Brick	11
28	1436AD	–	1449AD	34 10	108 57	Brick	9
29	1368AD	–	1644AD	35 28	110 26	Brick	3
30	1636AD	–	1911AD	35 28	110 26	Brick	10
31	1636AD	–	1911AD	34 16	108 57	Brick	3
32	1960.0	epoch		34 16	108 57	–	–
33	5300BC			34 31	113 22	Baked earth	2
34	480BC	–	220BC	34 45	113 39	Brick	8
35	206BC	–	24AD	34 41	112 26	Baked earth	8
36	206BC	–	220AD	34 45	113 39	Baked earth	5
37	9AD	–	23AD	34 41	112 26	Grave brick	33
38	25AD	–	220AD	34 41	112 26	Baked earth	3
39	220AD	–	300AD	34 41	112 26	Baked earth	54
40	265AD	–	316AD	34 41	112 26	Brick	29
41	386AD	–	534AD	34 43	112 47	Brick	38
42	562AD	–	565AD	34 41	112 26	Grave brick	34
43	581AD	–	907AD	34 41	112 26	Brick	12
44	618AD	–	907AD	34 41	112 26	Baked earth	11
45	960AD	–	1127AD	34 41	112 26	Brick	16
46	960AD	–	1127AD	34 10	113 28	Baked earth	14
47	1127AD	–	1279AD	34 10	113 28	Baked earth	9
48	960AD	–	1279AD	34 47	114 21	Brick	3
49	1368AD	–	1435AD	34 41	112 26	Brick	8
50	1436AD	–	1644AD	34 45	113 39	Brick	9
51	1436AD	–	1644AD	34 41	112 26	Brick	10
52	1821AD	–	1850AD	34 47	114 21	Grave brick	2
53	1636AD	–	1911AD	34 41	112 26	Brick	8
54	1960.0	epoch		34 41	112 26	–	–

Table 3-2 continued

Site No	Age		Λ (°,')	Φ (°,')	Sample type	N
55	5000BC	– 4000BC	30 21	112 12	Pottery	5
56	3000BC	– 2600BC	30 26	111 45	Baked earth	8
57	2200BC	– 1800BC	30 31	111 24	Baked earth	4
58	1600BC	– 1028BC	30 52	114 22	Pottery	7
59	1027BC	– 771BC	30 31	111 24	Tripod	4
60	770BC	– 220BC	30 13	115 05	Baked earth	5
61	480BC	– 220BC	30 21	112 12	Baked earth	3
62	206BC	– 220AD	30 31	112 12	Grave brick	9
63	420AD	– 589AD	30 31	112 12	Grave brick	3
64	4000BC		28 36	113 47	Pottery	4
65	3000BC	– 2000BC	29 39	111 45		1
66	206BC	– 8AD	28 12	112 59	Disc	4
67	25AD	– 220AD	28 12	112 59	Brick	4
68	220AD	– 265AD	28 35	112 20	Brick	3
69	265AD	– 420AD	28 12	112 59	Brick	3
70	960AD	– 1127AD	28 05	119 07	Kiln-wall	25
71	1271AD	– 1368AD	28 05	119 07	Brick	31
72	1127AD	– 1279AD	29 18	117 12	Brick	11
73	1368AD	– 1644AD	29 16	117 12	Brick	11

The age represents the duration of corresponding dynasty, i.e. the historical interval according to the Chinese chronology. Λ and Φ are north latitude and east longitude of sampling site respectively. N is number of samples used in this work.

The regions or provinces which the archaeological site with given site No. belongs to are given as follows.

01–15	Beijing region;	16–32	Xian region;
33–54	Luoyang region;	55–63	Hubei province;
64–69	Hunan province;	70, 71	Zhejiang province;
72, 73	Jiangxi province.		

5300BC to 1900AD). Details of the site localities are given in Table 3-2.

J_N is the intensity of NRM; J_V is the intensity of VRM, acquired by the specimen in the present geomagnetic field after two weeks storage in the laboratory. Both J_N and J_V are given in arbitrary units.

The direction and intensity of NRM were measured with astatic magnetometers. Tests on the stability of primary NRM, including a viscosity test, AF demagnetization at 50Hz up to 35mT, and thermal cleaning at 80 to 200°C were carried out and the results are presented in Tables 3-3 and 3-4. The average direction and the related statistical parameters (Fisher, 1953) for every site and every dynasty were calculated. The total intensity of the ancient geomagnetic field was determined by using the Thelliers method (Thellier and Thellier, 1959; Nagata et al., 1963). Some examples of NRM-TRM diagrams and the corresponding thermal demagnetization curves are

Table 3-3:   Percentage of viscous remanent magnetization

Site	Sample	J_N	J_V	$100J_V/J_N$
06	8	56.8846	0.2638	0.5
08	8	26.9954	0.5377	2.0
11	1	59.0162	0.9271	1.6
15	1	55.8758	1.8103	3.2
15	1	152.3745	9.6187	6.3
19	1	270.7540	10.6556	3.2
20	1	32.1279	2.0381	6.3
	3	131.9857	0.8302	0.6
23	3	83.4051	1.7591	2.1
35	1	35.7595	1.0644	3.0
37	2	71.9442	0.8486	1.2
38	2	110.9320	2.1724	2.0
39	5	38.7074	0.4637	1.2
40	2	92.5152	0.3226	0.3
41	1	102.3345	1.5583	1.5
	2	138.5665	2.1984	1.6
	3	140.2478	1.4893	1.1
42	2	131.5037	1.2590	1.0
43	1	65.3142	0.4064	0.6
44	1	85.3314	1.1841	1.4
45	3	39.1387	1.6451	4.2

J_N is the intensity of NRM; J_V intensity of VRM, acquired
by the specimen in the present geomagnetic field after two weeks
storage in the laboratory. Both J_N and J_V in arbitrary units.

shown in Figure 3-31.   Further details of measurements and data treatment are
given by Wei et al. (1980).

3.11.2. Inclination variations

Figure 3-32 shows the variation of inclination in Beijing, Xian, Luoyang regions and
some parts of Hubei province.   Inclination in the Beijing region, going back to
1000BC, ranged from 50° to 68° and the fluctuations appear to consist of a
superposition of cyclic variations of long (1800yr) and short (200yr – 300yr) period.

An archaeological site of Neolithic age in Jiangzhai (34°22'N, 109°12'E) in Lintong
county to the north-east of Xian was dated at 6495±185bp by the radiocarbon method
(Cai and Qui, 1979) and the ancient inclination was determined to be 48°.   As for the
period from 1500BC until 1900AD in the Xian region, the inclination ranged between
40° and 60° (see Figure 3-32).   The highest value occurred at about 900BC and the
lowest value at about 200AD.   And after 500AD the fluctuations appear cyclic with
apparent period of about 600yr with a minimum at about 1700AD as for the Beijing
region.

Table 3-4: Change of inclination after heating

Site	Sample	I_0 (°O.')	I_T (°T.')	T°C (°C)
06	10	45 56	47 37	80
08	6	58 27	59 00	100
11	1	52 23	51 58	80
19	3	54 05	52 12	80
20	1	28 47	28 07	100
	3	49 13	49 49	100
23	1	42 07	42 45	100
	2	48 48	47 14	100
	3	51 15	51 19	100
35	6	31 53	33 02	30
37	2	42 57	42 26	80
	3	34 32	34 57	80
38	1	43 01	44 13	100
39	1	65 31	65 02	80
	2	46 23	44 55	80
	3	44 16	45 01	100
	6	54 17	54 14	160
	7	42 13	42 46	100
	7	47 18	47 34	100
	10	46 18	62 53	100
40	1	50 10	49 59	100
	4	51 57	52 04	80
41	1	53 56	52 28	100
	2	49 27	51 04	100
	4	56 29	56 23	80
42	1	57 51	55 50	100
	4	54 16	54 46	100
	4	46 21	64 29	100
43	1	61 17	60 12	100
	2	57 09	57 33	100
44	2	62 26	61 31	100
45	1	57 50	57 45	100
	2	52 49	55 07	100

I_0 and I_T are the inclination before and after heating at temperature T.

It is interesting to compare the curves of variation of inclination for the Xian and Luoyang regions (Wei et al., 1981). It is clearly seen that the character of the variations is rather different although the distance between these two regions is only about 400km. The common feature is the very high inclination of ~60° attained at about 900BC and also at about 760AD in Xian and at about 800AD in Luoyang. The present value of inclination is about 50° for both regions.

The inclination in Hubei province, which is situated in southern China, was about 53° during the Shang Dynasty (1600BC to 1028BC). This value is much higher than the present value of 44°. The curves in Figure 3-32 demonstrate that the inclination attained high values in both middle and southern China, suggesting that the geomagnetic pole was formerly located close to East Asia.

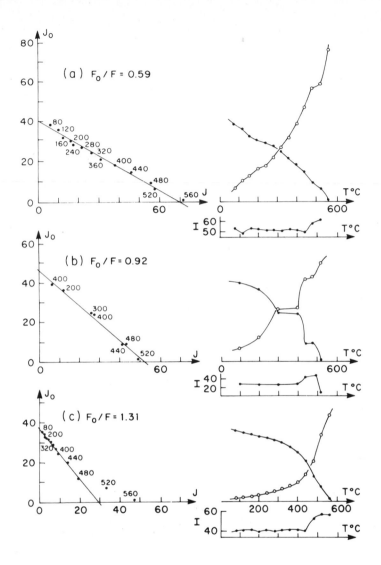

Figure 3-31:   NRM-TRM diagram and the corresponding thermal demagnetization curve.
The inclination at different temperatures is given beneath:
(a) baked earth. Neolithic period (~5300BC). County Mi, Henan province;
(b) brick, Qim Dynasty (221BC–207BC), Lintong, Shanxi province;
(c) brick, Three Kingdoms (220–280AD), Changsha, Hunan province.
J and J_o are given in arbitrary units.

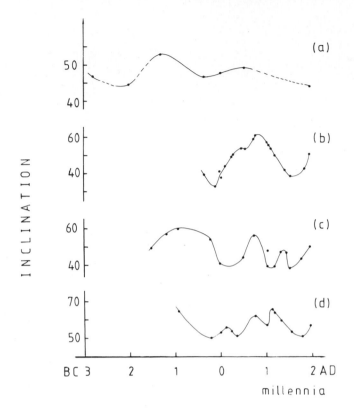

Figure 3-32:   Variation of inclination: (a) Hubei province;
(b) Henan province; (c) Shanxi province; (d) Beijing region.

3.11.3. Declination variations

The results for Luoyang are shown in Figure 3-33.   The declination was positive
(easterly) between 100BC and 350AD, the largest positive value of more than 11°
having occurred at the beginning of the first century AD.   After 350AD the declination
became negative (westerly), peaking at -23° at 500AD after which it decreased to
-3° for the 1960.0 epoch.

For the Xian region the results cover the period from 1500BC to 900BC.   Before
1200BC declination was positive and then changed sign.   Figure 3-34 shows a Bauer
diagram of declination plotted against inclination for the Luoyang region.   Both
clockwise and counter clockwise looping of the geomagnetic vector will be observed.
The regions investigated were located near the agonic line and any change in position
of this line will result in a change of sign of declination in the area swept out.

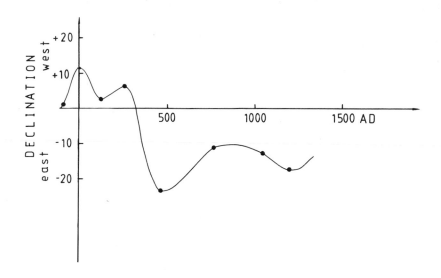

Figure 3-33: Variation of declination for Luoyang region.

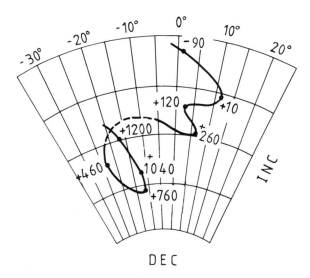

Figure 3-34: Secular variation for Luoyang region. The dates indicated are AD(+) and BC(−).

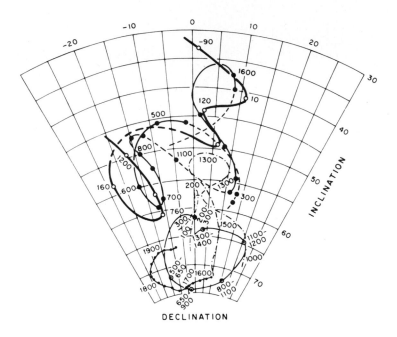

Figure 3-35:   Secular variation for Luoyan, London ,
Japan and Ukraine . The dates indicated
are BC(-) and AD(+).

The Bauer plot for Luoyang (Wei et al., 1981) is compared with Bauer plots for London (Aitken, 1970), Ukraine (Rusakov and Zagny, 1973), Japan (Hirooka, 1971) and other localities in Figure 3-35. It will be seen that the patterns of secular variation for Luoyang and Japan have many similar characteristics, especially for the interval 400AD to 1200AD. The Japanese curve shows a complex anti-clockwise loop for the interval of 500AD to 800AD while the Chinese curve has a similar loop for the interval of 460AD to 1200AD. This anti-clockwise loop appears progressively later for sites located towards the west, suggestive of westward drift.

3.11.4. Intensity variations

The total intensity of the ancient geomagnetic field in Beijing, Xian, and Luoyang regions and Hubey and Hunan provinces (Figure 3-36) has fluctuated during the last 7000yr attaining a peak of almost 160μT at about 500AD (Teng and Li, 1965; Sasajima, 1965).

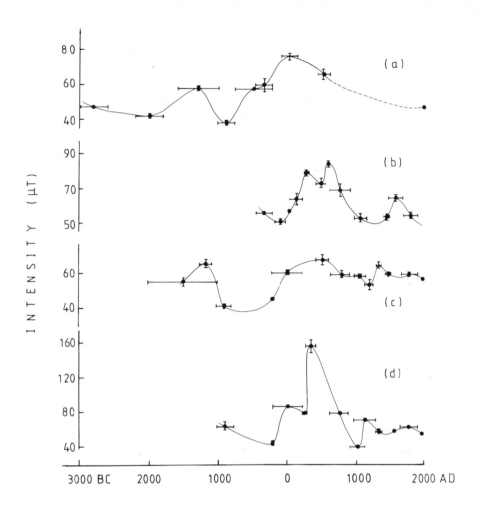

Figure 3-36:   Variation of total intensity: (a) Hubei province;
(b) Henan province; (c) Shanzi province; (d) Beijing region.

For comparison, the present intensity is about 55μT in Beijing. The difference between this maximum value and the observed minima of 42μT at about 200BC and 40μT at about 1000AD is ~90μT. The variations during the last thousand years look like damped oscillations with amplitudes less than 30μT and period ~500yr. The total intensity has also changed a great deal during the last 3500yr in the Xian region: there is a minimum which lasted from 100BC until 0AD, and afterwards quasi-cyclic changes with apparent periods of about 500yr - 800yr occurred.

For the Luoyang region, a maximum appeared at about 300AD and it is to be noted

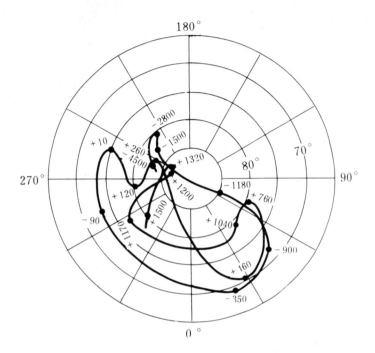

Figure 3-37: Virtual geomagnetic pole positions derived from archaeomagnetic data for China. Positive numbers label AD ages and negative numbers label BC ages.

that the character of total intensity variations are very different for Luoyang and Xian. We have already noted that the character of the inclination variations is different for these two regions.

3.11.5. VGP path for the last 6000 years

The positions of the VGP for different periods are presented in Figure 3-37. The movement of the northern geomagnetic pole traces out a number of loops centred on the far side (with respect to China) of the geographical pole through the last few thousands of years.

Both clockwise and anti-clockwise motion occurred, the VGP being located within

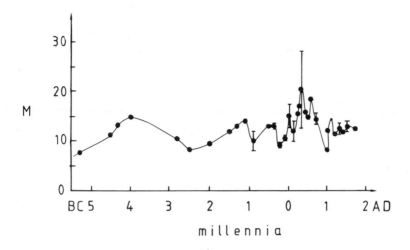

Figure 3-38: The averaged curve of variation of RDM. The standard errors are shown where the number of data averaged is more than three. M is given in $10Am^2$.

the sector from 150°W to 70°E the lowest VGP latitude being 68°N. For the period before Christ, the pole moved in a clockwise sense: westward movement was dominant but north-south motion of the pole is also evident for the interval from about 2000BC to 1000BC. After some wobble between 10AD and 250AD, the direction of motion was approximately eastward until 760AD when a backward turn occurred after which the motion was directed to the south-west for about 300yr and then rapidly to the west until about 1170AD when it turned to the north. For the time from 1170AD until 1500AD the VGP path traces out an anti-clockwise loop.

The speed of VGP motion appears to have slowed down when the sense of looping changed. Some recent results (Yukutake, 1979) indicate that the equatorial dipole has been drifting westwards since the 17th century, although the speed has been apparently slower since the beginning of the 19th century. That may be an omen of change in pole movement direction: the northern geomagnetic pole may move back to the east at some time in the near future.

A comparison of Chinese and Japanese VGP paths for the last 1700yr (Hirooka, 1971) indicates that they fall in the same longitude sector and that they show some similarities in shape, especially for the period from 250AD to 800AD. But the British VGP path (Aitken, 1974) is located in a totally different sector.

3.11.6. Changes of the geomagnetic moment

The reduced dipole moment (RDM), calculated from the total intensity data for the last 7300yr, is illustrated in Figure 3-38. It attained a maximum during the first few centuries AD. A comparison of Chinese data with world-wide data indicates that there is good agreement in general trend for the last three to four thousand years, but the Chinese RDMs are much higher than world-wide averages (Cox, 1968). This may be a result of either anomalous NDF intensity in this region and/or of the position of the geomagnetic centre which would have been located in the northern part of the eastern hemisphere.

3.12. RESULTS FROM JAPAN

By Kimio Hirooka,
Toyama University, Japan

3.12.1. Archaeomagnetic secular variation

According to Imamiti (1956), who catalogued historical records of instrumental measurements of the geomagnetic field made in Japan, observations of declination date back to 1613AD. The first such observation was made by Saris at Hirado, Kyushu by using a navigation compass. After Saris, eleven observers had recorded declination data by the time of the first geomagnetic survey of the islands of Japan (Knott and Tanakadate, 1889). No inclinations have been recorded prior to this survey.

Since no direct measurements of the geomagnetic field were made before 1613AD, the field direction prior to the 17th century may only be inferred from archaeomagnetic measurements of remanent magnetization of volcanic lava flows and baked clays. There are several pioneer works in which the direction of the geomagnetic field through historic time are described. Nagata (1943) and Kato and Nagata (1953) measured remanent directions of volcanic lava flows of which the dates of eruption were well recorded in old documents, e.g. the Aokigahara Lava (864AD) of Mt. Fuji, the An-ei Lava (1778AD) of Mt. Mihara and the Bunmei Lava (1471AD) and the An-ei Lava (1779AD) of Mt. Sakurajima. The first results of magnetic measurements on baked clays from archaeological sites were reported by Watanabe (1949) and Kobayashi et al. (1953).

It was Watanabe (1958, 1959) who initiated systematic archaeomagnetic studies in

Japan . He made many magnetic measurements of baked clays taken from ancient hearths, fireplaces and kilns and summarized the results obtained from 178 baked clay remains excavated at 55 archaeological sites, for the purpose of recovering the secular variation of the past geomagnetic declination and inclination. He obtained the secular variations for two different periods: first for the Jomon period (5600BP to 4400BP) and second for the last 1700 years. In the nineteen fifties however, the available archaeological information was not precise enough for the magnetic data to be arranged in their correct chronological order. Watanabe assumed, therefore, that inclination varied periodically between 40° and 60° , and adopted a sinusoidal variation curve with a period of 1000yr which satisfied the inclination values of the historical lava flows of 864 AD, 1778AD and 1779AD. For the Jomon period, he also assumed the same sinusoidal period. The secular variation curves of declination for both these periods were constructed by adjusting the ages so that the corresponding inclination values fitted the assumed sinusoidal curves as well as possible, and then the declination values were plotted against the adjusted ages. The chronology of the Jomon period is, as yet, not precise enough to distinguish age differences of a few hundred years, during which time the direction of the geomagnetic field might have changed considerably. Consequently, no new secular variation curves for the Jomon period have been produced since Watanabe's.

Momose et al. (1964) constructed secular variation curves for the period between the 7th and the 9th centuries AD from the measurements of roof tile kilns specially built for the construction of the old Buddhist temples in Nara, the old capital of Japan. Yukutake et al. (1964) obtained secular variation curves for the past 1500yr from a study of well dated lava flows and volcanic ash layers of Mt. Mihara, Oshima Island. Kawai et al. (1965) reported the results of an archaeomagnetic survey in south-western Japan and presented secular variation curves of declination and inclination for the past 1700yr.

Hirooka (1971) continued the study on baked clays of archaeological sites, mainly old kilns, distributed in south-western Japan to complete the archaeo-secular variation curves for the past 2000yr. Watanabe (1977) produced revised secular variation curves constructed on the basis of his newly obtained data.

All the secular variation curves mentioned above are illustrated in Figure 3-39. Although some minute differences are to be recognized among the curves, they show a very similar pattern on the whole. The geomagnetic field in Japan shows a characteristic secular variations with westerly declinations prevailing during the 700yr from the 5th to the 12th centuries AD, followed by easterly declinations until the end of the 18th century AD. Meanwhile, the inclination shows succesive increases and decreases at intervals of about 400yr to 600yr.

Figure 3-40 shows the archaeo-secular variation of the geomagnetic field directions during the last 2000yr in south-west Japan (Hirooka, 1971, 1977), obtained from

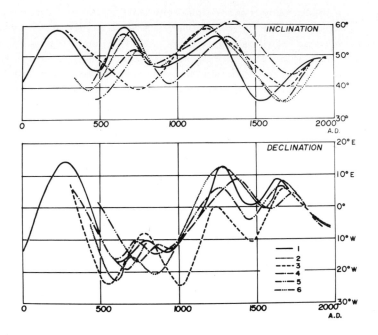

Figure 3-39: Declination and inclination secular variations recorded
in Japan after (1) Hirooka (1971), (2) Watanabe (1977), (3) Watanabe (1959),
(4) Kawai et al (1965), (5) Yukutake et al (1964) and (6) Momose et al (1964)

several very well dated kilns such as two roof-tile kilns and a kiln of Oribe-yaki
ceramics. At the time of the construction of the Heian Palace in Kyoto, many kilns,
including the two mentioned above, were specially built for the purpose of baking roof-
tiles for the palace which was completed in 794AD. Hence the kilns were dated at
790±10AD. On a potsherd, excavated from a kiln of Oribe-yaki ceramics (Motoyashiki
Kiln), the date "the 12th year of Keicho" (1607AD) was written, and as this kiln was
still in use at this date, its age is estimated to be 1620±20AD.

The secular variation curve shown in Figure 3-40 is now used for archaeomagnetic
age dating. Baked clays from many excavated kilns and hearths were measured, and
the ages determined by archaeomagnetic methods show good agreement with those
estimated by the archaeologists. In south-western Japan, the accuracy of magnetic
dating is now as good as 20yr for well baked and strongly magnetized clays.

After the systematic work of Watanabe (1959) and Hirooka (1971), furhter
archaeomagnetic measurements have been made at many excavated sites. These
newer data have been described in the excavation reports. Studies of Sue type kilns
from Sue Village (Suemura), Osaka Prefecture, have yielded comprehensive results

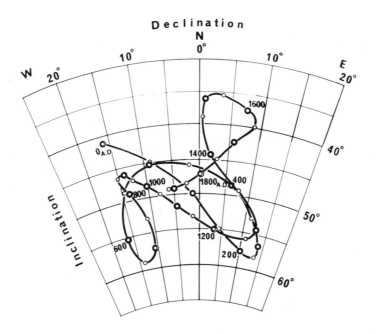

Figure 3-40:   Bauer plot of secular variations in the direction of the geomagnetic field through the last 2000 years for south-west Japan (Hirooka, 1977).

(Hirooka et al., 1970; Torii et al., 1976; Nakajima et al., 1977) These kilns are the oldest yet found in Japan and before that time earthenware was not baked in kilns. The technique of baking Sue ceramics was introduced from Korea in the middle of the 5th century AD when ceramic craftsmen immigrated into Sue Village and they continued their work for about 500 years. Sue ceramics are grey coloured, thin and hard. Several decades after the immigration, this new and advanced technique spread all over Japan.   Archaeologically well dated results obtained during the last five years are listed in Table 3-5, and it is apparent that regional differences of the direction of the geomagnetic field cannot be neglected. Japan is situated on an island arc which runs more than 1000km from north-east to south-west, so that the local field direction has varied from place to place. It is seen in the table that westerly declinations become pronounced in the 7th and 8th centuries AD at the sampling sites located in the east and north and this suggests that the isogonic lines ran in a north-west to south-east direction in that period. In the 17th and the 18th centuries AD, it is evident that the isogonic contours were aligned along a north-south direction and the tendency for easterly declinations increased towards the east. The same tendency is also recognized from the historical observations of 1643AD which were quoted by Imamiti (1956).

Table 3-5: Archaeomagnetic results from Japanese kilns obtained in the last five years in Japan

Kiln name lat.	Locality long. (°N) (°E)		N	D (°)	I (°)	α95 (°)	K	age (yr AD)	reference
Suemura TG19	34 29	135 30	21	−4.0	53.0	2.3	197.8	410–470	1
Suemura TG43-II	34 28	135 30	29	−5.4	44.6	2.6	106.7	460–510	1
Suemura TG51	34 28	135 30	6	−14.0	42.7	2.6	653.5	450–500	1
Dairenji	38 18	140 53	12	−9.5	49.9	3.8	135.2	500–550	3
Bungyo 1	36 19	136 22	9	−22.9	57.3	1.5	1150.2	560–600	2
Bungyo 5	36 19	136 22	9	−22.8	62.4	1.6	1005.4	560–600	2
Suemura TG65	34 28	135 30	10	−13.2	52.3	3.8	160.9	540–620	1
Suemura TG41-I	34 28	135 30	20	−10.9	57.2	1.9	311.0	540–620	1
Suemura TG10-II	34 29	135 29	21	−0.6	61.9	3.4	87.2	610–690	1
Suemura TG64	34 28	135 30	13	−9.9	64.5	5.6	56.1	610–690	1
Suemura TG223	34 30	135 29	13	−3.3	48.3	3.0	193.5	710–750	1
Suemura TG206	34 29	135 30	5	−11.6	55.1	4.1	357.3	710–750	1
Okayama 3	36 37	136 51	11	−15.6	53.9	1.0	1931.3	700–750	5
Kosugi RD 9	36 42	137 04	13	−9.4	55.8	1.5	781.8	700–750	6
Kosugi RD 2	36 42	137 04	13	−10.1	54.4	2.5	281.8	700–750	6
Kawarai I	34 54	136 32	9	−10.1	50.5	2.0	689.7	750–800	6
Kawarai II	34 54	136 32	9	−12.4	52.6	2.2	574.1	750–800	6
Kawarai III	34 54	136 32	11	−10.4	51.7	1.9	578.2	750–800	6
Tendai 1	36 35	138 52	13	−16.8	56.5	3.2	171.7	750–780	6
Sasayama 1	37 24	138 46	16	−22.9	57.2	2.4	248.2	770–810	6
Chinjuan 3	35 03	135 44	9	−10.4	48.3	2.2	277.6	780–800	2
Chinjuan 4	35 03	135 44	23	−13.4	49.5	1.4	496.1	780–800	2
Kattachi 4	33 01	130 29	10	−16.9	59.5	4.8	104.0	750–850	3
Kattachi 5	33 01	130 29	8	−11.0	58.5	4.4	159.9	750–850	3
Tozu 5	33 40	136 26	14	−13.9	47.5	2.9	194.8	780–850	3
Anyojishita 1	38 37	140 53	14	−16.1	51.1	1.2	1028.7	750–820	3
Anyojishita 2	38 37	140 53	14	−16.2	50.2	2.5	256.3	750–820	3
Anyojishita 3	38 37	140 53	14	−14.3	51.2	2.0	405.2	750–820	3
Orito 80	35 07	137 04	11	−12.2	43.2	4.3	114.2	780–820	3
Narumi 275	35 05	137 01	14	−12.7	45.4	3.2	152.6	780–820	3
Kurozasa 7	35 07	137 05	17	−14.4	50.8	2.1	300.0	800–830	2
Amanuma 1	35 37	139 19	17	−10.5	47.2	2.3	243.0	800–850	3
Amanuma 2	35 37	139 19	10	−15.7	50.7	3.0	270.0	850–900	3
Kitaoka 4	35 22	137 05	15	−12.9	50.9	1.7	537.3	800–900	6
Kitaoka 5	35 22	137 05	18	−13.4	56.6	2.0	302.3	800–900	6
Kitaoka 8	35 22	137 05	13	−14.5	50.7	1.8	561.6	980–1000	6
Kitaoka 14	35 22	137 05	11	−9.3	46.7	4.8	91.4	1000–1020	6
Kitaoka 7	35 22	137 05	19	−10.9	51.5	1.9	304.8	1020–1040	6
Kitaoka 15	35 22	137 05	15	−10.4	50.8	0.9	1985.1	1040–1060	6
Kurozasa 90	35 07	137 06	9	−18.9	48.3	1.2	1385.1	1000–1050	2
Gotenyama	35 37	139 19	10	−13.5	49.4	1.6	891.0	1050–1100	3
Kurozasa 21	35 07	137 07	10	−14.0	52.0	3.1	242.1	1050–1075	6

3.12.2. Variation of archaeointensity

Archaeomagnetic intensity variations in Japan have been investigated by several authors. Nagata et al (1963) deduced the pattern of variations over the last 5000yr by measuring volcanic lava flows. Sasajima (1965) and Sasajima and Maenaka (1966) studied archaological artifacts and hence obtained the pattern of variations for the past 2000yr. Kitazawa (1970) and Sakai (1980) extended the time span back to 7000BP.

Table 3-5, continued

Kiln name	Locality		N	D	I	α95	K	age AD	reference
Yawatayama 6	35 04	137 04	14	-18.1	65.2	3.2	158.7	1050-1100	6
Kitaoka 9	35 22	137 05	11	2.9	58.1	1.3	1287.4	1125-1175	6
Kitaoka 12	35 22	137 05	13	2.3	58.8	4.0	425.5	1175-1200	6
Kitaoka 16	35 22	137 05	11	-1.6	61.4	1.2	1482.9	1175-1200	6
Kitaoka 18	35 22	137 05	12	0.0	54.9	0.9	2373.7	1175-1200	6
Higashiyama 101	35 09	136 59	12	-3.0	55.9	3.9	123.6	1150-1175	2
Kamichosa 3	35 56	136 43	12	-0.5	58.6	2.6	277.6	1150-1200	2
Kamichosa 4	35 56	136 43	20	-2.6	58.7	1.6	403.4	1150-1200	2
Kamichosa 5	35 56	136 43	14	3.4	61.2	1.9	421.2	1150-1200	2
Kamichosa 6	35 56	136 43	19	6.9	59.8	1.6	468.8	1150-1200	2
Kitaoka 10	35 22	137 05	12	2.1	57.1	3.6	150.8	1280-1320	6
Kitaoka 19	35 22	137 05	14	6.5	58.9	1.0	1705.2	1300-1360	6
Hojuji 3	37 26	137 13	40	6.9	58.5	2.0	132.3	1300-1400	2
Toshiro	35 21	137 10	16	6.6	44.2	1.9	380.1	1530	2
Myodo	35 18	137 09	19	6.6	43.6	1.4	605.9	1500-1550	2
Kokutani 1	36 12	136 25	19	8.0	49.3	3.0	122.7	1600-1670	2
Kokutani 2	36 12	136 25	10	5.6	42.9	3.9	149.3	1675-1725	2
Anada 1	35 04	137 07	10	8.6	41.4	8.5	33.1	1651-1652	4
Anada 2	35 04	137 07	14	7.1	42.9	1.7	520.6	1600-1650	4
Tenjinmori 2	33 11	129 52	17	-0.6	40.2	2.1	79.7	1600-1700	2
Tenjinmori 8	33 11	129 52	16	-4.3	41.9	2.3	260.2	1600-1700	2
Himedana 1	34 38	133 19	23	6.1	34.8	2.1	201.1	1660-1680	2
Himedana 2	34 38	133 19	13	-1.0	39.7	3.1	178.6	1690-1710	2
Tanigama	33 11	129 54	36	-4.6	46.2	1.7	199.6	1840-1880	2
Yoshidaya	36 12	136 25	11	-1.6	47.3	2.3	384.9	1820-1840	2
Wakasugi 1	36 24	136 29	20	4.3	48.1	2.7	143.2	1820-1840	2

N, D, I, α95 and K are respectively the number of samples, declination,
inclination, Fisher's angle of confidence at 95% level and precision parameter.
References are as follows. 1: Nakajima et al.,(1977); 2: Hirooka, (1979);
3: Hirooka, (1980); 4: Hirooka, (1981); 5: Hirooka and Aoki, (1981); 6: this
contribution.

All these results show that the intensity of the geomagnetic field has been decreasing, with short period fluctuations, through the last 2000yr. Recently, a detailed study was carried out for the past 1300yr (Tokieda, 1981), and fairly cyclic fluctuations with a period of about 500yr were found, as shown in Figure 3-41, the amplitudes being as large as 20% of the total intensity. These intensity changes should provide a method of dating in the near future when sufficient data are accumulated to draw a precise secular variation curve.

3.12.3. Orientation of old Buddhist temples and ancient declination

The Chinese have been very much concerned with orientation of streets and buildings since ancient times. According to Imai (1942), magnetic compasses were accurate enough to detect the geomagnetic declination in the 11th century AD in China. Needham (1962) pointed out that the Chinese compass ("ssu-nan") was probably used as a device for navigation in the 1st or 2nd century AD. As many descriptions of "ssu-nan" were recorded in the 6th century AD, magnetic compasses

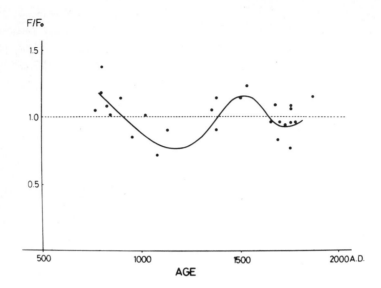

Figure 3-41:   Archaeointensity variation log covering the past
1300 years in Japan (Tokieda, 1980).

certainly existed at the time.

Buddhism was introduced into Japan in the 6th century AD at which time the system
and techniques of building Buddhist temples were imported from the Asian continent.
Many temples were constructed all over Japan from the 7th century AD onwards.
Buildings within the ancient Buddhist temples were arranged along central north-south
axial lines. In the most primitive type of plan, all the buildings of a temple were
arranged symmetrically about the north-south axis. The entrance gate was placed at
the south. To the north of the gate, the five-storied pagoda, the main hall and the
lecture hall were placed, in order, along the central axial line. Assuming that the
Chinese architects who built Buddhist temples used magnetic compasses for orienting
the buildings, the technique must have been imported to Japan with the introduction of
Buddhism.

From the orientation of the axial lines, ancient Japanese temples were divided into
two groups: (i) those in which the central axes are aligned very close to geographic
north and (ii) those in which the central axes deviate from north. Hirooka (1976)
(Figure 3-42) studied the relationship between the axial orientation of temples and the
age of their construcion and found that the temples belonging to the second group
show a clear tendency for the older temples constructed before the 12th century AD to

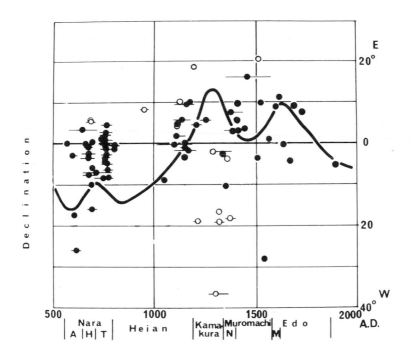

Figure 3-42: Declination inferred from the orientation of Buddhist temples. Solid and open circles respectively indicate temples built in flat and hilly areas. Periods are as follows: A=Asuka, H=Hakuho, T=Tempyo, N=Nanboku-cho and M=Azuchi-Momoyama.

have a westerly deflection of their central axes while the younger ones show widely dispersed directions. After the 11th century AD, they began to build Buddhist temples on the tops or flanks of mountains. Consequently it became difficult to find a sufficiently large area to systematically align the temple buildings which were then sporadically placed in small individual areas so that their desired orientations could not be realized. This might be the cause of the observed widely dispersed orientation of the younger temples. If we exclude temples which were built high up in the mountains, a clear tendency for the central axes to be deflected towards the east may be recognized. The directions of the central axes of temples are plotted against their age in Figure 3-42 together with the secular variation curve of archaeomagnetic declination for the past 1500yr for south-west Japan. It is obvious from this figure that the pattern of deviations of temple axes is similar to that of the declination variations. Hirooka thus concluded that the ancient Buddhist temples with deflected axes were orientated with a magnetic compass, and hence that the practical use of magnetic compasses in Japan can be traced back to the 7th century AD.

3.13. ARCHAEOMAGNETISM IN THE SOUTHWEST OF NORTH AMERICA

By Rob Sternberg
University of Arizona, USA.

3.13.1. Introduction

The south-western United States is an ideal region for archaeomagnetic studies. The magnificent cliff dwellings and pueblos, the excellent preservation of material culture in the arid enviromnment, and proximity to Mexico and the cultures of Mesoamerica have all served to encourage archaeological investigations. For a review of the archaeology of the south-west USA see Martin and Plog (1973) and of Mesoamerica see Weaver (1981). The availability of tree-ring dating (Dean, 1978; Schove, in Chapter 2 of this volume) provides the best prehistoric age control anywhere in the world. The abundance of well dated artifacts and features amenable to magnetic analysis have made this region a fruitful one for archaeomagnetic research.

3.13.2. Methods and techniques

Hearths and firepits once used for cooking or heating have been predominantly used in archaeomagnetic secular variation studies and are to be found in most living rooms at archaeological sites. They are typically circular in plan view with diameters of about 30cm and depths of about 10cm. Other archaeological features that have been less frequently sampled include plastered or unplastered burned floors and walls, small roasting pits or the larger communal hornos, and cremation pits. A typical NRM intensity for a hearth is between 0.01 and 10A/m. With the relatively modest sizes and remanences of hearths, problems of magnetic refraction (Aitken and Hawley, 1971; Dunlop and Zinn, 1980) are not likely to be encountered.

Preliminary evidence suggests that hearths may only carry a PTRM rather than a total TRM. Krause (1980) experimentally fired three hearths and deduced a maximum initial baking temperature of 450°C. Blocking temperature spectra for NRM and TRM show differences above 450°C (Sternberg, 1982). The stability of magnetization above this temperature suggests the presence of a high temperature VRM or a CRM acquired during firing in grains with blocking temperatures above 450°C. Any particular hearth was unlikely to have been used for more than 25 years. The PTRMs should still provide essentially a spot reading of the archaeomagnetic field.

Typically eight to twelve individually oriented samples have been collected from each feature investigated. A pillar was cut in the burned material and a non-magnetic brass or aluminium mould placed over it. Plaster of Paris was then poured around the pillar and allowed to harden. Levelling of the mould and taking an azimuth with a magnetic or sun compass gave fully orientated samples. Sampling techniques are described in more detail by Eighmy(1980). In the laboratory standard palaeomagnetic techniques were used. AF as opposed to thermal demagnetization was applied for cleaning since Curie temperatures of predominantly 578°C (Baumgartner, 1973) and low coercivities (Sternberg,1982) suggest that magnetite or titanomagnetite rather than haematite is the carrier of the remanence. Outlier tests were applied (Nichols, 1975; Krause, 1980; Sternberg, 1982; Wolfman, in press) to delete samples with aberrant directions possibly caused by differences in temperatures and atmosphere within the hearth during firing, mechanical disturbance since firing, or sampling problems. However, a cautious approach should be adopted in the treatment of outliers (Grubbs, 1969). The deletion of outliers from hearths fired experimentally in a known field has been found to improve the clustering of the sample magnetic directions, but the mean direction becomes less accurate (Krause, 1980).

3.13.3. Declination and inclination results

Results from three data sets covering the last 2000yr will be discussed.

1. Fifty-two independently dated prehistoric features from Arkansas and adjacent states yielded thirty-seven results for which α95 is less than 4° (Wolfman, in press).

2. As part of the Dolores archaeological programme in south-western Colorado, Hathaway et al. (in press) analyzed ninety-six features of which thirty-six were dated independently and yielded results with α95 values less than about 2.5° .

3. Sternberg and McGuire (1981), Eighmy et al. (1980) and Sternberg(1982) examined one hundred and fifty-eight features from eastern Arizona, western New Mexico, and south-western Colorado, of which seventy-three were independently dated and gave α95 values less than 10°.

The results are plotted in Figure 3-43 in which only the means have been shown for the sake of clarity. A 600yr oscillation having a peak-to-peak amplitude of about 20° with a minimum at 1100AD is apparent in declination. The inclination increases by about 20° from 800AD to 100AD but is fairly constant thereafter. The median α95s and interquartile ranges for all results from the three studies are respectively (i) 2.7° and 1.9° - 5.2°, (ii) 2.6° and 2.0° - 3.3° and (iii) 2.8° and 2.2° - 5.4° . These values reflect the nature of the baked clay material, sampling proficiency, and laboratory procedure including outlier analysis. Median age ranges for the independent dates are 15yr for Dolores where exellent tree-ring dating was available, 50yr for Arkansas where a combination of tree-ring, radiocarbon, and seriation dates was used, and 150yr for the Southwest where only radiocarbon dates and seriation

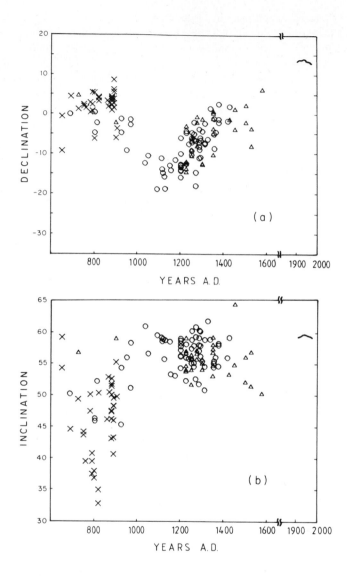

Figure 3-43: Mean declinations (a) and inclinations (b) for archaeomagnetic
features vs. time in degrees. Crosses are from Dolores; circles from
the Southwest; trangles from Arkansas. Curves are annual means from US
Geological Survey Observatory in Tucson, Arizona.

were available. Thus the scatter between the means shown in Figure 3-43 could be
due to the above uncertainities or to sampling errors, incomplete removal of
secondary components of magnetization, local or regional anomalies in the magnetic

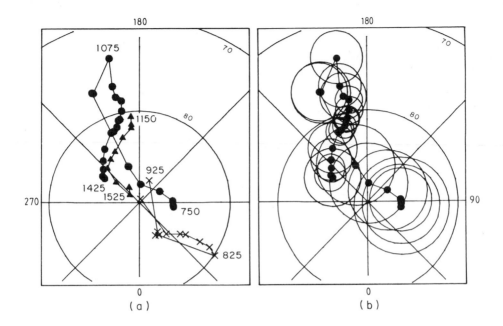

Figure 3-44:   (a) Interval mean virtual geomagnetic poles from Table 3-6:
Dolores (crosses), The Southwest (circles) and Arizona (triangles).
Stereographic projection with lines of latitude and longitude as indicated.
(b) The Southwestern curve with associated α95s.

field, or mechanical shifting of features since firing.   Secular variation curves were
derived by applying a moving window/ weighted Fisher statistics technique (Sternberg
and McGuire, 1981) to VGPs corresponding to the data sets of Figure 3-43.   For the
Arkansas data a constant 100yr window and an interval of 50yr were used. Thus
successive windows overlap each other. A 50yr window and 25yr interval were used for
the more precisely dated Dolores data. For the Southwestern data, overlapping 100yr
windows were applied from 1000AD to 1450AD where the data are more dense. The
results obtained were not sensitive to the particular windowing scheme chosen. The
Southwestern data were also weighted by the precision parameter k_i for each feature
so that total weight was $f_{ij} \cdot k_i$, where f_{ij} is the fractional overlap of the i-th feature with
the j-th window and a higher α95 cut off of 10° was used. Precision weighting should
reduce the impact of the less precise results but in fact it was found to produce little
change. An outlier test was used to discard features with anomalous directions. If a
feature   direction   had   less   than   a   0.5%   probability   of   being   consistent   with
contemporaneous features (McFadden, 1980), it was rejected. No outliers were found
in the Arkansas and Dolores data, but six were found among the seventy-three
Southwestern data. The results are given in Table 3-6.

Figure 3-44a shows the interval means and Figure 3-44b illustrates the interval

Table 3–6:   Mean virtual geomagnetic poles and α95
for Dolores, the Southwest and Arkansas.
Plat – pole latitude; Plong – pole longitude.

Time Interval (yr AD)		Plat (°)	Plong (°)	α95 (°)	Time interval (yr AD)		Plat (°)	Plong (°)	α95 (°)
Dolores					Southwest				
625	675	83.3	211.4	18.6					
650	700	89.6	167.5	9.8					
675	725	85.5	40.3	3.9					
700	750	84.4	52.7	3.0	700	800	86.2	82.6	5.3
725	775	83.8	55.3	2.6					
750	800	82.1	57.2	2.6	750	850	86.3	85.6	4.3
775	825	80.9	57.7	2.8					
800	850	80.0	55.0	5.5	800	900	86.3	92.8	3.4
825	875	85.9	31.0	2.2					
850	900	86.0	25.5	1.6	850	950	87.4	118.6	4.4
875	925	86.3	31.4	2.1					
900	950	87.4	156.7	5.0	900	1000	88.0	176.0	4.4
					950	1050	85.8	198.1	4.4
					1000	1050	77.0	204.1	3.1
					1025	1075	77.0	203.6	3.3
					1050	1100	74.0	192.2	2.9
Arkansas					1075	1125	77.0	193.5	2.4
					1100	1150	78.2	193.1	2.6
1100	1200	80.6	186.2	2.3	1125	1175	78.7	190.4	1.5
					1150	1200	79.8	191.3	1.5
1150	1250	81.4	186.5	1.8	1175	1225	80.7	193.7	1.4
					1200	1250	80.7	194.7	1.4
1200	1300	81.8	186.7	1.7	1225	1275	81.4	197.1	1.3
					1250	1300	81.7	200.4	1.4
1250	1350	82.7	195.1	1.6	1275	1325	81.8	202.4	1.5
					1300	1350	83.0	213.6	2.7
1300	1400	83.8	206.5	1.7	1325	1375	84.0	221.5	2.3
					1350	1400	84.6	228.0	1.5
1350	1450	84.3	214.8	2.1	1375	1425	84.9	235.2	1.8
					1400	1450	85.3	236.1	3.2
1400	1500	84.8	223.9	3.2					
1450	1550	86.6	231.0	3.5					
1500	1600	88.6	235.0	5.0					
1550	1650	88.0	216.1						

α95s for the Southwestern data. The assertion by Wolfman (1979) and Eighmy et al.
(1980) that the secular variation should be represented as some sort of band rather
than a simple line is hereby quantified. The interval α95s are generally smaller where
the data are more dense, and the smallest values about 1.5° are comparable to the
smallest feature α95s of about 1.0° .

3.13.4. Discussion of archaeodirections

The three VGP paths of Figure 3–44 show reasonable overall agreement and the
oldest point on the Dolores curve for 900AD – 950AD agrees nicely with the
Southwestern curve for this same period. Little secular variation is recorded by the

Dolores curve from 825AD to 925AD nor by the Southwestern curve from 700AD to 900AD but for 700AD to 900AD the Dolores VGPs are systematically offset from the Southwestern curve though the difference is probably not significant.   The possibility also exists that the Dolores curve may record a regional anomaly since it is based on data from a very restricted area.

The Arkansas and Southwestern curves show excellent agreement. The difference between any pair of points is not significant at the 95% confidence level but nevertheless there is a systematic offset between the two curves. Arkansas is at the same latitude, 35°N, as the Southwest but is about 17° of longitude or 1550km to the east. Thus only a small difference in geomagnetic directions for the two areas could be attributed to a non-dipole field (Shuey et al., 1970; Champion, 1980).

Along the Southwestern curve, the average rate of VGP motion between successive points ranges from 0.004 to 0.358 deg/yr, with a mean of 0.057± 0.015 deg/yr. The variations of speed of VGP motion along the Dolores and Southwestern curves appears to be a real characteristic of the archaeomagnetic secular variation.

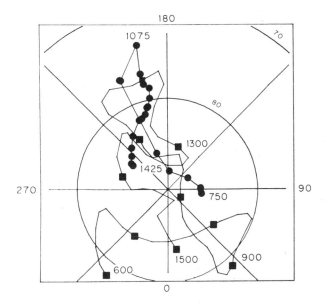

Figure 3-45:   Virtual geomagnetic poles from the Southwest: circles from Sternberg and McGuire (1981), squares from Dubois (1975a). Stereographic projection with lines of latitude and longitude as indicated. Dubois' points fall every hundred years, with some marked; data for Sternberg/McGuire curve in Table 3-6

Many Southwestern data accumulated by R.L. Dubois have never been adequately described in print, although some preliminary results, most of them not demagnetized, were plotted as declinations and inclinations with error-bars (Watanabe and Dubois, 1965): otherwise only a final VGP curve has been shown (Dubois, 1975a, b) and this is compared with the Sternberg/McGuire Southwestern VGP curve in Figure 3-45. The Sternberg/McGuire curve is smoother (note particularly the absence of the Dubois loop at ~1300AD and it runs about 100 years ahead of Dubois' curve for the interval 100AD to 1250AD. The source of these discrepancies is problematical because of the unavailability of Dubois' raw data. One possible reason is that Dubois simply drew his curve freehand (Wolfman, personal communication, 1980). The dangers of subjectively constructed curves are illustrated by the exact agreement of the Wolfman and Dubois curves for certain periods of time, and the greater detail shown in Wolfman's curve for other periods. The former would be impossible for independently constructed curves, and the latter is unlikely considering that dating in Arkansas is not as precise as in the Southwest. Wolfman's own curve does not show the close agreement with the Sternberg/McGuire curve illustrated above when his data are passed through one moving window.

For earlier periods of time, the relatively sparse data and imprecise radiocarbon dates generally prohibit determination of detailed secular variation curves. Parker (1976) gives undemagnetized results from four strata of archaic features in Utah which were radiocarbon dated from 4425BP to 7565BP. Champion (1980) describes a palaeomagnetic study on Holocene igneous rocks of the western USA, including thirty-six dated units but only one of these, however, dates after 500AD.

3. 13. 5. Archaeointensities - material and methods

Although several intensity studies on North, Central, and South American archaeological materials have been carried out few attempts have been made to analyze the results. The North and Central American data are here divided into three groups according to region and to accuracy of the dates. For discussion of the various chronologies see Taylor and Meighan (1978). Dating in Mesoamerica is based on radiocarbon which is not as precise as the tree-ring based chronologies of the Southwest.

Nagata et al. (1965), Bucha et al. (1970) and Lee (1975) used ceramics primarily but also studied baked clays (hearths), wattle-and-daub, and igneous rocks. There is some controversy surrounding the chronology of the Hohokam culture of the desert Southwest (Schiffer, 1982) because, although ceramics can be confidently assigned to particular phases which can be ordered chronologically, the absolute ages are still debatable. Bucha et al. (1970), Lee (1975), Sternberg and Butler (1978) and Sternberg (1982) studied Hohokam ceramics. Other Southwestern material can be dated more accurately and precisely because of the ultimate dependence in many cases on tree-ring dates. Lee (1975), Parker (1976) and Sternberg (1982) worked with such material. Finally, Schwarz and Christie (1967) reported results from some Canadian pottery dated by radiocarbon and by seriation and Champion (1980) obtained palaeointensities from igneous rocks dated mostly by radiocarbon.

Either the double heating technique (Thellier and Thellier, 1959; Thellier, 1977) or the Coe (1967a) variant of this technique was used. The palaeointensity was determined by linear regression from the now conventional $\Delta J_N - \Delta J_T$ diagram (Coe, 1967a). Some of Lee's (1975) palaeointensities were inferred using the AF and thermal demagnetization methods.

3.13.6. Discussion of archaeointensity results

The results, tabulated in the publications cited above were analyzed and adjustments were made to the dates of some samples by recalibrating radiocarbon dates (Damon et al., 1980; Klein, personal communication, 1982) and assigning age ranges where only central dates had been given. Unweighted averages and standard errors were calculated for multiple specimens from the same sample, or for contemporaneous samples from the same site when only single specimens were taken. For single specimens from single samples, no standard error was calculated. Finally, results were transformed to virtual axial dipole moments (VADM) which are considered to be as good as or better than the more commonly used virtual dipole moment (VDM) (Barton et al., 1981; Champion, 1980).

It should be noted that Southwestern pottery, even though made without the use of a potter's wheel, appears to have a significant TRM anisotropy due to preferential alignment of ferromagnetic grains during manufacture (Rogers et al., 1979). Sternberg (1982) found the average anisotropy in TRM susceptibility to be about 30%, but it can be larger than 100% in some samples. He used the easy-plane model (Aitken et al., 1981) to correct for this effect.

The palaeointensity results for the three groups of data are plotted in Figure 3-46. The considerable scatter in the data can be attributed largely to experimental error (e.g. see Coe, 1967b) and difficulty in interpretation of the experimental data (Sternberg, 1978; Coe et al., 1978). A common trend is to be observed in the Mesoamerican and Southwestern data sets: the field was high between 500BC and 1BC, yielding a VADM of $13 - 15 \times 10^{22} Am^2$. By 250AD the field reached a minimum between $7 - 9 \times 10^{22} Am^2$, and then rose to another high of $13 - 14 \times 10^{22} Am^2$ near 1000AD. The general congruence of these curves suggests that the non- dipole field was similar for the two regions despite the large separation of about 27° or 300km between the area studied by Champion (1980) and central Mexico.

The currently disputed Hohokam chronology (Schiffer, 1982) makes it risky at best, to plot data from this culture on an absolute time scale, though when plotted using the

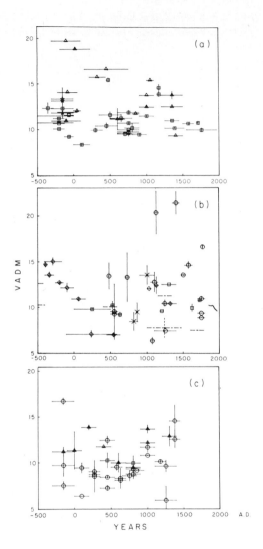

Figure 3-46: Archaeointensity results transformed to virtual axial dipole moments, units of
$10^{22} Am^2$. Age ranges and one standard error of mean VADMs shown. Results
from: (a) Mesoamerica, stars – Nagata et al. (1965), triangles – Bucha et al.(1970),
squares – Lee (1975); (b) North America, no symbol with dashed error bars
– Schwarz and Christie (1967), squares – Lee (1975), crosses – Parker (1976),
diamonds – Champion (1980), circles – Sternberg (1982); the curve is from annual
means in Nogales and Tucson, Arizona; (c) Hohokam culture, Southwest, triangles
– Bucha et al. (1970), squares – Lee (1975), circles – Sternberg (1982).

'long count' chronology (Figure 3-46c) they are not inconsistent with the other two data sets. The extreme alternative 'short count' chronology does not change the ages for the more recent ceramic phases, but compresses the time scale about 800 years by bringing the earliest phase from 300BC up to 500AD.

3.13.7. Conclusions

Although North American archaeomagnetic directional and intensity results are noisy, the directional variations have been recovered by using the moving window and weighted Fisherian statistics. Similar secular variation patterns are observed in the Southwest and Arkansas suggesting a fairly uniform non-dipole field over North America for the past 1500yr – 2000yr. The Southwestern declination and intensity data may show 600yr and 1000yr oscillations, respectively. Similar periodicities have been found in other regions and in global compilations of archaeomagnetic data (Burlatskaya, section 3.6 this volume; Champion, 1980). The rate of motion of the VGP varies by almost a factor of ten. The author wishes to acknowledge helpful discussions with Randy McGuire, and comments on the text by Mcguire and Bob Butler. This work was partially supported by National Science Foundation grant EAR-81-16196 and the State of Arizona.

3.14. RESULTS FROM PERU

By Ken Games,
University of Liverpool, UK

3.14.1. Introduction

Magnitude variations of the geomagnetic field in Peru during the last 2000yr were determined by a joint project involving the Universities of Oxford and Liverpool. In Liverpool, the field magnitude determinations were made by use of Shaw's (1974) method. In Oxford, the ceramics were dated by the thermoluminescence (TL) method.

3.14.2. Method

The method used to determine the magnitude of the geomagnetic field in this study was essentially that described by Shaw, and was carried out by Nigel Gunn. AF demagnetizing coils capable of producing a peak field of 170mT at 500Hz surround a two-axis specimen tumbler of the type described by Wilson and Lomax (1972). We believe that by using this system we have avoided the problem of samples acquiring rotational remanent magnetization (RRM) and gyromagnetic remanent magnetization (GRM) (Stephenson , 1980). The central tumbling mechanism can be interchanged for one modified to deliver a current to a set of Rubens coils (Rubens 1945) wound round the plastic cube which holds the specimen. Thus an ARM can be installed by applying a small, steady field to the specimen during AF treatment. The stages in the application of the method are as follows:

1. The specimen is AF demagnetized in steps of 10mT up to 170mT, all three components of the residual NRM being measured at each step. A specimen may be rejected at this stage if the direction of the NRM residual shows persistent change.

2. ARM(1), is installed in a suitable steady field, usually 675 μT, through the coercivity range 0 to 170mT. This is stepwise AF demagnetized and measured like the NRM, except that since the ARM direction is already known, only one component of magnetization need be measured.

3. The specimen is heated once to 700°C, and allowed to cool in a field of 50μT. The resulting TRM is then measured in the same way as the ARM(1).

4. ARM(2), is installed and measured, again in the same way as ARM(1).

5. Residual intensities of ARM(1) and ARM(2) are plotted against each other, with the peak demagnetizing field as parameter. If there is a field value **above which** the points fall close to a straight line of unit slope, this is identified as the coercivity range over which the remanence carrying coercivity of the specimen has not been altered by the laboratory heating. This usually occurs at high AF values: 100mT or higher. The ARM data are used for no other purpose. In particular they are not used as a 'correction' factor to influence the NRM/TRM ratio.

6. The NRM and TRM residual intensities are plotted together in the same way, except that here the origin is included as a data point, corresponding to the hypothetical demagnetization of the NRM and TRM in an infinitely strong alternating field. The NRM/TRM plot is examined for linearity in the unaltered coercivity region identified from the ARM data. The data are then edited, eliminating points which fail the ARM or the NRM/TRM linearity test.

7. If NRM/TRM linearity is observed over an acceptable range of AF values, the ancient field magnitude, B_A, is calculated from B_A = (NRM/TRM) x 50μT

3.14.3. Accuracy of archaeomagnetic determinations

In most cases only one specimen could be obtained from each sherd, and therefore the method, as applied here, incorporates no internal measure of its own errors. But after many trials of the method and comparisons with other results, experimental errors are subjectively estimated to be around $\pm 5\mu T$ or less on the desired field values. The results have in fact been categorized according to their credibility or reliability. Thus a 'good' result is one which on intrinsic experimental grounds, we are virtually certain is correct within $\pm 5\mu T$ limits, an 'acceptable' one is probably correct but with a slight risk of gross error not excluded by the measurements, while a 'suspect' result is considered equally to be either correct within $\pm 5\mu T$ or wildly wrong. But the final test of credibility is taken to be that the results should collectively conform to a curve consistent with plausible variations in the true field magnitude through historical time.

It is also worth pointing out here two possible sources of error which could both lead to the field magnitude results obtained in this study being too high. First, the TRM was given to these samples with a cooling time of approximately one hour. Since we would expect the cooling time in antiquity to be greater than this we can expect the TRM to be up to 10% lower than that acquired at the lower cooling rate (see Fox and Aitken, 1980). Second, no account was taken of shape or fabric anisotropy (Rogers et al. 1979) of the specimens. Since the TRM was normally installed perpendicular to the plane of the sherd, this could lead to the efficiency of the TRM acquisition being lower that that of the NRM, and so the archaeomagnitude value derived in the ordinary way would be too high.

3.14.4. Dating

The reliable dating of specimens is a major problem in resolving field magnitude variations. Although archaeological dating was available for all of the sherds used in this study, it was decided to apply the TL dating technique to as many samples as possible. The work was carried out by Andrew Murray at Oxford using the fine grain method first described by Zimmerman (1971). It was originally hoped that the archaeological stratigraphy of the collection site would serve as relative chronologies, with just a few TL dates obtained from convenient sherds to tie into the absolute time scale. This turned out to be impossible. In all, 20 TL dates were obtained. Potsherds from a single level did not always agree in either their TL dates or the archaeomagnitudes determined from them, while the TL dates from different levels showed no convincing correlation with the stratigraphic sequence. Hence all associative dates assigned to archaeomagnitudes are quoted with arbitrary errors of ± 300yr, and such results are distinguished in the final curve from ones which have individual TL dates or comparatively good historical dates (after 1300AD only).

3. 14. 5. Results

Figure 3-47 is a typical page of diagrams showing the full archaeomagnitude analysis and evaluation of the results for one specimen. Crosses represent data points at peak AF intervals of 10mT up to a maximum of 170mT (nearest the origin). A detailed analysis of all the results appears in Gunn and Murray (1980) and Gunn (1978).

The archaeomagnitude results which fall into the 'good' and 'acceptable' categories already described are shown in Figure 3-48. Results which were merely acceptable and had no firm date are not included. The results may be divided into three categories:

1. Good results, with TL or other firm dates. The curve is constrained to pass close to these data points;

2. Acceptable results with TL or other firm dates. The curve is adjusted to pass fairly close to each of these data points, but a single such point is not considered sufficient grounds for a major deviation of the curve;

3. Good results with uncertain dates. The curve is constrained to approach these field magnitude values at some time not unreasonably far from the estimated dates.

3. 14. 6. Conclusions

The main features of the curve in Figure 3-48, such as the field maxima approaching 60 μT (more than double the present field strength), are clear enough to stand comparison with similar information for other parts of the world, as shown by other contributions to this Chapter. Unlike the overall decline in field magnitude during this period, which seems to be a world-wide phenomenon, these non-rapid changes do not correlate well between different areas. This is probably due in part to the failure of the data for some areas to resolve the more rapid fluctuations, but short term secular variations of large amplitude in geomagnetic field components other than the axial dipole must also contribute to the effect. This raises the possibility that at times in the past when the field was stronger than it is now, these harmonics may have made up a bigger fraction of the total. Because of their rapid variations they would tend to be averaged out of palaeomagnetic data. Thus the ancient geomagnetic field would appear to have been predominantly dipolar throughout a geological period even if this was not its typical instantaneous condition.

There have been several attempts to modify Shaw's method, and while these procedures may be effective in special circumstances, they lack some of the theoretical assurance and simplicity of the original method. For this reason we feel happier in taking the straightforward approach, even though it means a higher rate of

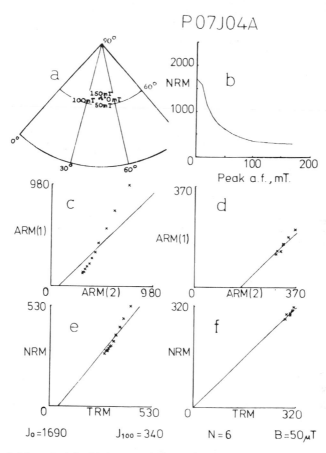

Figure 3-47:   Archaeomagnetic analysis of specimen PO7J04A:
(a) direction, (b) intensity of NRM during AF demagnetization, (c) full
and (d) edited ARM1/ARM2 residual intensities, (e) full and (f) edited
NRM/TRM residual intensity data. J_0=total NRM intensity,
J_{100}= NRM intensity after AF demagnetization in 100mT.

All intensities are in units of $\mu Am^2 kg^{-1}$. N – number of points,
B = ancient field magnitude derived from (f)

rejection of specimens.   Undoubtedly the most serious sources of error, unfortunately
only recognized after the work had been done, were the effects of anisotropy and
cooling rate. Here again, though, since the magnitude of these effects is not known

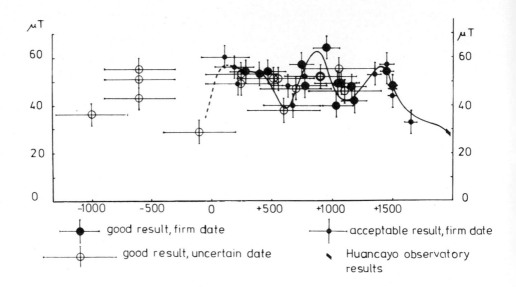

Figure 3-48:   Geomagnetic field magnitude vs. time for Peru. Estimated total
errors to one standard deviation. Arbitrary limits on the field magnitude
values are shown at ±5μT

precisely for each specimen, we do not feel justified in making any correction to the
field magnitudes.

Finally the TL dating has been invaluable to this study, and ideally one would like a
TL date for every specimen used if the maximum amount of information is to be
obtained from the sherds.

3. 15. ARCHAEOMAGNETIC RESULTS FROM AUSTRALIA

By Mike Barbetti, University of Sydney, Australia

Ancient hearths are fairly common in the flat, semi-arid, inland region of south-eastern Australia. They range from about 0.2m to 1.5m in diameter and usually contain hand-sized lumps of sandy sediment, rock or broken termite nest, spread in a more or less level fashion on top of charcoal and ash.

Palaeomagnetic studies began in 1970, and it was soon discovered that the lumps were generally preserved in the positions in which they had originally cooled. This meant they could be used for directional as well as field strength studies, and it also meant that charcoal trapped under the lumps was particularly reliable for radiocarbon dating.

A series of hearths in lunette sediments around L. Mungo contained evidence of one or possibly two geomagnetic excursions between 31000bp and 25000bp (Barbetti & McElhinny, 1976; Coe, 1977).

The first palaeomagnetic directions from Holocene hearths showed a pronounced change in inclination around 500yr ago, suggesting a shift of the dipole axis (Barbetti, 1977). New data confirm the change; they and previously published data for the last 3000 years are presented in Figure 3-49. Comparison of hearth and lake sediment secular variation data indicates the latter were smoothed by diagenesis, and that ancient carbon was incorporated at deposition (Barton & Barbetti, 1982).

A comprehensive series of Holocene palaeomagnetic field strength measurements is being compiled for south-eastern Australia. Results from baked sediments around two burnt tree stumps were published by Barbetti et al (1977); they are included in Figure 3-50, along with results from numerous other tree-stumps and hearths (Barbetti et al, in prep.). The data show a weak field around 6000bp, followed by a steady gradual rise to a maximum between 4000bp and 3000bp. Thereafter, the field strength generally remains high, although the character of the record seems to change to one of rapid fluctuation, with peaks at about 2000bp, 1200bp and 400bp. Results have also been obtained from sediments baked by two lava flows (Barbetti & Sheard, 1981); their ages, though probably Holocene, are not yet precisely known, and the data are therefore omitted from Figure 3-50.

The probable errors in the radiocarbon time-scale during the late Pleistocene and Holocene, which depend strongly on geomagnetic strength variation, are discussed by Barbetti & Flude (1979) and Barbetti (1980).

The archaeological implications, arising from firing temperatures and hearthstone movement as determined by the palaeomagnetic measurements, are discussed by Clark and Barbetti (1982).

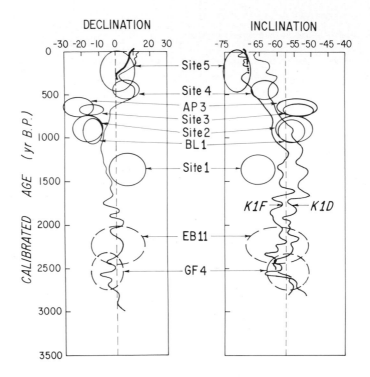

Figure 3–49:   Geomagnetic direction in south–eastern Australia for the last 3000
years, with all the data adjusted to the latitude of Toolangi Magnetic
Observatory (37°32'S, 145°28'E – near Melbourne). Dashed lines
are the ADF direction, short bold lines are compiled from historical
measurements. Ellipses are 95% confidence limits for the true values,
calculated from Fisher (1953) statistics and radiocarbon calibration
according to Klein et al. (1982). Results for sites 1 to 5 and GF4 are from
Barbetti (1977). Other results are from work in progress. Thin lines are
sediment results from L. Keilambete, adjusted for radiocarbon errors arising
from the incorporation of old carbon at the time of deposition (Barton and
Barbetti, 1982)

This work has been supported by the Australian National University, the Universities
of Adelaide and Sydney, and the Australian Research Grants Scheme. The cheerful
cooperation of many prehistorians and colleagues is also gratefully acknowledged.

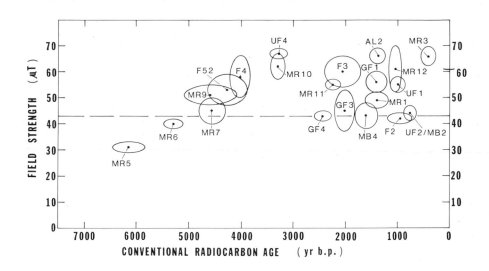

Figure 3-50:   Geomagnetic field intensity in south-eastern Australia over the
last 6000bp. The dashed line at 43μT is the field expected
at the latitude of L.Mungo (33.7°S) for a geocentric axial dipole
source of $8.10^{22}Am^2$. The short bold line at 61 μT is the historic
variation at Toolangi Magnetic Observatory. Ellipses are 95% confidence limits
for the true value based on the standard error of the site mean strength
multiplied by the appropriate value of Student's t-test, and twice the normal
radiocarbon errors. Results from sites MR7 and MR9 are from Barbetti et al.
(1977). Other results are from Barbetti et al (in preparation)

3. 16. ANALYSIS OF GLOBAL INTENSITIES FOR THE PAST 50000 YEARS

By Mike McElhinny,
Australian National University, Canberra

3. 16. 1. Introduction

The earliest archaeomagnetic intensity measurements showed that the Earth's dipole moment was very much higher about 2000 years ago than its present value and probably much lower between 5000 and 6000 years ago (Bucha, 1967, 1969; Smith, 1967; Cox, 1968). This was taken to indicate that the variations might be sinusoidal with periodicity of between 8000yr and 9000yr and with maximum and minimum respectively about 1. 5 and 0. 5 times the present dipole moment.

Work in archaeomagnetism increased over the succeeding ten years and many new results have been produced. Burlatskaya and Nachasova (1977) produced a catalogue of world-wide data but it has many shortcomings, with much duplication and omission. The data listed are heavily biassed towards those obtained in the Soviet Union and in recent years many more intensity values have become available from different parts of the world which extend that global coverage of the catalogue quite considerably. However, even when these new results are added the resulting global data are still very concentrated in the European region (0-90°E) to the extent that they outweigh data from the rest of the world by a factor of nearly two to one.

In this section 1175 archaeomagnetic intensity results have been analysed covering the past 12000yr and the overall picture that emerges is compared with the few results that are available between 15000bp and 50000bp.

3. 16. 2. The Period 0-12000bp

It is convenient to divide these data into two sets, one being the European region (defined here as the whole of the northern hemisphere between 0° and 90°E) and the other the rest of the world combined. This enables a comparison to be made between the main concentration of data from one eighth of the Earth's surface (746 results) with what amounts to a random sampling both in time and space of the rest of the world (427 results).

From the European region as defined above, the data come from Bulgaria and Yugoslavia (Kovacheva, 1980), Czechoslovakia (Bucha, 1967), the Ukraine and Moldavia (Rusakov & Zagny, 1973) and other parts of Europe (UK and France) and the Soviet Union as summarized in Nachasova (1972) and Burlatskaya & Nachasova (1977). Results are also available from India (Athavale, 1966), Egypt (Athavale,

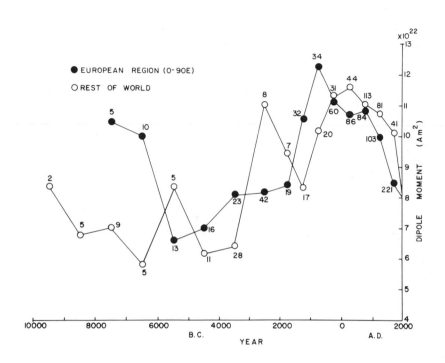

Figure 3-51: Comparison of average dipole moments
from archaomagnetic data from the European region (defined as the northern
hemisphere from 0–90°E) and the Rest of the World combined.
The numbers are the number of data averaged in each 500–year–interval
(to 2000BC) or 1000–year–interval (prior to 2000BC).
Data listed in Table 3–7.

1969; Games, 1980) and Greece (Walton, 1979).

From the rest of the world the data are fairly widely spaced with 23 results from Australia (Barbetti et al., 1977; 1982) and 155 results from Japan (Nagata et al., 1963; Kitazawa, 1970; Kinoshita, 1970; Hirooka, 1971; Domen, 1977; Tanaka, 1978) and Mongolia and China (Burlatskaya et al., 1975; Burlatskaya & Nachasova, 1977). In the American hemisphere there are 117 results from Bolivia and Peru (Nagata et al., 1965; Kitazawa & Kobayashi, 1968; Lee, 1975; Walton, 1977; Gunn & Murray, 1980), and from North America and Hawaii 132 results (Schwarz & Christie, 1967; Bucha et al., 1970; Coe, 1967; Doell & Cox, 1972; Lee, 1975; Sternberg & Butler, 1978; Coe et al., 1978; Champion, 1980).

Table 3-7:   Average VDM's (VADM) for 500yr intervals (to 2000BC)
and 1000yr intervals (from 2000BC to 10000BC)
from archaeomagnetic intensity data for the European
Region (0°-90°E, northern hemisphere)
and the Rest of the World calculated separately.
N is the number of points for each interval,
dipole moments are in units of $10^{22}Am^2$,
and SD is the estimated standard deviation

Time Interval (yr AD or BC)		European Region VDM(VADM)	N	SD(%)	Rest of the World VDM(VADM)	N	SD(%)
1980-1500	AD	8.45	221	1.17(13.8)	10.11	41	1.79(17.7)
1500-1000	AD	10.02	103	1.51(15.1)	10.74	81	2.16(20.1)
1000-500	AD	10.85	84	1.71(15.8)	11.05	113	2.13(19.3)
500-0	AD	10.68	86	1.63(15.3)	11.58	44	2.79(24.1)
0-500	BC	11.20	60	2.27(20.3)	11.30	31	2.90(25.7)
500-1000	BC	12.33	34	2.59(21.0)	10.14	20	1.66(16.4)
1000-1500	BC	10.57	32	2.98(28.2)	8.29	17	2.76(33.3)
1500-2000	BC	8.35	19	1.82(21.8)	9.45	7	2.99(31.6)
2000-3000	BC	8.18	42	1.36(16.6)	11.03	8	2.69(24.4)
3000-4000	BC	8.08	23	1.69(20.9)	6.40	28	1.59(24.8)
4000-5000	BC	6.98	16	2.01(28.8)	6.17	11	1.86(30.1)
5000-6000	BC	6.59	13	1.34(20.3)	8.34	5	0.96(11.5)
6000-7000	BC	10.01	10	1.22(12.2)	5.81	5	0.87(15.0)
7000-8000	BC	10.46	5	1.08(10.3)	7.05	9	1.95(27.7)
7000-9000	BC	-	-	-	6.76	5	1.34(19.8)
9000-10000	BC	-	-	-	8.36	2	0.06

All the results have been carefully examined to determine if radiocarbon ages have been calibrated. If not these were corrected using the calibration curve of Clark (1975). For radiocarbon ages older than 6500bp there is no calibration available, so it has been assumed that true ages and radiocarbon ages converge again at 10000bp. Calibration was made by using linear interpolation beyond radiocarbon ages of 6500yr. The method of analysis discussed below involves averaging data over 1000 year intervals in this time bracket so that even a very crude calibration is quite adequate for the purpose.

Barton et al. (1979) have summarized the various ways of representing archaeomagnetic intensity information so as to allow comparison of data from different geographic regions. Where possible VDMs have been calculated making use of the measured inclinations.   When inclinations are not available the VADM has been calculated. The advantage of the VDM representation (Smith, 1967) is that dipole wobble does not introduce a scatter in them. Barton et al. (1979) have pointed out that errors in VADMs due to tilting of the dipole axis will tend to cancel out when they are averaged from widely separated regions over short periods of time.   However the standard deviations of VDMs will be less if dipole wobble has occurred. In the analysis here both VDMs and VADMs have been combined and they occur in about equal numbers.

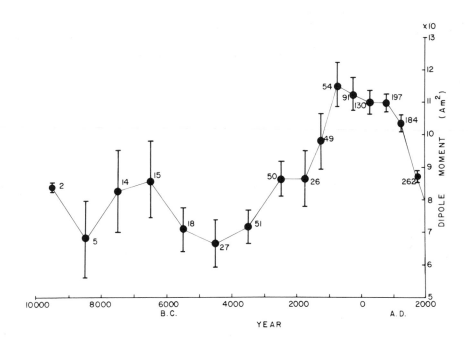

Figure 3–52:   Global mean dipole moments with 95%
errors for 500–year–intervals (to 2000BC) or 1000–year–intervals (prior
to 2000BC) from archaomagnetic data.
Data listed in Table 3–8.

     The results are summarized in Table 3–7 and Table 3–8.   Table 3–7 gives the
separate data for the European region and the rest of the world. 500yr intervals are
used back to 2000BC but for earlier times 1000yr intervals are used where the data
become more sparse. These means are illustrated in Figure 3–51. Both sets of data
show the same broad trend of a high around 0 to 1000BC and a low around 5000BC,
but prior to that the data are too few to draw any conclusions. Indeed there seems no
clear evidence for any sinusoidal variation, a point that was emphasized by Barton et
al. (1979). If the two data sets are combined into a global average, the variation is
listed in Table 3–8 and illustrated in Figure 3–52.   The overall data suggest a
maximum in the dipole moment occurred around 2500BP (500BC) and this is preceded
by a minimum around 6500BP (4500BC).   This is consistent with independent
evidence from radiocarbon data as summarized by Barton et al. (1979).

     Because the data are not uniformly distributed with time the best estimate of the
average dipole moment over the past 10000yr as shown at the bottom of Table 3–8 has

Table 3–8:   Global average VDMs (VADM) using all the data of Table 3–7.

Time interval	VDM (VADM)	N	SD (%)	95% limits
1980–1500AD	8.706	262	1.421	0.172
1500–1000	10.337	184	1.852	0.267
[1980–1000AD	9.379	446	1.800 (19.2)	0.166]
1000–500	10.966	197	1.957	0.271
500–0	10.984	130	2.128	0.365
[1000–0	10.973	327	2.023 (18.4)	0.218]
0–500BC	11.231	91	2.485	0.507
500–1000	11.581	54	2.506	0.665
[0–1000BC	11.338	145	2.488 (21.9)	0.404]
1000–1500	9.778	49	3.080	0.858
1500–2000	8.649	26	2.188	0.837
[1000–2000BC	9.387	75	2.839 (30.2)	0.640]
2000–3000	8.639	50	1.923 (22.3)	0.530
3000–4000	7.159	50	1.824 (25.5)	0.498
4000–5000	6.649	27	1.954 (29.4)	0.733
5000–6000	7.077	18	1.461 (20.6)	0.672
6000–7000	8.609	15	2.318 (26.9)	1.167
7000–8000	8.269	14	2.362 (28.6)	1.231
8000–9000	6.76	5	1.342	1.170
9000–10000	8.36	2	0.057	0.079
10 means (0–1000bp)	8.75	10	1.58 (18.0)	0.97

been calculated by averaging the ten 1000yr means. This gives a value of 8.75 x 10^{22}Am^2.

3.16.3. The Period 15000–50000bp

Barbetti & Flude (1979) have noted that the dipole moment appears to have been much lower in the period 10000bp to 50000bp than it is today. Since there are no data between 12000bp and 15000bp the discussion will be limited here to the period 15000bp and 50000bp. Four of these are the unusually high values associated with the Lake Mungo Excursion (Barbetti & McElhinny, 1976). Since these values are associated with wide departures from the average geomagnetic field direction at that time, they have not been included in the discussion that follows. All the 18 values are shown in Figure 3–53 and compared with the variation deduced for the past 10000yr. Excluding the Excursion all 14 values are very low and their mean is 4.44 x 10^{22}Am^2, only one half the average value of 8.75 x 10^{22}Am^2 for the past 10000 yr.

Barbetti (1980) has summarized radiocarbon and comparative thermoluminescence or Uranium series (^{230}Th/^{234}Th) ages between 40000bp and 10000bp. Atmospheric

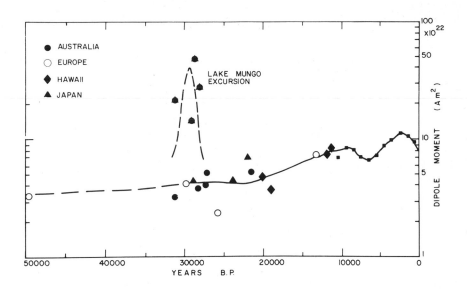

Figure 3-53:   Archaomagnetic dipole moments in the
time range 10000bp to 50000bp.

^{14}C concentrations deduced from these comparisons are almost all much higher than their present value implying a prolonged period of reduced dipole moment. Therefore these age comparisons confirm quite independently the results from the 14 intensity values.

3.17. SOME SPECIFIC COMMENTS

By Daniel Wolfman,
Arkansas Archaeological Survey, USA

Although the results presented in this chapter provide a fairly complete review of the details of secular variation of geomagnetic intensity obtained from baked clay samples and from adobe bricks, attention should be drawn to the availability of a few archaeointensity results from other geographical areas; namely from India, (Athavale, 1966), Egypt (Athavale, 1969), and Peru (Nagata et al. 1965; Kitazawa and Kobayashi 1968). Declination and inclination SV data additional to those described in this chapter may be found in Aitken (1958), Aitken et al., (1963), Aitken and Hawley (1966, 1967) and Thellier (1981) for Western Europe; by Kawai et al. (1972a), Dollfuss and Hesse (1977) and Becker (1979) for the Middle East and by Wolfman (1973, 1983) for Mexico and Central America. For future work, the Near East and northwest India should present excellent opportunities for the development of long directional SV sequences. In the former region, civilization was well developed by 3000BC and well stratified, well-baked hearths and ovens exist in profusion at earlier stratified village sites. Similarly, large cities existed in northwest India before 2000BC.

In order to expand the global coverage of archaeomagnetic results, the investigation of sites occasionally occupied by semi-nomadic groups must be considered. Wolfman (1982), using samples from 28 well-baked hearths in two stratigraphic units in the Modoc Rock Shelter in Central USA, have constructed a fairly complete record of directional changes for most of the period from 9000bp to 7000bp from one such site. It is virtually certain that other suitable material, from similar contexts, exist elsewhere in the world and that from these, archaeomagnetic direction and to lesser extent intensity records will eventually be recovered for much of the early Holocene.

A key question arising from the results presented in this chapter concerns the accuracy with which past geomagnetic field intensities can be determined from baked clay samples. Apart from experimental and other random errors, systematic errors can be caused by differences in natural and laboratory cooling rates, mineralogical changes during laboratory heatings, fabric and shape anisotropy and stress effects during drilling. In particular, Walton (Section 3.3) has drawn attention to mineralogical changes which may take place during Thellier-type experiments in such a way that the straight-line criterion of acceptability is apparently satisfied from a relatively low temperature. It should also be noted that the Shaw technique may be subject to similar criticism. Therefore we are prompted to ask which, if any, archaeointensity results in the published literature can now be considered acceptable. In this respect the situation may not be as bad as it first appears because it seems that Walton's published intensity determinations are probably systematically too high and since the errors do not appear to be random (as suggested by the first results from a new method being applied by Walton to Greek pottery) it might be possible to devise a correction to be applied to the earlier published results. Next it is suggested that the fairly close agreement found by Games (Section 3.6) between archaeointensities obtained from sherds and from adobe bricks from Egypt is encouraging since two very different types of remanent magnetism are involved.

In a general way, the determination of archaeodirections from baked clays would appear to be more straightforward than the determination of archaeointensity. However, the very stringent requirements of precision and accuracy required for archaeological dating mean that particular attention must be paid to the relative efficiencies of thermal as opposed to AF demagnetization and to criteria for the selection of the optimal level of demagnetization. It is apparent that there is little agreement on which demagnetization techniques yield the best results. In Europe, thermal demagnetization has generally been used while AF techniques have been preferred in USA. Wei (Section 3.11) reports having used both demagnetization techniques on baked material collected in China but has not compared the results obtained on the same material using the two methods. The preferred procedure in Japanese archaeomagnetic work has been to use NRM measurements (Hirooka, 1971; Asami et al., 1972). Further work on the nature of secondary components of RM especially of those due to weathering and to the possible use of chemical demagnetization to remove them is most desirable if more accuracy is to be obtained.

Future work, which will require more extensive cooperation between geophysicists and archaeologists, can be expected to provide a great deal of information of considerable interest to both disciplines.

3.18. DISCUSSION

By the Editors

3.18.1. Presentation of intensity results

As a direct result of laboratory experiment, a value is obtained for the strength of the ancient magnetic field F_A which existed at a particular 'instant' of time, that is the time when a particular piece of pottery was baked and subsequently cooled down, or when a particular fireplace was last used. Therefore, it is very important that this experimental result, F_A, should be reported, together, of course, with the relevant age and geographical coordinates of the site. Yet all too often, this basic result has not been reported: in the literature archaeointensities are to be found expressed indirectly in at least six different ways, largely out of desire to compare coeval results from different regions.

a. As the ratio F_A/F_P of the measured ancient field strength (F_A) to the present day value (F_P).

b. As the ratio F_A/F_D of the measured ancient field strength (F_A) to that (F_D) at the same site due to an axial geocentric dipole of present-day moment ($M = 8 \times 10^{22} Am^2$).

c. As F_E, the field strength at the equator corresponding to the measured F_A, calculated, assuming that F_A originates from a geocentric dipole. If the ancient inclination was measured, the geocentric dipole may be assumed to be inclined rather than axial.

d. As a reduced dipole moment (RDM) which is the moment of a geocentric dipole, inclined at the present angle, which would account for the observed ancient field strength (F_A) at the site. The RDM is calculated when the ancient inclination is not known.

e. As a virtual dipole moment (VDM), which is the moment of the inclined geocentric dipole corresponding to the measured ancient field strength (F_A) and the measured ancient inclination (I_A) (assuming zero declination).

$$VDM = 0.5 \, F_A R^3 (1 + 3 \cos^2 I_A)^{1/2} \tag{26}$$

f. As a virtual axial dipole moment (VADM), which is the moment of the axial geocentric dipole corresponding to the measured ancient field strength (F_A) and the site latitude (L) (independent of ancient inclination).

$$VADM = F_A R^3 (4 - 3 \cos^2 L)^{-1/2} \tag{27}$$

In equations 26 and 27, R is the Earth's radius.

It is unfortunate that such a diversity of choices appear in the literature, although with our present level of confidence in archaeointensity curves, the differences are probably not significant.

Apart from (a) and (b), all these representations of the ancient field strength F_A are based on the premise that the measured variations in archaeointensity are due to variations in the strength and orientation of the main dipole. The latter is sometimes referred to as 'dipole wobble'. This premise is clearly false when we are considering 'spot' values of the geomagnetic field (and it must be stressed that archaeomagnetic studies yield 'spot' values) because observations of the geomagnetic field over last four centuries or so indicate quite clearly that changes in the configuration of the non-dipole field have played a very important role, possibly the dominant role, in producing secular variations with characteristic times of the order of up to at least a few centuries. Moreover, lake sediment results strongly suggest that the non-dipole field has played a very important part in creating secular variations in direction with characteristic times up to a few thousand years.

Since the non-dipole field makes an important contribution to the strength and orientations of the geomagnetic field at any 'instant' of time, values of F_E, RDM, VDM and RADM calculated from a single spot value of F_A are meaningless. Clearly, the pattern of archaeointensity variations will vary from region to region and differences are to be expected around any given parallel of latitude as well as from pole to equator.

Representations (a) and (b) are designed to relate the measured value of F_A to some base value. This is reasonable so long as the base value is clearly stated. The use of the axial-dipole field value, F_D is preferable to the actual present field value F_P at any given locality since the latter value is of no fundamental importance.

The separation of the contributions of the main dipole and non-dipole parts of the field requires careful analysis of a substantial data set. In order to facilitate such analyses (e.g. Barton et al. 1980) it is important that the basic experimental result should be clearly stated for each archaeointensity determination.

3.18.2. Reliability of archaeointensity determinations

The newer archaeointensity techniques described in the chapter indicate strongly that adequate precautions have often not been taken in the past. This situation has arisen partly out of ignorance of complications such as anisotropy and cooling rate effects, but largely because there has been a tradition of it. For this reason it is probably misleading to use many of the early results in any analysis of archaeointensity data. Nevertheless, because of lack of samples, we still need to use non-ideal results. Therefore, it would be useful if a simple reliability scheme (e.g. 1 = ideal; 2 = non-ideal but consistent; 3 = still worth reporting; see also Games 3.14) could be generally accepted.   This would facilitate subsequent filtering of good from questionable data.

As of now it would be better to use only a small number of the more reliable intensity data which have accurate ages in any overall analysis.   The incorporation of second-rate data will inevitably lead to wrong rather than incomplete conclusions.

3.18.3. Rate of change of F

The question of whether very rapid rates of change of the geomagnetic field (faster than or equal to the maximum instantaneous rates observed for modern isoporic foci) have persisted for many hundreds of years remains unproven.   If such rapid rates of change have in fact occurred, they must almost certainly be a manifestation of the non-dipole field, and should therefore be associated with large changes in direction. In general, this seems not to be the case, although changes in inclination and intensity are often correlated, and many of the archaeointensity data sets which appear to support large rapid changes in F do not include archaeoinclination information.   We note that problems with archaeointensity determinations discussed in sections, 3.2, 3.3 and 3.14 of this Chapter, tend to an overestimation of the ancient field intensity.   For any estimate of rate of change of the field parameters to be reliable, accurate age determination is critical.   Errors in dating can inevitably lead to wildly wrong estimates of rate of change as discussed in section 3.8.

3.18.4. Regional agreement of intensity determinations

There is good regional agreement in some parts of the globe, e.g.   Arkansas and southwestern USA (Sternberg, section 3.13), Greece, Bulgaria and Yugoslavia, Paris (Kovacheva section 3.6), Thomas section 3.8, many parts of the USSR (Burlatskaya et al. section 3.10).   Other parts do not show good internal agreement, e.g. China (Wei, section 3.11).

3.18.5. Long-range agreement of intensities

There is general agreement that the dipole moment showed relatively little variation and that it was somewhat weaker than the present-day moment prior to 4500BP. Thereafter it grew to a maximum around 2000BP – 2500BP and then declined to its present day values.   The support for Cox's (1969b) 8000 year sinusoid becomes weaker as the global coverage of data improves.

3. 18. 6. Drift of non-dipole field

Phase differences of inclination records from different sites imply westward drift at approximately present day rates Burlatskaya's (section 3. 10) analysis indicates that a simple westward drift model is too simplistic: for longitude 20°W – 60°E it is 0. 33 deg/yr and 100°E it is 0. 12 deg/yr and Wei (section 3. 11) provides some evidence of eastward drift.   That is to say, the drift rate and sense are functions of both time and longitude.   The sense of drift is often inferred from the sense of movement around loops described by Bauer or VGP plots (Skiles, 1970).   An excellent illustration of the limitation of the applicability of this rule is provided by the counter-clockwise VGP loops for Japan for 500AD – 800AD and for China for 460AD – 12200AD from which eastward drift is implied.   In fact the phase differences exhibited by the declination and inclination logs indicate westward drift.

3. 18. 7. Periodicities

Many authors report periodicities in directional results of the order of 500 – 600 years.   Periodicities reported in intensity data are also commonly around 500 – 600 years and around 1000 years.   Most estimates are simply based on counting cycles in the time domain, which, given the quality of most data sets, is quite adequate, though Burlakskaya reports the results of spectral analyses of many data sets. Although there is no clear cut pattern, these trends are found globally and for different intervals during the last 8000 years.   This implies that the same fundamental type of mechanism is responsible for the SV globally and throughout the Holocene.   For many published curves there is more high frequency information towards the recent end of the record – an effect which we must attribute to the higher resolution of both archaeomagnetic and chronological information in young records.

3. 18. 8. Relationship between intensity and inclination

It is common to find steep inclinations coinciding with high field strengths, e. g. Kovacheva Fig. 3-8, Burlatskaya Fig. 3-23a, b, Aitken et al.   section 3. 9.   This could be a consequence of "radial dipoles" in the northern hemisphere dominantly pointing downwards or possibly of dipole wobble.

REFERENCES

Ade-Hall, J.M., Khan, M.A., Dagley, P. and Wilson, R.L. A detailed opaque, petrological and magnetic investigation of a single lava flow from Scotland — III. Invetigation into the possibility of obtaining the intensity of the ambient magnetic field at the time of cooling of the flow. *Geophys. J. R. astr. Soc.*, 1968, *16*, 401–415.

Aitken, M.J. Magnetic dating — I. *Archaeometry*, 1958, *1*, 16–20.

Aitken, M.J. Dating by archaeomagnetic and thermoluminescent methods. *Phil. Trans. R. Soc.*, 1970, *269A*, 77–88.

Aitken, M.J. *Physics and Archaeology*. Clarendon Press, Oxford, 1974. (2nd ed.).

Aitken, M.J. Archaeological Involvements in Physics. *Physics Letters*, 1978, *40C*, 277–351.

Aitken, M.J., Alcock, P.A., Bussel, G.D. and Shaw, C.J. Archaeomagnetic determination of the past geomagnetic intensity using ancient ceramics: allowance for anisotropy. *Archaeometry*, 1981, *23*, 53–64.

Aitken, M.J., Hawley, H.N. and Weaver, G.H. Magnetic dating: further archaeomagnetic measurements in Britain. *Archaeometry*, 1963, *6*, 76–80.

Aitken, M.J. and Hawley, H.N. Magnetic dating — III. *Archaeometry*, 1966, *9*, 187–197.

Aitken, M.J. and Hawley, H.N. Archaeomagnetic measurements in Britain IV. *Archaeometry*, 1967, *10*, 129–134.

Aitken, M.J. and Hawley, H.N. Archaeomagnetism: evidence for magnetic refraction in kiln structures. *Archaeometry*, 1971, *13*, 83–85.

Ajuushayev, G. Burlatskaya, S.P., Nachasova, I.E., Dodon, G. and Balbar, I. Archaeomagnetic definitions of the geomagnetic field for the territory of Mongolia. *Geomagnetism i Aeronomia*, 1976, *16*, 542–548.

Asami, E., Tokieda, K. and Kishi, T. Archaeomagnetic study of kilns in San-in and Kyushu, Japan. *Memoirs Faculty of Literature and Science. Shimane University, Natural Science*, 1972, *5*, 18–22.

Athavale, R.N. Intensity of the geomagnetic field in India over the past 4000 years. *Nature*, 1966, *210*, 1310–1312.

Athavale, R.N. Intensity of the geomagnetic field in prehistoric Egypt. *Earth Planet. Sci. Lett.*, 1969, *6*, 221–224.

Barbetti, M. Measurements of recent geomagnetic secular variation in southeastern Australia and the question of dipole wobble. *Earth Planet. Sci. Lett.*, 1977, *36*, 207–218.

Barbetti, M. Geomagnetic strength over the last 50000 years and changes in atmospheric ^{14}C concentration; emerging trends. *Radiocarbon*, 1980, *22*, 192–199.

Barbetti, M.F. and Flude, K. Geomagnetic variations during the Late Pleistocene period and changes in the radiocarbon time-scale. *Nature*, 1979, *279*, 202–205.

Barbetti, M.F. and McElhinny, M.W. The Lake Mungo geomagnetic excursion. *Phil. Trans. R. Soc.*, 1976, *A281*, 515–542.

Barbetti, M. and Sheard, M.J. Palaeomagnetic results from Mounts Gambier and Schank. *J. Geol. Soc. Aust.*, 1981, *28*, 385–394.

Barbetti, M.F., McElhinny, M.W., Edwards, D.J. and Schmidt, P.W. Weathering processes in baked sediments and their effects on archeomagnetic field intensity measurements. *Phys. Earth. Planet. Inter.*, 1977, *13*, 346–354.

Barbetti, M., Schmidt, P.W., Edwards, D.J. and McElhinny, M.W. Geomagnetic strength in southeastern Australia for the last 7000 years. 1982. (in preparation).

Barton, C.E. and Barbetti, M. Recent geomagnetic secular variation from lake sediments, ancient fireplaces and historical measurements in southeastern Australia. *Earth Planet. Sci. Lett.*, 1982, *59*, 375–387.

Barton, C.E., Bowler, J.M. and Polach, H.A. Magnetic stratigraphy, sedimentology and ^{14}C ages of three Australian maars, *unpublished manuscript, available on request.*, 1981.

Barton, C.E., McElhinny, M.W. and Edwards, D.J. Laboratory studies of depositional DRM. *Geophys. J. R. astr. Soc.*, 1980, *61*, 355–377.

Barton, C.E., Merrill, R.T. and Barbetti, M. Intensity of the Earth's magnetic field over the last 10000 years. *Phys. Earth. Planet. Inter.*, 1979, *20*, 96–110.

Baumgartner, E.P. *Magnetic Properties of Archaeomagnetic Materials*. Doctoral dissertation, University of Oklahoma, 1973.

Becker, H. Archaeomagnetic investigations in Anatolia from prehistoric and Hittite series (First preliminary results). *Archaeo-Physika*, 1979, *10*, 382–387. In: Proceedings of the 18th International Symposium on Archaeometry and Archeological Prospection.

Benkova, N.P., Cherevko, T.N., Burlatskaya, S.P. and Kovacheva, M. The ancient geomagnetic field in Bulgaria from the beginning of our e till nowadays. *Bulgarian Geophysical Journal*, 1977, *3*, 89–95.

Benkova, N.P., Burlatskaya, S.P. and Cherevko, T.N. The comparison of the models of the main geomagnetic field with archaeomagnetic data. *Izv. A.N. SSSR, Fizika Zemli*, 1979, *3*, 76–88.

Braginsky, S.I. Spherical analysis of the geomagnetic field upon the angular data and extrapolated $°g_1$. *Geomagnetism i Aeronomia*, 1969, *9*, 777–778.

Braginsky, S.I. The analytical description of the geomagnetic field of the previous epochs and the determination of the spectrum of the magnetic waves in the Earth's core. *Geomagnetism i Aeronomia*, 1972, *12*, 1092–1105.

Braginsky, S.I. The analytical description of the geomagnetic field of the previous epochs and the determination of the spectrum of the magnetic waves in the Earth's core, II. *Geomagnetism i Aeronomia*, 1974, *14*, 522–529.

Braginsky, S.I. and Burlatskaya, S.P. The comparison of the archaeomagnetic data with analytical representa of the geomagnetic field for the last 350 years. *Izv. A.N. SSSR, Fizika Zemli*, 1972, *1*, 95–99.

Bucha, V. Intensity of the Earth's magnetic field during archaeological times in Czechoslovakia, *Archaeometry*, 1967, *10*, 12–22.

Bucha, V. Changes in the Earth's magnetic moment and radiocarbon dating. *Nature*, 1969, *224*, 681–683.

Bucha, V. Influence of the Earth's magnetic field on radiocarbon dating. In Olsson, I.U. (Ed.), *Nobel Symposium*. Vol. 12: *Radiocarbon, Variations and Absolute Chronology*. John Wiley and Sons, New York, 1970.

Bucha, V., Taylor, R.E., Berger, R. and Haury, E.W. Geomagnetic intensity changes during the past 8500 years in the western hemisphere. *Science*, 1970, *1968*, 111–114.

Burakov, K.S. and Nachasova, I.E. The method and the results of the investigation of the geomagnetic field in Hiva from the middle of the sixteenth century. *Izv. A.N. SSSR, Fizika Zemli*, 1978, *11*, 93–99.

Burlatskaya, S.P. *Archaeomagnetism: Investigation of the Geomagnetic Field for Former Epochs*. Nauka, Moscow, 1965. 127 pp.

Burlatskaya, S.P. The variation of the geomagnetic field's intensity for the last 8500 years upon the world archaeomagnetic data. *Geomagnetism i Aeronomia*, 1970, *10*, 694–699.

Burlatskaya, S.P. The secular variations of the geomagnetic field upon archaeomagnetic and palaeomagnetic data. *Geomagnetism i Aeronomia*, 1972, *7*, 662–675.

Burlatskaya, S.P. Characteristics of the spectrum of secular geomagnetic field variations in the past 8500 years. *Geomagnetism i Aeronomia*, 1978, *18*, 621–623.

Burlatskaya, S.P. The geomagnetic field's variations for the last 8000 years. In *The second all-USSR Congress: 'The Main Geomagnetic Field, Rock Magnetism and Palaeomagentism'*. Tbilisi University, 1981.

Burlatskaya, S.P. and Braginsky, S.I. The comparison of archaeomagnetic data with the analytical representation of the geomagnetic field for the last 2000 years. *Archaeometry*, 1978, *20*, 73–83.

Burlatskaya, S.P. and Nachasova, I.E. About reliability of the secular variation periods of the geomagnetic field upon archaeomagnetic data. *Geomagnetism i Aeronomia*, 1978, *18*, 724–727.

Burlatskaya, S.P. and Voronov, U.N. The change of the geomagnetic field's intensity in the past in the territory of Abchazia. *Geomagnetism i Aeronomia*, 1974, *17*, 130–132.

Burlatskaya, S.P., Nechayeva, T.B. and Petrova, G.N. The characteristics of the geomagnetic field's secular variations upon world data. *Izv. A.N. SSSR, Fizika Zemli*, 1968, *12*, 62–70.

Burlatskaya, S.P., Nechayeva, T.B. and Petrova, G.N. Some archaeomagnetic data indicative of the westward drift of the geomagnetic field. *Archaeometry*, 1969, *11*, 115–130.

Burlatskaya, S.P., Kruczyk, J. Malkowski, Z., Nechayeva, T.B. and Petrova, G.N. The geomagnetic field in the territory of south Poland for the last 1000 years. *Acta Geophysica Polonica*, 1969, *17*, 165–179.

Burlatskaya, S.P., Nachasova, I.Ye. and Burakov, K.S. New determinations of the parameters of the ancient geomagnetic field for Mongolia, Soviet central Asia and Abkhazia. *Geomag. Aeron.*, 1975, *16*, 447–450.

Burlatskaya, S.P., Nachasova, I.E. and Burakov, K.S. New determinations of the ancient geomagnetic field's parameters in Mongolia. *Geomagnetism i Aeronomia*, 1976, *16*, 914–918.

Burlatskaya, S.P., Nachasova, I.E. and Tschelidze, Z.A. New determinations of the geomagnetic field's intensity in Georgia for the last 2000 years. In *The second all-USSR congress: 'The Main Geomagnetic Field, Rock Magnetism and Palaeomagnetism'*. Tbilisi University, 1981.

Burlatskaya, S.P. and Nachasova, I.Ye. *Archeomagnetic determinations of the geomagnetic field elements* (Tech. Rep.). Sov. Geophys. Comm. Acad. Sci. U.S.R.R., Moscow, 1977. 112 pp.

Burlatskaya, S.P., Nechayeva, T.B. and Petrova, G.N. Geomagnetic field intensity in the last 2000 years, according to global data. *Geomagnetism i Aeronomia*, 1970, *10*, 693–695.

Cai, L.Z. and Qui, S.H. Statistical analyses of ^{14}C dating data. *Archaeology*, 1979, *6*, 554–561.

Champion, D.E. *Holocene geomagnetic secular variation in the western United States: implications for the global geomagetic field* (Open file report 80–824). U.S. Geological Survey, 1980. 314 pp.

Clark, R.M. A calibration curve for radiocarbon dates. *Antiquity,,* 1975, *50*, 61–63.

Clark, P. and Barbetti, M. Fires, hearths and palaeomagnetism. In Ambrose, W. and Duerden P. (Eds.), *Archaeometry: An Australian perspective*. Australian National University Press, Canberra, 1982.

Coe, R.S. Palaeointensities of the Earth's magnetic field determined from Tertiary and Quaternary rocks. *J. geophys. Res*, 1967a, *72*, 3247–3262.

Coe, R.S. The determination of palaeointensities of the Earth's magnetic field with emphasis on mechanisms which could cause non-ideal behavior in Thellier's method. *J. Geomag. Geoelectr.*, 1967b, *19*, 157–179.

Coe, R.S. Source models to account for the Lake Mungo palaeomagnetic excursion and their implications. *Nature*, 1977, *269*, 48–51.

Coe, R.S., Gromme, S. and Mankinen, E.A. Geomagnetic palaeointensities from radiocarbon dated lava flows on Hawaii and the question of the Pacific nondipole low. *J. geophys. Res*, 1978, *83*, 1740–1756.

Cox, R.M. Lengths of geomagnetic polarity intervals, *J. geophys. Res*, 1968, *73*, 3247–3260.

Cox, A. A paleomagnetic study of secular variation in New Zealand. *Earth Planet. Sci. Lett.*, 1969, *6*, 257–267.

Creer, K.M. Geomagnetic secular variation during the last 25000 yr: an interpretation of data obtained from rapidly deposited sediments, *Geophys.J. R. astr. Soc.*, 1977, *45*, 91–109.

Creer, K.M., Readman, P.W. and Papamarinopoulos,S. Geomagnetic secular variations in Greece through the last 6000 years obtained from lake sediment studies. *Geophys.J. R. astr. Soc.*, 1981, *66*, 193–219.

Damon, P.E., Lerman, J.C., Long, A., Bannister, B., Klein, J. and Linick, T. Report on the workshop on the calibration of the radiocarbon dating time scale. *Radiocarbon*, 1980, *22*, 1740–1756.

Dean, J. Tree-ring dating in archeology. In *Anthropological Papers*. Vol. 99: *Miscellaneous collected papers 19-24*. University of Utah, 1978.

Dodson, M.H. and McClelland-Brown, E. Magnetic blocking temperatures of single-domain grains during slow cooling. *J. geophys. Res*, 1980, *85*, 2625–2637.

Doell, R.R., and Cox, A. The Pacific geomagnetic secular variation anomaly and the question of lateral uniformity in the lower mantle. In Robertson, E.C. (Ed.), *The Nature of the Solid Earth*. MaGraw-Hill, New York, 1972.

Dolfuss, G. and Hesse, A. Les structures de combustion du Tope Djaffarabad, Periodes I A III. *Cahiers de la Delegation Archeologique Francaise en Iran*, 1977, *7*, 11–47.

Domen, H. A single heating method of paleomagnetic field intensity determination applied to old roof tiles and rocks, *Phys. Earth Planet. Interiors.*, 1977, *13*, 315–318.

Dubois, R.L. Secular variation in southwestern United States as suggested by archaeomagnetic studies. In R.M.M. Fisher, M.Fuller, V.A. Schmidt and P.J. Wasilewski (Eds.), *Takesi Nagata Conference - Magnetic Fields: Past and Present*. Goddard Space Flight Center, Greenbelt, Maryland, 1975.

Dubois, R.L. Secular variation from archaeomagnetic data. In *Abstract with programs, South Central Section, 9th Ann. Meeting*. Geological Society of America, 1975.

Dunlop, D.J. and West, G.F. An experimental evaluation of single domain theories. *Rev. Geophys. Space Phys.*, 1969, *7*, 709–757.

Dunlop, D.J. and Zinn, M.B. Archeomagnetism of a 19th century pottery kiln near Jordan, Ontario. *Can. J. Earth Sci.*, 1980, *17*, 1275–1285.

Eighmy, J.L. *Archaeomagnetism: a Handbook for the Archeologist.* United States Department of Interior, Heritage Conservation and Recreation Service, 1980. 104 pp.

Eighmy, J.L., Sternberg, R.S. and Butler, R.F. Archaeomagnetic dating in the American Southwest. *American Antiquity*, 1980, *45*, 507–517.

Fisher, R.A. Dispersion on a sphere. *Proc. Roy. Soc. London*, 1953, *A217*, 295–305.

Folgheraiter, G. Ricerche sull' inclinazione magnetica all' epoca etrusca. *Rendiconti della R. Accademia dei Lincei*, 1896, *V*, 293–300.

Folgheraiter, G. Sulla forza coercitiva dei vasi etruschi. *Rendiconti della R. Accademia dei Lincei*, 1897, *VI*, 64–70.

Folgheraiter, G. La magnetizzazione dell' agrilla colla cottura in relazione colle ipotesi sulla fabbricazione del vasellame nero etrusco. *Rendiconti della R. Accademia dei Linei*, 1897, *VI*, 368–376.

Folgheraiter, G. Sur les variations seculaires de l'inclination magnetique dans l'antiquite. *Archives des Sciences Physiques et Naturelles*, 1899, *8*, 5–16.

Fox, J.M.W. *Archaeomagnetic Measurement of the Ancient Geomagnetic Field Intensity.* Master's thesis, Oxford University, 1979.

Fox, J.M.W. and Aitken, M.J. Cooling rate dependance of thermoremanent magnetisation. *Nature*, 1980, *283*, 462–463.

Fritsche, H. Die Elemente des Erdmagnetismus fuer die Epochen 1600, 1650, 1700, 1780, 1842 und 1885 und ihre saecular Aenderungen. St. Petersburg, 1889.

Games, K.P. The magnitude of the palaeomagnetic field: a new, non–thermal, non–detrital method using sun–dried bricks. *Geophys. J. R. astr. Soc.*, 1977, *48*, 315–329.

Games, K.P. The magnitude of the archaeomagnetic field in Egypt between 3000 and 0 BC. *Geophys. J. R. astr. Soc.*, 1980, *63*, 45–56.

Gheradi, S. Sul magnetismo polare de palazzi ed altri edifizi in Torino. *Il Nuovo Cimento*, 1862, *16*, 384–404.

Grubbs, F.E. Procedures for detecting outlying observations in samples. *Technometrics*, 1969, *11*, 1–21.

Gunn, N. *Archaeomagnetic Field Strengths from Peru.* Doctoral dissertation, University of Liverpool, 1978.

Gunn, N.M. and Murray, A.S. Geomagnetic field magnitude variations in Peru derived from archaeological ceramics dated by thermoluminescence. *Geophys. J. R. astr. Soc.*, 1980, *62*, 345–366.

Hathaway, J.H., Eighmy, J.L. and Kane, A.E. Preliminary modification of the Southwest virtual geomagnetic pole path AD 700 to AD 900. in press. Dolores Archeological Program Results submitted for publication.

Hirooka, K. Archaeomagnetic study for the past 2000 years in southwest Japan. *Mem. Fac. Sci. Kyoto Univ. Ser. Geol. & Min.*, 1971, *38*, 167–207.

Hirooka, K. The direction of horizontal axial lines of ancient Buddhist temples and archaeological terrestrial magnetism. *J. Archaeological Soc. Nippon*, 1976, *62*, 49–63.

Hirooka, K. Recent trend in archaeomagnetic and palaeomagnetic studies in Quaternary research. *Quaternary Research*, 1977, *15*, 200–203.

Hirooka, K. Dating of ancient kilns with archaeomagnetic method. In S.Narsaki (Ed.), *Ceramic Art of the World.* Vol. 2: *Japanese Ancient Period.* Shogakukan, Tokyo, 1979. (in Japanese).

Hirooka, K. Age dating by using the secular variation of the geomagnetic field direction. In The Editorial Committee of Kobunkazai (Eds.), *Natural Scientific Approaches in Archaeology and Art History.* The Japan Society for the Promotion of Science, Tokyo, 1980. (in Japanese).

Hirooka, K. Archaeomagnetism and age dating. In M. Akamatsu, M.Ando, T.Takimoto and M.Miyaishi (Eds.), *History of Earthenwares.* Annals of Seto City, 1981. (in Japanese).

Hirooka, K., Nakajima, T., Tokieda, K. and Kawai, N. On the results of the measurements of thermoremanent magnetization of the old kilns of Suemura. In H. Nakamura (Ed.), *Osaka Prefectural Board of Education. Suemura.* Osaka: , 1970. (in Japanese).

Hirooka, K. and Aoki, M. Archaeomagnetic measurements on Kiln Okayama 3 in Hirazakura. In Oyabe Municipal Board of Education (Ed.), *Research Report of Buried Cultural Properties of Oyabe City*. Vol. 5: *Kiln Hirazakura-Okayama 3.*, 1981. (in Japanese).

Imai, I. A historical note of the magnetic compass in China. *J. Shanghai Sci. Inst.*, 1942, *12*, 147-166.

Imamiti, S. Secular variation of the magnetic declination in Japan. *Memoirs of Kakioka Magnetic Observatory*, 1956, *7*, 49-55.

Kato, Y. and Nagata, T. On the secular variation in geomagnetic declination in the historic of Japan. In *Proc. 7th Pacific Congress.*, 1953.

Kawai, N., Hirooka,K., Sasajima,S., Yaskawa,K., Ito,H. and Kume,S., Archaeomagnetic studies in southwestern Japan. *Ann. Geophys.*, 1965, *21*, 574-575.

Kawai, N., Hirooka, K. and Nakajima, T. Archaeomagnetism in Iran. *Nature*, 1972, *236*, 223-225.

Kinoshita, H. List of archeomagnetic and paleomagnetic results, 1950-1970 in Japan. *J. Geomag. Geoelect.*, 1970, *22*, 507-545.

Kitazawa, K. Intensity of the geomagnetic field in Japan for the past 10000 years. *J. geophys. Res*, 1970, *75*, 7403-7411.

Kitazawa,K and Kobayashi,K. Intensity variation of the geomagnetic field during the past 4000 years in South America. *J. Geomag. Geoelect*, 1968, *20*, 7-19.

Klein, J., Lerman, J.C., Damon, P.E. and Ralph, E.K. Calibration of radiocarbon dates: Tables based on the consensus data of the workshop on calibrating the radiocarbon time scale. *Radiocarbon*, 1982, *24*, 103-150.

Knott, C.G. and Tanakadate, A, A magnetic survey of all Japan. *J. Coll. Sci.*, *Imperial Univ Japan*, 1889, *2*, 163-262.

Kobayashi, K., Matuzasaki, H. and Momose, K. On the direction of thermoremanent magnetization of hearths excavated at Hiraide Ruin. *Kagaku*, 1953, *23*(23), 85-86. (in Japanese).

Kolomiitseva, G.I. and Pushkov, A.N. The analytical model of the main geomagnetic field for the 2000 years. In *The Time-Place Structure of the Geomagnetic Field*. Nauka, Moscow, 1974.

Kolomiitseva, G.I. and Pushkov, A,.N. The analysis of the geomagnetic field for the 2000 years. In *The Analysis of the Time-Place Structure of the Geomagnetic Field*. Nauka, Moscow, 1975.

Kono, M. Palaeointensity determination by a modified Thellier method. *Phys. Earth. Planet. Inter.*, 1978, *13*, 305-314.

Kono, M. and Ueno, N. Palaeointensity determination by a modified Thellier method. *Phys. Earth. Planet. Inter.*, 1977, *13*, 305-314.

Kovacheva, M. Archaeomagnetic investigations in Bulgaria: field intensity determinations. *Phys. Earth. Planet. Inter.*, 1977, *13*, 355-359.

Kovacheva, M. Summarised results of the archaeomagnetic investigation of the geomagnetic field variation for the last 8000 years in south-eastern Europe. *Geophys.J. R. astr. Soc.*, 1980, *61*, 57-64.

Kovacheva, M. and Veljovitch, D. Geomagnetic field variation in south-eastern Europe between 6500 and 100 years BC. *Earth Planet. Sci. Lett.*, 1977, *37*(1), 131-137.

Kovacheva,M., Benkova.N.P., Cherevko,T.N., and Burlatskaya S.P. The ancient geomagnetic field in Bulgaria 15-20c. *Bulgarian Geophysical Journal*, 1977, *3*, 88-92.

Krause, G.J. *An Experimental Approach towards defining Archaeomagnetic Dating Techniques*. Doctoral dissertation, Colorado State University, Fort Collins, 1980.

Lawley, E.A. The intensity of the Tertiary geomagnetic field in Iceland during Neogene polarity transitions and systematic deviations. *Earth Planet. Sci. Lett.*, 1970, *10*, 145-149.

Leaton, B.R., Malin, S.R.C. and Evans, M.I. An analytical representation of the estimated geomagnetic field and its secular change for epoch 1965. *J. Geomag. Geoelectr.*, 1965, *17*, 187-194.

Lee, S.S. *Secular variation of the Intensity of the Geomagnetic Field during the past 3000 years in North, Central and South America*. Doctoral dissertation, University of Oklahoma, Norman, 1975. 422 pp.

Levi, S. The effect of magnetite particle size on palaeointensity determination of the geomagnetic field. *Phys. Earth. Planet. Inter.*, 1977, *13*, 245-259.

Lowrie, W. and Fuller, M. Effect of annealing on coercive force and remanent magnetization in magnetite. *J. geophys. Res*, 1969, *74*, 2698-2710.

Martin, P.S. and Plog, F. *The Archaeology in Arizona: a study of the Southwest Region*.

Doubleday/Natural History Press, Garden City, 1973. 422 pp.

McFadden, P.L. Determination of the angle in a Fisher distribution which will be exceeded with a given probability. *Geophys. J. R. astr. Soc.*, 1980, *60*, 391-396.

Mercanton, P. Etude des properties magnetiques des poteries lacustres. *Bulletin de la Societe Vaudoise des Sciences Naturelles*, 1902, *XXXVIII*, 335-346.

Mercanton, P. Stabilite d'aimanation des poteries lacustres. *Bulletin de la Societe Vaudoise des Sciences Naturelles*, 1910, *XLVI*, lxx. Series 5.

Mercanton, P. Etat magnetique de terres cuites prehistoriques. *Bulletin de la Societe Vaudoise des Sciences Naturelles*, 1918a, *52*, 9-15. Series 5.

Mercanton, P. Etat magnetique de quelques terres cuites prehistoriques. *Comptes Rendus de l'Academie des Sciences*, 1918b, *166*, 681-685.

Mirzachanov, R.G. The ancient magnetic field in Azerbaijan. In *The Main Geomagnetic Field*. Institut Fiziki Zemli AN SSSR, 1970.

Momose, K., Kobayashi, K., Tsuboi, K. and Tanaka, M. Archaeomagnetism during the Old Tomb and the Nara periods. pp. 33-38.

Nachasova, I. Ye. Magnetic field in the Moscow Area from 1480 to 1840. *Geomag. Aeron.*, 1972, *12*, 277-280.

Nagata, T. The natural remanent magnetism of volcanic rocks and its relation to geomagnetic phenomena. *Bull. Earthquake Res. Inst., Tokyo Univ.*, 1943, *21*, 1-196.

Nagata T. *Rock Magnetism*. Mazuren Company Ltd., Tokyo, 1961. 346 pp.

Nagata, T., Arai, Y. and Mamose, K. Secular variation of the geomagnetic total force druing the last 5000 years. *GJR*, 1963, *68*, 5277-5281.

Nagata, T., Kobayashi, K. and Schwarz, E.J. Archeomagnetic intensity studies of South and Central America. *J. Geomag. Geoelect.*, 1965, *17*, 399-405.

Nakajima, T., Torii, M., Natsuhara, N. and Kawai, N. On the results of measurements of thermoremanent magnetization of Sue type old kilns in Toga Area. In *Research Report of Buried Cultural Properties of Osaka Prefecture*. Osaka Prefectural Board of Education, 1977. (in Japanese).

Nechayeva, T.B. The obtaining and analysis of the inclination and intensity variation of the Ukraine geomagnetic field. In *The Materials of the 8th Conference of the Main Geomagnetic Field and Palaeomagnetism*. Naukova Dumka, 1970.

Needham, J.I. Science and Civilisation in China. In . Cambridge University Press, LOndon, 1962.

Neel, L. Theorie du trainage magnetique des ferromagnetiques au grains fins avec applications aux terres cuites. *Annales de Geophysique*, 1949, *5*, 99-136.

Neel, L. Some theoretical aspects of rock magnetism. *Advances in Physics, a Quarterly Supplement of the Philosophical Magazine*, 1955, *4*, 191-243.

Nichols, R.F. *Archeomagnetic study of Anasazi-Related Sediments of Chaco Canyon, New Mexico*. Doctoral dissertation, University of Oklahoma, Norman, 1975.

Nodia, M.Z., Apakidze, A.M. and Chelidze, Z.A. The results of the geomagnetic field's investigations in the territory of Georgian SSR by archaeomagnetic method. In *The Materials of the 8th Conference of the Main Geomagnetic Field and Palaeomagnetism*. Naukova Dumka, 1970.

Nodia, M.Z., Apakidze, A.M. and Chelidze, Z.A. Some findings of an archaeomagnetic study of the declination of the geomagnetic field in the territory of Georgia. *Bulletin of the Academy of Sciences of the Georgian SSR*, 1970, *58*(3), 565-568.

Smith, P.J. The intensity of the ancient geomagnetic field: a review and analysis. *Geophys. J. Roy. astron. Soc.*, 1967, *12*, 321-362.

Parker R.A. *Archeomagnetic Secular Variation*. Doctoral dissertation, University of Utah, Salt Lake City, 1976. 250 pp.

Pentecost, J.L. and Wright, C.H. Preferred orientation in ceramic materials due to forming techniques. In W.M. Mueller, G.R. Mallet and M.J. Fay (Eds.), *Advances in X-ray Analysis*. Plenum Press, New York, 1964.

Petrova, G.N. and Burlatskaya, S.P. Modern conceptions about secular variations. In *The Problems of the Investigations of the Paleosecular Variations of the Geomagnetic Field*. Institut Vulaknologii AN SSSR, Vladivostok, 1979.

Pozzi, J.P. and Thellier, E. Sur la perfectionements recents apportes aux magnetometres a tres haute sensibilite utilises en mineralogie magnetique et en paleomagnetisme. *Comptes Rendus de l'Academie des Sciences*, 1963, *257*, 1037–1040.

Smith, R.W. *The Magnetization and Stable Remanence of Haematite*. Doctoral dissertation, University of Pittsburgh, Pa., 1967.

Rimbert, F. Contribution a l'etude de l'action de champs alternatives sur les aimanations remanentes des roches, application geophysiques. *Revue de l'Institut Francais de Petrole*, 1959, *14*, 17–54 & 123–155.

Rogers, J., Fox, J.M.W. and Aitken, M.J. Magnetic anisotropy in ancient pottery. *Nature*, 1979, *277*, 644–646.

Roquet, J. Sur les remanences des oxydes de fer et leur interet en geomagnetisme. *Annales de Geophysique*, 1959, *10*, 226–247 & 282–325.

Rubens, S.M. Cube-surface coil for producing an uniform magnetic field. *J. Sci. Instr.*, 1945, *16*, 243.

Rusakov, O.M. and Zagny, G.F. The virtual pole's position upon archaeomagnetic investigation's data for the Ukraine and Moldavia. In *The Materials of the 8th Conference of the Main Geomagnetic Field and Palaeomagnetism*. Naukova Dumka, 1970.

Rusakov, O.M. and Zagny, G.F. Intensity of the geomagnetic field in the Ukraine and Moldavia during the past 6000 years. *Archaeometry*, 1973, *15*, 275–285.

Sachs, A., Absolute dating from Mesopotamian records. *Phil. Trans. R. Soc.*, 1970, *A269*, 19–22.

Sakai, H, Variation of the geomagnetic field intensity deduced from archaeological objects. *Rock Magnetism and Paleogeophysics*, 1980, *7*, 61–67.

Sasajima, S., Geomagnetic secular variation revealed in baked earth in west Japan (Part 2): Change of the field intensity. *J. Geomag. Geoelectr.*, 1965, *17*, 413–416.

Sasajima, S., and Maenaka, K. Intensity studies of the archaeosecular variation in west Japan, with special reference of the hypothesis of the dipole axis rotation. *Mem. Coll. Sci., Univ. Kyoto, Ser. B, Geology and Mineralogy*, 1966, *23*, 53–67.

Schiffer, M.B. Hohokam chronology: an essay on history and method. In R.H. McGuire and M.B.Schiffer (Eds.), *Hohokam and Patayan*. Academic Press, New York, 1982.

Schmidbauer, E. and Veitch, R.J. Anhysteretic remanent magnetization of small multidomain Fe_3O_4 particles dispersed in various concentrations in a non-magnetic matrix. *J. Geophys.*, 1980, *48*, 148–152.

Schwarz, E.J. and Christie, K.W. Original remanent magnetization of Ontario potsherds. *J. geophys. Res*, 1967, *72*, 3263, 3267.

Shaw, J. A new method for determining the magnitude of the palaeomagnetic field. *Geophys.J. R. astr. Soc.*, 1974, *39*, 133–141.

Shaw, J. Rapid changes in the magnitude of geomagnetic field. *Geophys.J. R. astr. Soc.*, 1979, *58*, 107–116.

Shuey, R.T., Cole, E.R. and Mikulich, M.J. Geographic correction of archeomagnetic data. *J. Geomag. Geoelectr.*, 1970, *22*, 485–489.

Sigalas, I., Gangas, N.H.J. and Danon, J. Weathering model in palaeomagnetic field intensity measurements on ancient fired clays. *Phys. Earth. Planet. Inter.*, 1978, *16*, P15–P19.

Skiles, D.D. A method of inferring the direction of drift of the geomagnetic field from paleomagnetic data. *J. Geomag. Geoelectr.*, 1970, *22*, 441–461.

Stacey, F.D. The Physical Theory of Rock Magnetism. *Adv. Phys*, 1963, *12*, 45–133.

Stephenson, A. Gyromagnetism and the remanence acquired by a rotating rock in an alternating field. *Nature*, 1980, *284*, 48–49.

Sternberg, R.S. *Archaeomagnetic Secular Variation of Direction and Paleointensity in the American Southwest*. Doctoral dissertation, University of Arizona, Tucson, 1982.

Sternberg, R.S. and Butler, R.F. An archaeomagnetic paleointensity study of some Hohokam potsherds from Snaketown, Arizona. *Geophys. Res. Lett.*, 1978, *5*, 101–104.

Sternberg R.S. and McGuire, R.H. Archaeomagnetic secular variation in the American Southwest. *EOS*, 1981, *62*, 652.

Tanaka, H. Geomagnetic paleointensities during the past 30000 years in Japan. *Rock Magnetism and Palaeogeophysics*, 1978, *5*, 95–97.

Tarchov, E.N. The geomagnetic field in Leningrad upon archaeomagnetic data. *Geomagn. i aeronomia*,

1963, *3*, 728–734.

Tarchov, E.N. Some results of the archaeomagnetic investigations in the western part of the USSR. *Geomagn. i aeronomia*, 1965, *5*, 135–140.

Tarchov, E.N. *The Geomagnetic Inclination in the European part of the USSR and Siberia upon Archaeomagnetic Data*. Doctoral dissertation, IZMIR AN SSSR, 1967.

Tarchov, E.N. The inclination in Central Russia upon archaeomagnetic data. In *The Materials of the 8 Conference of the Main Geomagentic Field and Palaeomagnetism*. Naukova Dumka, Kiev, 1970.

Tarchov, E.N. and Ivanov, N.V. Secular variations of the geomagnetic field's inclination in the territory of Litovskaja SSR upon archaeomagnetic data. *Geomagn. i aeronomia*, 1965, Vol. *5*.

Taylor, R.E. and Meighan, C.W. (Editors). *Chronologies in New World*. Academic Press, New York, 1978. 587 pp.

Teng, H.H. and Li, T.C. The geomagnetic field in Peking region and its secular variation during the last 2000 years. *Acta Geophysica Sinica*, 1965, *14*, 181–196.

Thellier, E. Magnetometre insensible aux champs magnetiques troubles des grandes villes. *Comptes Rendus de l'Academie des Sciences*, 1933, *197*, 232–234.

Thellier, E. Aimantation des bricques et inclination du champ magnetique terrestere. *Annales de l'institut de Physique du Globe*, 1936, *14*, 65–70.

Thellier, E. On the magnetization of bricks and pottery, *Ann. Phys.*, 1938, *16*, 157–302.

Thellier, E. Proprietes magnetiques des terres cuites et des roches. *Journal de Physique et la Radium*, 1951, *12*, 205–218.

Thellier, E. Early research in the intensity of the ancient geomagnetic field. *Phys. Earth. Planet. Inter.*, 1977, *13*, 241–244.

Thellier, E. Sur la direction du champ magnetique terrestre en France durant les deux derniers millenaires. *Phys. Earth. Planet. Inter.*, 1981, *24*, 89–132.

Thellier, E. and Rimbert, F. Sur l'analyse d'aimantations fossiles par action de champs magnetiques alternatifs. *Comptes Rendus de l'Academie des Sciences*, 1954, *239*, 1399–1401.

Thellier, E. and Rimbert, F. Sur l'utilisation en paleomagnetisme, de la desaimantation par champs alternatifs. *Comptes Rendus de l'Academie des Sciences*, 1955, *240*, 1404–1406.

Thellier, E. and Thellier, O. Sur l'intensite du champ magnetique terrestre dans le passe historique et geologique. *Annales de Geophysique*, 1959, *15*, 295–376.

Thomas, R.C. *Archaeomagnetism of Greek Pottery and Cretan Kilns*. Doctoral dissertation, University of Edinburgh, 1981.

Tokieda, K. Age dating of old earthenwares by archaeointensity method. In The Editorial Committee of Kobunkazai (Ed.), *Natural Scientific Approaches in Archaeology and Art History*. The Japan Society for the Promotion of Science, Tokyo, 1980. (in Japanese).

Torii, M., Nakajima, T., Asai, I., Koide, K., Natsuhara, N. and Kawai, N. Archaeomagnetic research of Sue type kilns in Ono area. In *Suemura I. Research Report of Buried Cultural Properties of Osaka Prefecture*. Osaka Prefectural Board of Education, 1976. (in Japanese).

Walton, D. Archaeomagnetic intensity measurements using a SQUID magnetometer, *Archaeometry*, 1977, *19*, 192–200.

Walton, D. Geomagnetic intensities in Athens between 2000 BC and AD 400, *Nature*, 1979, *277*, 643–644.

Walton, D. Time–temperature relations in the magnetisation of assemblies of single domain grains. *Nature*, 1980, *286*, 245–247.

Walton, D. Errors and resolution of thermal techniques for obtaining the geomagnetic intensity. *Nature*, 1982, *295*, 512–515.

Watanabe, N. On the remanent magnetism of pottery and baked earth from prehistoric Japan. *Proc. 2nd Joint Meeting Eight Acad. Soc. Japan*, 1949, pp. 149–154.

Watanabe, N. Secular variation in the direction of geomagnetism as the standard scale for geomagneto-chronology in Japan. *Nature*, 1958, *182*, 383–384.

Watanabe, N. The direction of remanent magnetism of baked earth and its application to chronology for anthropology and archaeology in Japan: An introduction to geomagneto–chronology. *J. Faculty Sci., Univ. Tokyo,* , 1959, *5*, 1–188.

Watanabe, N. Magnetic dating in archaeology. *Kagaku no Ryoiki*, 1977, *31*, 683–691.

Watanabe, N. and Dubois, R.L. Some results of an archaeomagnetic study on the secular variation in the southwest of North America. *J. Geomag. Geoelectr.*, 1965, *17*, 395–397.

Weaver, M.P.   *The Aztecs, Maya and their Predecessors: Archeology of Mesoamerica.*   Academic Press, New York, 1981.   2nd edition.

Wei, Q.Y., Li, D.J., Cao, G.Y., Zhang, W.S. and Wang, S.P. Archaeomagnetic research of Jiangzhai relic, Neolithic Epoch. *Acta Geophysica Sinica*, 1980, *23*, 403–414.

Wei, Q.Y., Li, D.J., Cao, G.Y., Zhang, W.S. and Wang, S.P. Secular variation of the direction of the ancient geomagentic field for Luoyang region, China. *Phys. Earth. Planet. Inter.*, 1981, *25*, 107–112.

Weinberg, B.P.   The Catalogue of the magnetic measurements in the USSR and neighbouring territories from 1556 until 1926. (in Russian).

Wilson, R.L. and Lomax, R. Magnetic remanence related to slow rotation of ferromagnetic material in alternanting magnetic fields. *Geophys.J. R. astr. Soc.*, 1972, *30*, 295–303.

Wolfman, D. *A Re-evaluation of Mesoamerican Chronology: A.D. 1-1200.* Doctoral dissertation, University of Colorado, Boulder, 1973.   293 pp.

Wolfman, D. Archaeomagnetic dating in Arkansas. *Archaeo-Physika*, 1979, *10*, 522–533.

Wolfman, D. Geomagnetic dating methods in archeology. In M.B. Schiffer (Ed.), *Advances in Archaeological Method and Theory*. Academic Press, New York, 1983.   (in press).

Wolfman, D. Archeomagnetic dating in Arkansas and the border areas of adjacent states. In M. Jeter and N. Trubowitz (Eds.), *Arkansas Archeology in Review*. , in press.

Yukutake,T. Review of the geomagnetic secular variations on the historical time scale. *Phys. Earth. Planet. Inter.*, 1979, *20*, 83–95.

Yukutake, T., Sawada, M. and Yabu, T. Magnetization of ash-fall tuffs of Oshima Volcano, Izu, I. *J. Geomag. Geoelectr.*, 1964, *16*, 178–182.

Yukutake, T. and Tachinaka, H. Separation of the earth's magnetic field into the drifting and standing parts. *Bull. Earthquake Res. Inst.*, 1969, *47*, 65–97.

Zagny, G.F. The structure of the geomagnetic field's archaeosecular variations in the Ukraine and Moldavia for the last 5500 years. *Geofiz. jurnal*, 1981, *3*, 60–66.

Zimmerman,D.W. Thermoluminescent dating using fine grains from pottery. *Archaeometry*, 1971, *13*, 29–52.

4.
Palaeomagnetism of Unconsolidated Sediments

4.1. INTRODUCTORY COMMENTS

By the Editors

4.1.1. Introduction

Although unconsolidated sedimentary rocks were among the first to be investigated palaeomagnetically (McNish and Johnson, 1938; Ising, 1943b; Johnson et al., 1948), it was not until Mackereth (1971) demonstrated that changes in the direction of the geomagnetic field through post-Glacial time (~ the last 10000 years) were recorded in the bottom sediments of Lake Windermere, that the rapid development of this area of palaeomagnetic investigation began.

The earlier workers, recognizing the crucial importance of establishing a reliable time-frame, chose to work on varved sediments. Unfortunately it turned out that the high energy environment of deposition caused significant errors in the alignment of the magnetic particles which carry the remanent magnetization: water current velocities of 5 to 30cm/s can cause misalignments of the order of 10°. Also, the presence of flat or elongated grains can cause the recorded NRM inclinations to be too shallow and the rolling of grains on sloping surfaces of deposition can cause an additional deviation called the bedding error (King, 1955).

In contrast, Mackereth (1971) chose to study the NRM of the relatively homogenous muds deposited in the low energy environment of post-Glacial lakes. Some years earlier, he had developed (Mackereth, 1958), a pneumatically powered corer which permitted the collection of physically undisturbed samples of lake-bottom sediments. It was largely this invention which allowed him to recover the first high quality sedimentary record of secular variations in geomagnetic declination which he reported in 1971. The basic principles of the method may be illustrated by pointing to the analogy between muds deposited on the floors of lakes and cassettes of magnetic tape as outlined in Table 4-1.

Table 4-1:   Analogy between magnetic tape recording and play-back
and palaeomagnetic studies of unconsolidated sediments

function	process occurring in tape recorder	equivalent palaeomagentic process
the signal	sound waves (for example)	geomagnetic secular variations
the recording process	instrument recording head, amplifier, filters etc.	acquisition of PDRM by the sediment
the record	contained in a magnetic tape	contained in a core of sediment
recording errors	due (for example) to fluctuations in speed of tape transport during recording,	due (for example) to variations in deposition rate, physical disturbance of sediment
play-back	play-back of tape: record converted via instrument electronics and loudspeakers to facsimile of initial signal	laboratory magnetometer measurements of PDRM + computer data processing and plotting
reproduction errors	due (for example) to fluctuations in speed of tape transport during playback or faulty or low quality electronics	due (for example) to inaccuracies in transformation from depth to time scale resulting from errors in C14 dates or instability of remanent magnetization
result	HI-FI !	LOW-FI !

4.1.2. Techniques of coring unconsolidated sediments

4.1.2.1 The Mackereth corer

The main advantage of the pneumatic corer developed by Mackereth (1958) is that it can be made to penetrate the bottom sediments at rates which are both slow and controllable. Thus the structure of the cored sediment is subjected to minimal disturbance.

Cores of 6m length and 55mm diameter can be taken in plastic pipes using small boat (~4m) and two or three men. Core recovery rates are typically about 3 per day from 80m depth and about 6 per day from 30m depth. For operation of the extended 12m Mackereth corer it is desirable to have two small boats to allow greater flexibility in transporting the corer across the lake and for lowering and retrieving it. A rubber flotation bag is normally used to extract it from the lake bottom. The maximum depth of operation is limited by air pressures required to equal the hydrostatic pressure at the lake bottom. The air hose currently used will withstand a pressure of ~30 bars. This pressure should exceed the hydrostatic pressure by ~5 bars for reliable operation of the corer thus limiting the depth of water to ~250m. Normally, however, operation is limited to ~100m by the mass and volume of the air-hose which it is practicable to carry in a small boat. By milling a groove along the core tubes they can be prevented from twisting during penetration by means of a tongue fixed to the anchor drum. The corer may be orientated while on the bottom by photographing a compass and spirit levels housed in a water-tight box.

4.1.2.2 Gravity-piston corers

The Kullenberg (1947) type of corer relies on heavy weights to be driven into the bottom sediments and has to be winched out. This means that a rather large vessel is required. Cores of 15m to 20m length obtained using Alpine type gravity-piston corers as operated from the research vessel 'Limnos' by Canada Centre for Inland Waters have been used for palaeomagentic studies of the bottom sediments of the Great Lakes of North America (Creer et al, 1976a,b; Mothersill, 1979,1981). Gravity-piston corers were also used to collect sediments from the Black Sea on which palaeomagnetic studies were carried out by Creer (1974). All these lakes are too deep to permit the use of Mackereth type corers.

Good quality declination records are difficult to obtain from gravity-piston cores because the corer often 'corkskrews' its way into the bottom sediments causing the cored sediment to become twisted and/or sheared into short sections between which relative azimuthal orientation is lost. This unwanted effect can be reduced, though not eliminated, by giving the cable time to untwist before the corer is dropped to the bottom.

4.1.2.3 Livingstone Corer

Livingston (1955) introduced a piston sampler which is pushed into the sediment manually, using a long rod. Cores are taken by making successive drives of about 1m into the sediment. Operation in water depths greater than 15-20m is difficult due to flexibility of the long rod. The Livingstone corer must be operated from a fixed platform and as such a frozen lake surface is suitable. The technique and other considerations involved are discussed in section 4.3.2.

4.1.3. Measurements on unopened cores

Low field susceptibility and the horizontal component of the NRM may be measured using 'Digico' equipment in which the sediment cores are pushed through hollow, cylindrically shaped detector heads, measurements being made at discrete intervals (Molyneux et al., 1972; Molyneux and Thompson, 1973). The remanent magnetization of unopened cores may be AF cleaned using rotating demagnetizing fields.

In principle, a three component cryogenic magnetometer is an ideal instrument for measuring the complete remanent magnetization vector: in practice results from only one such instrument have been reported in the literature (Dodson et al., 1974).

The resolution of the signal derived from long core instruments is limited by the fact that a finite length of core influences the sensing device. The resolution may be improved by deconvolution, but this does not get over the difficulty that core tubes are rarely completely filled with sediment and transverse cracks and gaps are commonly found in most cores.

4.1.4. Detailed magnetic measurements

Mackereth type and gravity-piston type cores are normally collected in long plastic liner tubes in which they can be kept permanently. The long cores are then cut into sections of ~1.5m length for convenience of handling and this is often done in the field. With modified Livingstone corers (see section 4.4.2), sections of 1m to 3m length are recovered. These are extruded and wrapped in plastic film for storage prior to sub-sampling. On arrival at the laboratory the core sections may be X-rayed to identify the more important lithological features (which may be of use for within-lake correlation), to observe the structure of the sediment (if it is layered we may deduce whether penetration has been vertical and whether there has been any distortion) and to detect any gaps in the core.

For sub-sampling, the core sections are split into halves along their length using a thin wire or a strong thread to cut the sediment. Sub-samples are then taken from at least one of the halves of each section in small plastic boxes (~18 x 18 x 18mm cubes or as available) which are pushed into the sediment, taking care to preserve orientation by aligning them with the side of the core tube and with the normal to the sediment face.

The remanent magnetization and magnetic susceptibility of the sub-samples are then measured on standard fluxgate or cryogenic magnetometers.

4.1.5. Data analysis

The procedure adopted by different research groups varies somewhat and the reader is referred to the original research papers for details. The essential stages generally followed are outlined in Figure 4-1.

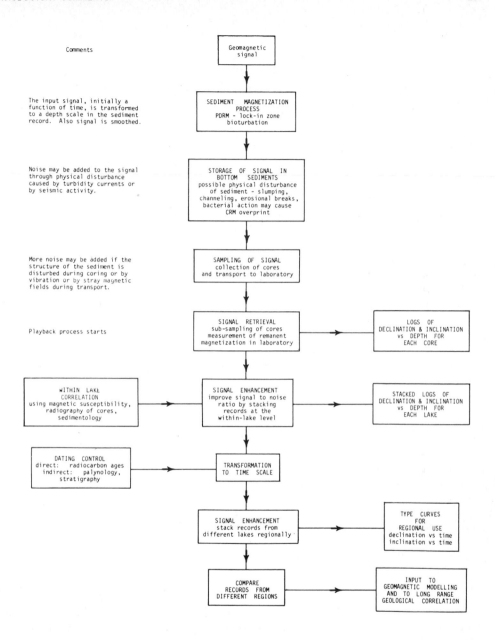

Figure 4-1:   Flow chart illustrating stages in the palaeomagnetic study of
lake sediments.

4. 1. 6. Objectives

Geologists and geographers are primarily interested in using limnomagnetic declination and inclination logs to help them date sediments and to determine variations in rates of deposition or to attempt to deduce the sources of origin of the sediments by comparing their basic magnetic properties with those of rocks exposed in possible source areas.   There is already a substantial volume of literature on the latter subject, e. g. Thompson et al. , (1978); Thompson et al. , (1979); Thompson and Morton, (1979); Thompson et al. (1980).   Geophysicists on the other hand are primarily interested in determining the spectral content and the regional and global patterns of the longer period geomagnetic secular variations in order to gain a better understanding of the physics of the Earth's core.

4. 2. PALAEOMAGNETIC RESULTS FROM RECENT EUROPEAN SEDIMENTS

By Gillian M. Turner,
Victoria University of Wellington, New Zealand

4. 2. 1. Introduction

The first palaeomagnetic study on Recent European sediments was reported by Ising (1943a), working with varved clay sequences from Sweden and using the existing varve chronology. He established, as McNish and Johnson (1938) had in a parallel study of New England varves, that the sediments carried a strong remanent magnetization, the declination of which gave a reliable recording of the magnetic field direction at the time of deposition, but the inclination of which was appreciably shallow. This "inclination error" was further investigated by Griffiths (1955), also working on Swedish varve sequences, and by King (1955) in laboratory redeposition experiments. Griffiths was able to draw up a secular variation record for the period 1100BC – 750AD (3050BP – 1200BP), which can be readily correlated with the more recent results shown below.

Although laboratory experiments on the acquisition of detrital remanence (DRM) in sediments continued through the 1960's, no further palaeosecular variation results were reported until 1971.

Current interest in the palaeomagnetism of rapidly deposited sediments was initiated by Mackereth's (1971) publication of measurements of the horizontal component of remanence in a sediment core spanning late and post–Glacial times from Lake Windermere, England. He reported large scale declination swings with an amplitude of about 20°, which, with the radiocarbon age control then available, seemed to have a regular period of about 2700 years, and predicted their use as a secondary dating method for sediment sections unsuitable for radiocarbon techniques.

In a subsequent, more detailed study of several Windermere cores, although no formal spectral analysis was performed, Creer et al. (1972) reported (i) a half cycle of approximately 11000 years, possibly associated with the period of precession of the equinoxes, 25800 years, (ii) anticlockwise loops of the magnetic vector approximately 5000 years in duration, and (iii) shorter period loops in both senses, including the prominent 2000 – 3000 year declination swings noted by Mackereth.

Following these first Windermere results a large number of studies have been conducted across Europe on rapidly deposited sediments from a variety of environments: lake, marine and continental shelf sediments and cave deposits. Unfortunately at present no satisfactory method has been developed to retrieve palaeointensities from DRM carrying sediments (Turner and Thompson, 1979). Therefore this account is restricted to directional results only.

The density and geographic range of the results across Europe have enabled investigation of the spatial and temporal changes in the geomagnetic field, extending the direct knowledge we have for the past few hundred years from observatory measurements. Thus it has been possible (i) to evaluate different models for the sources of the secular variation field on a time scale of several thousand years, and (ii) to build up secular variation master curves for certain restricted regions which can in turn be used to correlate and date other sections and cores in the area (Thompson and Turner, 1979).

However, before spatial and temporal differences between records can be quantified it is essential first, to establish detailed and accurate chronological control, and second, to substantiate both the palaeomagnetic signature and dating control by within–site and between–site correlations. Some of the studies discussed below do not meet both these criteria, so may be less reliable.

In a marine environment it is often possible to assume a uniform deposition rate, and a few well estimated age determinations are sufficient to confirm (or cast doubt on) this assumption. However, in a lake, activity in the drainage basin can have a marked effect on sedimentation on comparatively short time scales, and it is always necessary to build up a detailed age–depth profile.

Within–site and within–lake correlations serve to verify the internal consistency of the sediments in a study, while between lake correlations rule out local disturbances, such as slumping and current effects, and identify dating discrepancies.

4.2.2. European Results

Publications reporting most of the European secular variation results from recent sediments are listed in Table 4-2, and their locations shown in Fig. 4-2.   This is not intended to be an exhaustive list, but contains only the main reports of results, and excludes some records shorter than a few thousand years in length. The quality and density of both palaeomagnetic and chronological control vary widely between studies, and not all studies are suitable for detailed comparison.

Table 4-2:   Summary of results from recent European sediments

	LOCATION		REFERENCE	AGE(yr)	COMMENTS
1	Britain	54.5°N, 3.8°W	Mackereth (1971) Creer et al. (1972) Thompson (1973) Turner & Thompson (1981)	 10000 – 0	L. Windermere, dec. only L. Windermere L.Neagh 10 cores:L. Lomond L.Windermere, L.Geirionydd
2	Switzerland	47.5°N, 8.8°E 46.4°N, 6.3°E	Thompson & Kelts (1974) Creer et al. (1980)	13000 – 0 15000 – 0	L. Zurich, 1 core Lac de Joux, 1 core
3	Sweden	~63°N, 17°E	Griffiths (1955)	3050 –1200	Varved clays
4	Poland	~53°N, 15-22°E	Creer et al. (1979)	5500 – 0	17cores, 4 lakes scattered, poor dating
5	Finland	~64°N, 30°E	Stober & Thompson (1977)	5500 – 0	2 cores: L. Vuokonjarvi, independent dating unreliable

Figures 4-3 and 4-4 show the declination and inclination magnetograms of a series of well dated records lying approximately on a NW-SE line across Europe. It is important to note that differences in the labelling of major declination swings have arisen in the literature. The system of Stober and Thompson (1977) and Turner and Thompson (1979, 1981) is used here. Also, although in some publications radiocarbon dates have been calibrated, and time-scales have been presented in terms of calendar years (years BP), as calibration is only possible for the last 7000 years, all the records shown here are plotted in conventional ^{14}C years (years bp). Since the two scales diverge by as much as 1000 years at times, the choice of scale can influence interpretations markedly, and should always be clearly stated.

The best constrained record is that of Turner and Thompson (1981) for Britain. The records of 10 cores from three different lakes were dated and stacked to form single magnetograms of declination and inclination against time. Independent suites of radiocarbon age determinations were obtained on one core from each lake and transferred to the others from the same lake by means of magnetic susceptibility and lithological correlations. A single, well defined pollen horizon, the elm decline (5200BP – 5300BP) was identified in one core from each lake, and confirmed the correlations and date at this level. The time scales assigned to each core were thus completely independent of their magnetic remanence records. The records were interpolated to

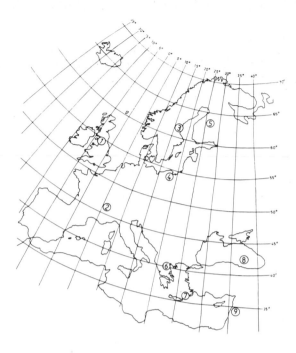

Figure 4-2: Location of the European studies listed in Table 4-2 and discussed in text

equal time intervals and stacked. With 10 estimates at each time horizon it was also possible to calculate 95% confidence limits for each point on the composite curve, using Fisher statistics. At this level all the major features and those sub-features identified in Figures 4-3 and 4-4 were considered significant and therefore thought to reflect real features of the local geomagnetic field (Turner and Thompson, 1982). The British record correlates well with archaeomagnetic results for the last two millennia (Aitken, 1970; Thellier, 1981), and with observatory data for the last 400 years, giving support for the reliability of the entire record.

The same averaging technique was used by Creer et al., (1981) on four cores from Lake Trikhonis, Greece. Although radiocarbon dates on the material were all contaminated by ancient carbon, the record was dated by correlation with the archaeomagnetic record for SE Europe (Kovacheva, 1980), which contains all the major features, but fewer details, and confirmed by pollen analyses. Although two other lakes were reported in this study their records were not long enough to allow combination of all three into a single curve.

Creer et al. (1980) reported a 15000bp year record in a single core from Lac de Joux,

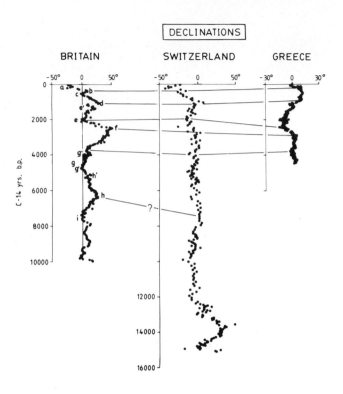

Figure 4-3: Declination log for Britain (Turner and Thompson, 1981;
Turner and Thompson 1982), relative declination logs for
Switzerland (Creer et al., 1980) and Greece (Creer et al., 1981).
Time-scale in in uncalibrated radicarbon years. Each declination
scale is obtained by dividing the scale of inclination
changes in Figure 4-4 by the cosine of the inclination at the site
of a geocentric axial dipole. This makes the angular displacements
equivalent between the logs and between Figures 4-3 and 4-4
(see Verosub et al., 1980).

Switzerland, showing large scale swings in both declination and inclination beyond the
range of the British record.   Unfortunately the two other cores from the site cover only
the late-Glacial and contain no post-Glacial record. In an earlier paper however
Thompson and Kelts (1974) reported a core of similar length from Lake Zurich, dated by
radiocarbon. Although the upper part of the declination record has a linear trend
superimposed on it, probably due to rotation of the corer during penetration, the main
declination and inclination swings and their dates provide valuable confirmation of those
at Lac de Joux.

The independent dating controls on the records from Poland, Finland and the Lebanon

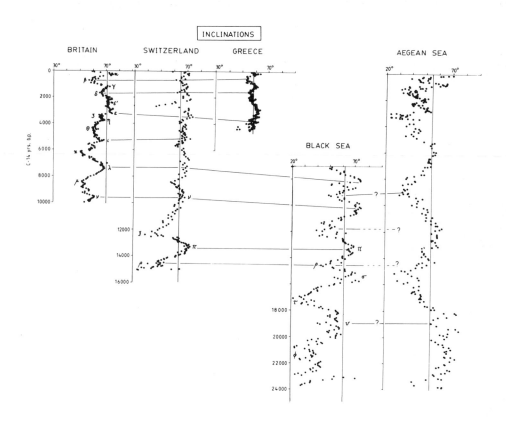

Figure 4-4:   Inclination logs for Britain (Turner and Thompson, 1981; 1982),
Switzerland (Creer et al., 1981), the Black Sea (Creer, 1974)
and the Aegean Sea (Opdyke et al., 1972). The time-scale is in
uncalibrated radiocarbon years. The vertical line on each log
shows the geocentric dipole field inclination at the site.
The Black Sea and Aegean Sea data have been smoothed using
a 3 point moving average.

are not very precise, but the magnetic variations in the Finnish and Lebanese data are
well defined and can be correlated broadly with the records mentioned above.

The cores from the Aegean and Black Seas carry long inclination records but, as they
were collected in several sections, no declination information. The Black Sea record is
from a single 11m core, and contains seven radiocarbon age estimates along its length.
Although, by comparison with the Swiss record, the time scale derived from these seems
reasonable for the period 15000bp - 7000bp, the possibility of systematic errors in the
two dates below this level cannot be ruled out.   The Aegean record is made up from a
4.8m core spanning 3500bp - 0bp, and a 6m core from 27000bp - 3500bp (the
lowermost 3000 years are not shown in Figure 4-4). Three levels in the longer core have

been dated at about 3500bp, 8000bp and 24000bp, each by two or more radiocarbon estimates. The time scale is based upon the uniform sedimentation rate suggested by these three dates.

4.2.3. Discussion

4.2.3.1 Long term secular variations (>2000 years).

The tie-lines drawn on Figures 4-3 and 4-4 demonstrate that all the major declination and inclination features 2000 years or more in duration can be traced across the 35° of longitude spanned by the records for the last 15000 years. The same basic pattern of post-Glacial features has also been seen in lake sediment cores obtained by R. Thompson and the author from both Iceland and Israel, extending the range of correlation to 60° in longitude and 30° in latitude. Correlation of the Aegean Sea record poses the only problem. Creer (1974) suggested that, due to the slower sedimentation rate, fewer features are resolved in the Aegean core, two swings in the Black Sea inclination record corresponding to each in the Aegean. Such a systematic loss of alternate swings seems unlikely, but a poor resolution, together with a deviation from the uniform sedimentation rate assumed in deriving the Aegean time scale, may account for the apparent lack of correlation.

There have been several reports of reversed directions of magnetization from Scandinavian late and post-Glacial sediments dated between about 13000bp and 11000bp (Morner et al., 1971; Noel and Tarling, 1975; Noel, 1975). The interpretation of these as a geomagnetic reversal or event has been strongly criticized (Thompson and Berglund, 1976; Thompson, 1976), due to lack of within-site repeatability and the likelihood of sedimentological disturbances during the rapidly changing climatic conditions prevailing around the time of deposition. The contemporaneous records shown here, from homogeneous sediments deposited in calm conditions, show no such aberrant directions, so we are led to conclude that normal amplitude secular variations have persisted undisturbed for at least 25000 years.

There is no evidence of an overall inclination error, as was found in the coarser grained varved sediments, rather the inclination logs are generally centred on, or just below, the geocentric dipole field direction at each site, with the major swings being predominantly to lower values. The earlier inclination swings recorded in the Aegean and Black Seas and the lower inclination and declination swings from Switzerland appear to have larger amplitudes than the later swings, although this effect may be caused by sedimentological differences.

The variations of 25000 and 20000 years duration suggested by Creer et al., (1972) and Opdyke et al., (1972) cannot be verified without substantially longer data sets.

4.2.3.2 Short term secular variations (<2000 years).

Whether shorter scale features are discernible in a record depends on the sedimentation rate, sampling interval, the time taken for the sediment to acquire a stable remanence and inherent scatter. When several records have been stacked it will also depend on the accuracy of the chronological matching. Thus the frequency limit above which the geomagnetic signal cannot be distinguished from sedimentological and measurement noise will vary between sediments and localities.

Some sub-features within, and progressive changes in the shapes of, the main swings can however be traced across Figures 4-3 and 4-4. The rounded easterly declination extremities d and f in the Greek record are sharply pointed in Britain. The westerly extremities e and g both carry short period, low amplitude sub-swings (e′, g′ and g′′) in all the records. Similarly ε′ can be traced across the inclination records.

The parallelism of the tie-lines between Figures 4-3 and 4-4 demonstrates that the phase relation between declination and inclination is maintained between Greece and Britain. Detailed inspection of the ages of the features and common sub-features shows a general 100-300 year lag of the British record behind the Greek one.

4.2.3.3 Time series analysis.

As we have distinguished long and short term secular variations it is natural to proceed with a more formal investigation of their frequency content.

Thompson (1973) reported a period of 2700 years from spectral analysis of the declination record from Lake Windermere then available. Opdyke et al., (1972) had previously deduced a period of 6000 years from direct inspection of the Aegean Sea inclination results. Similarly Creer (1974) derived a period of 2800 years from his Black Sea core, which he assumed was the second harmonic of Opdyke et al.'s 6000 year period, since each swing in the Aegean record seemed to correspond to two from the Black Sea. These reports led to the general acceptance of long period secular variation as a cyclic phenomenon (Thompson, 1975). Some authors have used this assumption of periodic secular variation of either declination or inclination between two horizons of known age to assign "magnetic ages" to a sequence (e.g. Creer et al., 1976a). Subsequent results and analyses of both declination and inclination from the same records have not however supported this idea of a simple cyclicity at a single period, but present a more complex picture (Turner and Thompson 1981).

Of the records shown, detailed spectral analysis has been performed only on the British results. Declination and inclination records were analysed both separately and together as a complex number series. In the latter method the predominant sense of looping of the magnetic vector at a particular frequency can be deduced from the balance of power between the positive and negative parts of the transform (Denham, 1975). The newer, more detailed dating control had the effect of removing the periodicity of the declination features, 'stretching' the older swings and 'compressing' the more recent ones, so they have a much broader frequency spectrum, with significant peaks at approximately 5000, 2000 and 1000 calendar years, corresponding roughly to the lengths of the individual swings. The complex representation spectrum was strongly biased towards clockwise looping, but contained some prominent peaks corresponding to anti-clockwise loops also. When the lower and upper halves of the record were analysed separately, two quite different spectra were obtained, indicating that the frequency content changes with time. Although the different dating schemes used by Thompson (1973) and Turner and Thompson (1981) were principally responsible for the difference in results, the fact that the former used the conventional ^{14}C time scale, while the latter used the calibrated scale (calendar years) contributed also. Thus, at present the European results yield no evidence for periodically varying long term features in the geomagnetic field.

4.2.3.4 Source of the long term secular variations.

Proposed sources of the long term secular variations have included (a) changes in the orientation and intensity of the main dipole, (b) movement, such as westward drift, of the sources of the non-dipole field, and (c) changes in the intensities of the non-dipole sources, e.g. periodic oscillation, pulsation or growth and decay. Typical time scales for each are not known, but calculations show that low order non-dipole disturbances may survive as long as 7000 years (Yukutake, 1968). Evidence of each of these processes can be seen in worldwide observatory data for the past few centuries (Yukutake, 1979). Thus it is unlikely that any one will adequately explain the palaeosecular variations of the past several thousand years.

However, the geographic range over which long term features have been correlated , and the persistent 100-300 year lag of the British records behind those from Greece do imply an element of westward drift, rather than stationary sources changing only in intensity. The predominantly clockwise looping of the magnetic vector observed in the British results is also suggestive of westward drift (Runcorn, 1959; Skiles, 1970). However, if some features have survived long enough to circle the globe several times, others must have changed sufficiently quickly and/or drifted at different rates, to alter the appearance of successive swings at a locality and to provide the short term differences observed between the same swing at different localities.

The earlier suggestion of Opdyke et al. (1972) of circular motion of the geomagnetic dipole axis about the rotation axis to explain the approximately 6000 year long swings in the Aegean, cannot be ruled out, although they suggested an anti-clockwise rotation, whereas the historically observed sense of rotation is clockwise. It is possible in this way that both dipole and non-dipole field may contribute to an overall picture of westward drift.

4.2.4. Summary

1. Recent rapidly deposited sediments have provided several continuous detailed records of the secular variations of the geomagnetic field in Europe, spanning up to 27000 years. Of these, five carry sufficient dating control to enable study of the temporal and spatial changes across the continent.

2. All the records display swings in both declination and inclination, of typical amplitudes expected during a time of settled polarity. Long term features (>2000 years) and some sub-features (<2000 years) can be correlated across all records; changes in the shapes of the main swings occur between records. There is a general lag of about 0.1 degree of longitude per year between east and west.

3. Although early results suggested periodically recurring features in both declination and inclination , subsequent results have provided no support for the idea.

4. These results are consistent with a generally westward drifting source model, but from the European data alone, it is impossible to separate effects of the main dipole and non-dipole features.

4.3. LATE QUATERNARY SECULAR VARIATIONS RECORDED IN CENTRAL NORTH AMERICAN WET LAKE SEDIMENTS

By Steve P. Lund
University of Southern California, Los Angeles, USA.
and Subir K. Banerjee
University of Minnesota, Minneapolis, USA.

4.3.1. Introduction

This contribution summarizes our current understanding of the palaeomagnetic secular variation from central North America over the last 20000 years, based on quantitative estimates of the field variability recorded at three locations: Lake St. Croix, Minnesota (45.0°N , 267.2°E); Kylen Lake, Minnesota (47.3°N, 268.2°E); and Anderson Pond, Tennessee (36.0°N, 274.5°E) – see Figure 4-5. These three small lakes were chosen for study because the sediments could be radiocarbon dated to provide high-resolution age control. The palaeomagnetic results from these three lakes have been used to characterize some distinctive features of the regional palaeosecular variation. These features provide important insight into the long term behaviour of the geomagnetic field.

4.3.2. Field and laboratory procedures

Oriented vertical cores of sediment, 5 or 10cm in diameter, were recovered from the three lakes using modified Livingstone piston-corers (Wright, 1967). The coring procedure consisted of recovering successive segments of 1m or 3m length. These were then extruded smoothly with no physical distortion and wrapped with plastic film and aluminium foil to minimize dessication.

Two replicate cores were recovered from each lake in order to test the reproducibility of the palaeomagnetic results. Also, the individual core segments of the two replicate cores were staggered vertically so that the core segment boundaries of one core would correspond to the middle of core segments from the other core. This offset provided an independent check of the declination orientation since the variation in declination between two contiguous segments of one core should be the same variation seen in the single replicate core segment. Two 19m cores from Lake St. Croix, two 8m cores from Kylen Lake, and two 7m cores from Anderson Pond were collected in this manner. In all three lakes, the distance between the two replicate cores was less than two metres.

The cores from Lake St. Croix and Anderson Pond consist entirely of visually uniform, slightly silty mud. The cores from Kylen Lake consist of an organic rich gyttja in the upper six metres which grades smoothly into a muddy silt in the lower two metres. The coring procedures and core sedimentology are described in more detail by Lund (1981).

A palaeomagnetic record was recovered from each core by first discretely sampling the sediment at 3 to 10cm intervals with a small hand corer of square cross-section, and then measuring the two to four specimens (20 x 20 x 18mm) collected at each sampling horizon in a cryogenic rock magnetometer. Typically, the multiple specimens from three horizons per metre of each core were step-wise demagnetized in peak alternating magnetic fields of 2.5, 5, 10, 20, 40, 60, and 80mT in order to pick a field suitable for cleaning the remaining specimens. A 10mT field was most suitable for cleaning the Lake St. Croix and Kylen Lake specimens; a 20mT field was most suitable for cleaning the Anderson Pond specimen. (Note 1mT = 10 Oe.)

Detailed rock magnetic analyses of the sediments in each core (Lund, 1981) indicate that (i) detrital magnetite is the NRM carrier and the only magnetic phase present, (ii) the variability of the NRM in each lake is due to subtle changes in the magnetite grain size associated with small changes in sediment influx, and (iii) the NRM is a DRM that forms during or soon after deposition (Verosub, 1977). The vector results, summarized below, indicate that there is no inclination error associated with this DRM mechanism.

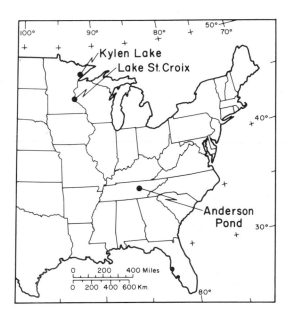

Figure 4–5:   Location map of lakes sampled for palaeomagnetic study of
the late Quaternary magnetic field variability.

4.3.3. Vector results

The final palaeomagnetic record from each core was determined by vector-averaging
the cleaned palaeomagnetic results from the two to four specimens collected at each
sampling horizon. The results from the replicate cores in each lake were then compared
with one another to estimate whether any systematic errors were present in the data, due
to misorientation of individual core segments in the coring process.

During coring, each core segment was oriented to magnetic north and the core
segment was assumed to be vertical. If individual core segments were misoriented with
respect to these coordinates then systematic errors in inclination (I) and/or declination
(D) occurred.   Lund (1981) has considered these potential errors in detail. Vertical
deviations of individual core segments, greater than the typical three degree dispersion
of the individual scalar inclination measurements, were not detected.

There were, however, occasional azimuthal misorientations of some core segments
with respect to magnetic north. These were detected by comparing the declination results
from the replicate cores. The azimuthal offsets in declination between individual core

segments were corrected by overlapping the declination variations of the offset core segments from the two replicate cores to produce a single composite declination record. We consider this procedure to be the most reliable and unbiased method for correcting occasional azimuthal misorientations. However, the reliability of each lake's composite declination record can only be assessed by comparing the results with other independent records.

The vector results from each lake are summarized in Figure 4-6. The data sets LSCSUM(I, D, t), KLMSUM(I, D, t), and APTSUM(I, D, t) represent the stacked mean of all the palaeomagnetic data from both cores of Lake St. Croix, Kylen Lake, and Anderson Pond respectively. The data are plotted versus radiocarbon age on the basis of 25 radiocarbon dates from the three lakes. In each lake, the palaeomagnetic results from the two replicate cores were always statistically identical.

The composite data sets LSCSUM, KLMSUM, and APTSUM have been optimally smoothed with regard to depth using the cross-validation technique described by Clark and Thompson (1978). The resultant vector data sets LSCMOD (I, D, t), KLMMOD(I, D, t), and APTMOD(I, D, t) represent our best estimate of the palaeomagnetic record in each lake, and are therefore our preferred models for the actual geomagnetic field variability observed at each site. These smoothed data sets are also shown in Figure 4-6. The heavy lines represent the best estimate of the field variability and the lighter lines enclose the region of 95% confidence that is estimated by the lighter cross-validation technique.

The palaeomagnetic results from the three lakes can be compared with each other and with previous independent estimates of the palaeomagnetic field variability to estimate their reliability.   Comparisons of I, D, and their phase relationship should be quantitatively identical (within the resolution of typical palaeomagnetic data) for sites within a few geographic degrees (1°~110km) of each other (Thompson and Barraclough, 1981; Champion, 1980). Sites within ~2000km of each other should have, at least, similar results.

The historic magnetic field variation at Lake St. Croix, measured locally for the last 1000yr and estimated by spherical harmonic analysis for the last 400yr (Barraclough, 1974), is accurately reproduced in the lake's palaeomagnetic record. Most important, the palaeomagnetic inclination is statistically identical to the actual field inclination, indicating that there is no systematic inclination error caused by the DRM mechanism at Lake St. Croix.

The palaeomagnetic records from Lake St. Croix and Kylen Lake, which should be identical except for a 2° difference in axial-dipole inclination, overlap in the interval 4000bp to 9000bp. Within this interval, the scalar amplitudes and phase relationships of the vector data agree closely and several points of correlation are shown in Figure 4-6. The only disagreement between these two data sets is a 4° difference in the site mean inclinations. The source of this difference is unknown.

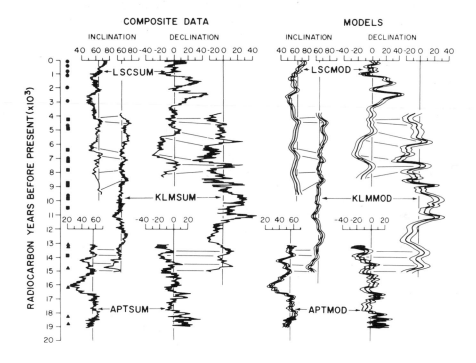

Figure 4-6:   Summary of palaeomagnetic results from three North American
lakes. The composite inclination and declination records
from two replicate cores in each lake are shown on the left.
LSCSUM is from Lake St. Croix, KLMSUM is from Kylen Lake,
and APTSUM is from Anderson Pond. The time domain is
provided by 9 radiocarbon dates from Lake St. Croix
(circles), 10 radiocarbon dates from Kylen Lake (squares),
and 6 radiocarbon dates from Anderson Pond (triangles). An
optimally smoothed smoothed model for each composite data set
is shown at right. LSCMOD is from Lake St. Croix, KLMMOD is
from Kylen Lake, and APTMOD is from Anderson Pond. The model
data sets can be considered the best estimate of the
palaeomagnetic field variability at each site. The preferred
correlations between the various data sets are noted.

The palaeomagnetic results from Kylen Lake and Anderson Pond overlap in the interval
13000bp to 15000bp. Within this interval, the vector waveforms appear to correlate
closely between the two widely separated lakes. Some points of preferred correlation are
indicated in Figure 4-6. Close correlations have also been noted (Lund, 1981) between
the southwest USA archaeomagnetic record of Dubois (1975) and the Lake St. Croix
record, as well as between the radiocarbon-dated lava-flow results of Champion (1980)
and both the Lake St. Croix and Kylen Lake records.

All of these correlations suggest that there is no systematic error in any of the three lake palaeomagnetic records. We can therefore assume that each record represents the actual palaeomagnetic field variability observed at the sampling site during the last 20000yr.

4.3.4. Analysis of secular variation

The palaeomagnetic variability estimated from Lake St. Croix, Kylen Lake, and Anderson Pond characterize the regional reliability of the late Quaternary palaeomagnetic field from central North America. We will attempt to analyze this variability through parametric statistical study, spectral analysis, and waveform description.

Equi-spaced palaeomagnetic time series of various lengths and sampling intervals were generated from the three lake palaeomagnetic records. The resultant vector time series fall into two categories: (i) those derived from the data in individual cores, such as TS3B from Lake St. Croix core 75B, TS15KB from Kylen Lake core 76B, and TS1A1 from Anderson Pond core 76A (all in Table 4-3); and (ii) those derived from the smoothed model data, such as TSLSCM from Lake St. Croix LSCMOD, TSKLMM from Kylen Lake KLMMOD, and TSAPTM from Anderson Pond APTMOD (all in Table 4-3).

The scalar and vector statistical properties associated with each time series were determined in order to estimate the probability distribution and dispersion of the palaeomagnetic field variability at each site. The statistical study also estimated the similarity of the observed palaeomagnetic field variability to that expected from an ADF model. The representative results of two time series from each lake are presented in Table 4-3.

The inclination and declination means, the resultant field vectors, and the resultant VGPs from each lake are all within 3° of that expected by the ADF model. The vector statistics, however, are still often significantly different from the axial-dipole expectation. In Lake St. Croix and Anderson Pond, these deviations away from the axial-dipole orientation are far-sided, a fact that is consistent with other statistical estimates of the late Tertiary geomagnetic field behaviour (McElhinny and Merrill, 1975; Cox, 1975).

Multiple x^2 tests were run on each time series following the usage of Baag and Helsley (1974), (Lund, 1981). The results in Table 4-3 describe what proportion of the x^2 tests passed, and whether this is considered significant. The tests were used to estimate (i) whether the scalar time series were Gaussian in distribution, and (ii) whether the field vector of VGP time series were Fisherian in distribution. The results suggest that the inclinations from near 45°N latitude and all the declinations tend towards a Gaussian distribution. The inclinations from Anderson Pond (36°N) , however, differ significantly from a Gaussian distribution since the data are skewed markedly to low values. This type of inclination behaviour has also been noted by Cox (1975) from palaeosecular variation

Table 4-3: Summary of scalar and vector results

DATA* SET	LENGTH (Years)	N	SAMPLE MEAN	ST. DEV.	INCLINATION SKEW.	KURT	X^2 TESTS (Gauss.)	SAMPLE MEAN	ST. DEV.	DECLINATION SKEW.	KURT	X^2 TESTS (Gauss.)
TS3B	9350	73(I)	60.9	5.5	.17	3.19	yes(4/4)	0.9	16.6	.38	2.74	yes(3/4)
	8340	64(D)										
TSLSCM	9520	145(I)	60.3	4.7	.29	3.35	no(2/4)	0.6	15.2	.63	3.10	no(0/4)
	8330	127(D)										
TS15KB	9740	190	66.5	4.5	0.23	3.04	yes(4/4)	-0.7	16.4	.17	2.69	yes(4/4)
TSKLMM	10240	156	66.8	4.2	.08	2.96	yes(4/4)	-2.3	14.9	.43	2.70	yes(3/4)
TS1A1	5910	141	52.4	9.1	-.85	3.19	no(0/4)	-0.2	11.3	.49	3.06	yes(3/4)
TSAPTM	6030	115	51.7	3.6	-1.06	3.51	no(0.4)	1.8	10.3	.05	2.58	yes(4/4)

DATA* SET	LENGTH	N	I_R	D_R	FIELD VECTOR α_{95}	S_f	X^2 TESTS (Fisher.)	LAT.$_R$	LON.$_R$	V.G.P. α_{95}	S_p	X^2 TESTS (Fisher.)
TS3B	8340	64	61.5	1.3	2.2	10.0	no(10/15)	87.9	65.7	3.1	14.1	no(2/15)
TSLSCM	8330	127	60.8	0.9	1.4	9.2	no(1/15)	87.0	77.0	2.0	12.9	no(0/15)
TSLSCM	9740	190	67.4	0.5	1.0	7.9	no(0/15)	88.0	271.2	1.5	12.2	no(0/15)
TSKLMM	10240	156	67.5	-1.1	1.0	7.3	no(0/15)	87.8	258.3	1.6	11.3	no(0/15)
TS1A1	5910	141	53.0	-0.5	1.6	11.2	no(8/15)	88.4	105.1	1.8	12.3	yes(14/15)
TSAPTM	6030	115	52.1	1.6	1.7	10.6	no(6/15)	87.1	65.2	1.8	11.3	no(9/15)

N = number of data points in time series, I_R and D_R are the vector mean inclination and declination, S_f = field vector dispersion, S_p = V.G.P. dispersion.

*TS3B and TSLSCM are Lake St. Croix time series, TS15KB and TSKLMM are Kylen Lake time series, TS1A1 and TSAPTM are Anderson Pond time series.

Table 4-4: Summary of scalar spectral analysis

LAKE ST. CROIX		KYLEN LAKE	
INCLINATION	DECLINATION	INCLINATION	DECLINATION
0.205 ± 0.047* mcpy	0.110 ± 0.031 mcpy	0.110 ± 0.047 mcpy	0.158 ± 0.063 mcpy
4880 < $^{6340}_{3960}$ *years	9060 < $^{12680}_{7050}$ years	9060 < $^{15850}_{6340}$ years	6340 < $^{10570}_{4530}$ years
0.425 ± 0.032 mcpy	0.362 ± 0.047 mcpy	0.362 ± 0.047 mcpy	0.456 ± 0.063 mcpy
2350 < $^{2540}_{2190}$ years	2760 < $^{3170}_{2440}$ years	2760 < $^{3170}_{2440}$ years	2190 < $^{2540}_{1920}$ years
0.599 ± 0.031 mcpy	--------------------	0.568 ± 0.063 mcpy	--------------------
1670 < $^{1760}_{1590}$ years	--------------------	1760 < $^{1980}_{1580}$ years	--------------------
0.870 ± 0.050 mcpy	0.833 ± 0.027 mcpy	0.820 ± 0.079 mcpy	0.962 ± 0.077 mcpy
1150 < $^{1220}_{1090}$ years	1200 < $^{1240}_{1150}$ years	1220 < $^{1350}_{1110}$ years	1040 < $^{1130}_{960}$ years
1.370 ± 0.071 mcpy	1.176 ± 0.053 mcpy	1.250 ± 0.087 mcpy	1.299 ± 0.079 mcpy
730 < $^{770}_{690}$ years	850 < $^{890}_{800}$ years	800 < $^{860}_{750}$ years	770 < $^{820}_{730}$ years
--------------------	1.754 ± 0.088 mcpy	1.613 ± 0.074 mcpy	1.613 ± 0.074 mcpy
--------------------	570 < $^{600}_{550}$ years	620 < $^{650}_{590}$ years	620 < $^{650}_{590}$ years

*Results are tabulated as frequencies (millicycles per year (mcpy)) and wavelengths (years).

studies of Hawaiian Island lavas. Both the field vector and VGP distributions at each site differ significantly from a Fisherian distribution.

The dispersion of the field vectors and VGP's in each time-series were estimated without assuming any particular probability distribution based on the equations of Cox (1970). The resultant dispersions are consistent with the estimated late Quaternary palaeomagnetic field dispersion, but they are systematically 3° lower than the estimated late Tertiary palaeomagnetic field dispersion (McElhinny and Merrill, 1975).

The frequency dependence of the scalar dispersions was estimated through spectral analysis of the individual core time-series from Lake St. Croix and Kylen Lake. The spectral estimates for each lake, summarized in Table 4-4, were determined by summing the results of multiple spectral analysis that used either periodogram or maximum-entropy spectral techniques (Lund, 1981). The results from the two independent lakes are almost identical even though they occur in two different (but overlapping) time intervals. The slight variations in frequency can be easily attributed to (i) slight errors in the radiocarbon ages, (ii) variability in the real-time/radiocarbon-time transfer function, or (iii) real time-dependent frequency shifts in the palaeomagnetic spectrum. The inclination and declination spectra within each lake are also very similar. Only the 1700yr wavelength in inclination has no counterpart in declination.

The spectral peaks fall into two distinct groups. The first group consists of the longest wavelength in each spectrum, typically 9000yr. This peak, which normally accounts for 60% of the spectral power, is difficult to resolve because it is near the wavelength of the typical data set. Even so, the power associated with this low frequency is quite significant and may be due to some form of dipole variation. There is no spectral evidence for unresolved longer wavelengths in the data sets that might appear as trends in the scalar data.

The second group consists of a fundamental wavelength at 2400yr with multiples at wavelengths of 1200yr, 800yr, and 600yr. The fundamental wavelength of 2400yr is noteworthy because it is near the estimated time for the non-dipole field to circle the Earth due to westward drift. It can be argued that these observed wavelengths simply combine to describe the irregular spatial variability of the non-dipole field that is recorded as the non-dipole field drifts past each site. This irregular variability probably changes continuously in time. Therefore, the spectral estimates from two separate time intervals may contain (i) different fundamental wavelengths if the drift rate changes and (ii) different proportions of the higher frequency components as the spatial waveform changes.

This cause for the higher-frequency variability can be considered further by filtering the palaeomagnetic data to look at the higher-frequency waveforms in more detail. The data from Lake St. Croix core 75B have been filtered to remove all wavelengths longer than 5000yr (Figure 4-7). The filtered inclination and declination waveforms do appear

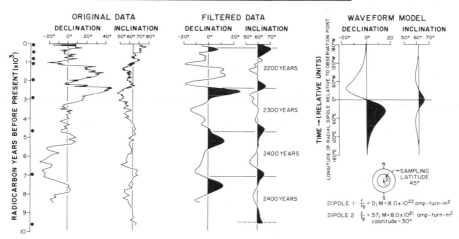

Figure 4-7: Palaeomagnetic field variation recorded in Lake St. Croix 75B
and a simulation model to explain its waveform variability.
The original data from core 75B are shown at left. The time domain is
determined from 9 radiocarbon dates (circles). The same data, filtered
to remove all wavelengths longer than 5000yr and shorter than 120yr,
are shown at centre. The highly distinctive waveform that remains has been
shaded to accentuate the waveform and its phase relationship. A simple model
for this waveform, determined by drifting a small radial-dipole underneath
the sampling site, is shown at the right. It is quite evident that the waveform
of the simple simulation model and the filtered data are very similar.

to show some cyclicity. More important, the inclination and declination waveforms appear to have a distinct, repeatable morphology. This waveform morphology is exactly what a mid-northerly latitude site might expect to see if a non-dipole focus, simulated by a radial-dipole near the core-mantle interface, were to drift latitudinally past the site. The waveform calculated from such a simulation model is shown in Figure 4-7 for comparison. The model waveforms and phase relationship are almost identical to the observed waveforms from Lake St. Croix. It therefore appears that latitudinal drift of non-dipole foci, as simulated by drifting radial-dipoles, can account for almost all the late Quaternary non-dipole field variability from North America.

Another way to assess the importance of latitudinal drift of the non-dipole field is to consider the historically observed looping behaviour of site field vectors and VGPs. Runcorn (1959) and Skiles (1970) have suggested that counter-clockwise looping in the vector data is indicative of eastward drift of the non-dipole field, while clockwise looping in the vector data is indicative of westward drift. Dodson (1979) details circumstances under which this correlation breaks down.

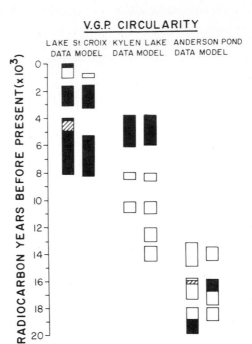

Figure 4-8:   VGP circularity recorded in three North American lakes. The periods
of definable clockwise (black regions) and counterclockwise (white regions)
looping behaviour in replicate cores from each lake are shown above. Hachured
regions indicate disagreement between the replicate cores. The definable
looping behaviour in the model data is also shown. It appears that extended
periods of coherent looping behaviour are present. Clockwise looping
predominates for the last 8000yr, but counter-clockwise looping predominates
between 8000bp and 20000bp.

The VGP paths for Lake St. Croix, Kylen Lake, and Anderson Pond have been
analyzed for this looping behaviour and the results are shown in Figure 4-8. Periods of
definable looping behaviour are boxed-in, where black regions indicate clockwise
looping, white regions indicate counter-clockwise looping, and hatched regions indicate
that the replicate cores gave opposite results. Loops were counted if more than half a
circle could be discerned. The α95 was used to estimate which loops were statistically
significant. The looping behaviour of the time-series model from each lake is also shown
in order to assess the effects of noisy data in the analysis.

These results are very similar within comparable time intervals and suggest that the
looping behaviour of the VGP is definitely not random. It appears that clockwise looping
has predominated in North America over the last 8000 years except for a brief period of

counter-clockwise looping between about 500bp and 2000bp. The looping was predominantly counter-clockwise between 8000bp and 20000bp. The palaeomagnetic records from Great Britain over the last 11000yr (Creer et al., 1972; Turner and Thompson, 1979) show almost identical results. The simplest interpretation of these results is that westward drift has predominated for the last 8000yr at mid-northerly latitudes, but that eastward drift predominated between 8000bp and perhaps 20000bp. This interpretation is now under further study through detailed modelling studies. Independent of the direction of drift, it seems clear that latitudinal drift of the non-dipole field at mid-northerly latitudes has dominated the late Quaternary palaeomagnetic field variability.

4.3.5. Conclusions

The palaeomagnetic results summarized here represent our current best estimate of the late Quaternary palaeomagnetic field variability from central North America. The results correspond closely wherever comparison is possible and suggest that no systematic errors due to unresolved non-geomagnetic sources remain in the data.

The palaeomagnetic time-series generated from these records have permitted us to quantitatively describe the central North American palaeomagnetic field variability for the last 20000yr. The assessment is based on parametric statistical study, spectral analysis, waveform morphology, and vector circularity studies of the various scalar and vector time-series. Six major conclusions about the late Quaternary magnetic field variability are suggested from this work.

First, the statistical properties of all data sets greater than 6000yr in length resemble closely the expected statistical properties of the ADF model. There are, however, some small asymmetries in the data that are consistent with other independent statistical studies of the late Tertiary palaeomagnetic field.

Second, the palaeomagnetic spectrum appears to be band-limited to wavelengths not much beyond 10000yr in length. Also, the greatest spectral power resides at the long-wavelength end of the spectrum and the spectral power generally decreases with increasing wavelength.

Third, the power spectrum seems to naturally separate into a single large-amplitude waveform with a period of about 9000yr, and a group of lower-amplitude waveforms with wavelengths of about 2400, 1200, 800, and 600yr. All of these wavelengths appear in both the scalar inclination and declination spectra. It is tempting to associate the 9000yr wavelength with some form of dipole variability and to associate the 2400yr wavelength and its multiples at 1200, 800, and 600yr with the non-dipole field variability.

Fourth, the observed lower-wavelength group of spectral peaks could be due predominantly to the latitudinal drift of the non-dipole field past each sampling site. The similarity of the 2400yr wavelength to the period of historic westward drift, the similarity of the filtered waveform to a simple model of latitudinal non-dipole field drift, and the observed coherence in VGP circularity all argue for this correlation.

Fifth, the VGP cirularity suggests that westward drift of the palaeomagnetic field at mid-northerly latitudes has predominated for the last 8000yr, but that eastward drift may have predominated from 8000 to 20000yr ago. This correlation is based on the simplest interpretation of the VGP looping and is not a unique interpretation. More detailed study of the looping behaviour at several world-wide sites is necessary to provide a unique answer.

Finally, the overall late Quaternary palaeomagnetic field behaviour from central North America is entirely consistent with the historically observed geomagnetic field variability. There is no evidence for excursions or other abnormalities in the data that cannot be readily explained by correlation to the present day field behaviour.

4.3.6. Acknowledgements

H.E. Wright Jr. and Paul and Hazel Delcourt were instrumental in site location and core recovery. Discussions with H.E.Wright Jr., Ken Hoffman, Shaul Levi, and John King are appreciated. Laboratory assistance was provided by Dave Bogdan, Jim Marvin, and John King.

This work has been funded in part by the following grants from the National Science Foundation : EAR-8008833 (S.P.L.), EAR-7919986 (S.K.B.) and EAR-8019466 (S.K.B.).

4.4. RESULTS FROM THE GREAT LAKES

By John Mothersill
Lakeland University, Thunder Bay, Canada.

4.4.1. Introduction

In recent years palaeomagnetic studies have been carried out on the Late Quaternary sedimentary sequences in Lake Erie (Creer et al., 1976a) in Lake Michigan (Creer et al., 1976b; Dodson et al., 1977; Vitorello and Van der Voo, 1977), in Lake Superior (Mothersill, 1979) and in Lake Huron (Mothersill, 1981; Mothersill and Brown, in press). An overall study of the Great Lakes area was carried out by Creer and Tucholka, 1982a). Additional palaeomagnetic studies are presently underway in Batchawana and Nipigon Bays, Lake Superior (Mothersill) and in Lake Ontario and Lake Erie (Carmichael and Mothersill).

The objective of most of the early palaeomagnetic studies was to determine whether high resolution palaeomagnetic curves could be obtained for the Late Quaternary stratigraphic sequence. Subsequent objectives of these studies were to utilize palaeomagnetic curves for time-parallel correlation of the Late Quaternary sequence within basins and from basin to basin within the Great Lakes area; to tie the palaeomagnetic curves into an absolute time scale and the established post-Glacial lake phases; and to determine rates and/or changes in rates of sedimentation within the Great Lakes area.

The Late Quaternary history of the Great Lakes has been reviewed by Hough (1958, 1962 and 1966), Prest(1970) and Terasme et al. (1972). Glacial ice receded from Lake Erie about 13600bp, and from Lake Ontario and southern Lakes Michigan and Huron about 12900bp. Glacial ice remained in the vicinity of Lake Erie until 12700bp and in the vicinity of Lake Ontario until 12200bp, Lake Huron until 11800bp and Lake Michigan until 11600bp resulting in the deposition of glacial outwash deposits prior to the deposition of the homogeneous post-Glacial lacustrine silty clays as noted by Mothersill (1981) for the Lake Huron sequence. Glacial varve deposition which began in the Lake Superior basin when glacial ice began to retreat about 11000bp continued until 9000bp when glacial ice finally retreated to the north of the Hudson Bay divide (Saarnisto, 1975).

4.4.2. Inclination results

In general, palaeoinclination curves show a more consistent and better defined pattern of oscillation swings resulting from the secular variations of the Earth's magnetic field than the palaeodeclination curves. The reason for this is the possibility of rotation of the coring device during the coring operation or the core later breaking into segments, with relative rotation of these segments occurring during the recovery operation. Both these common occurrences would affect the palaeodeclination but not the palaeoinclination values and tend to decrease the usefulness of the palaeodeclination log.

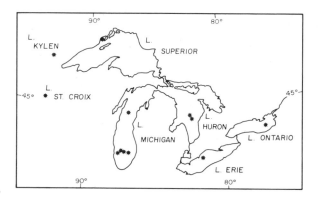

Figure 4-9:   Location map of cores utilized for palaeomagnetic studies,
Great Lakes area.

The sketch map (Figure 4-9) shows the location of cores that have been utilized for palaeomagnetic determinations in the Great Lakes area using the methodology outlined in section 4.1. The palaeoinclination and palaeodeclination logs for these cores have been plotted as a function of depth (Figures 4-10 and 4-11) Also noted on these logs are pertinent stratigraphic and palynological contacts: –

i. the boundary of the oak and pine pollen zones dated independently at 8500bp for Lake Erie core 14794;

ii. the top of glacial outwash deposits in the Lake Ontario C-80-10 and Lake Huron C-78-1 and C-78-2 cores;

iii. the diastem and dry zone marker horizon for the two Lake Huron core suggested by Mothersill (1981) as marking the two Lake Stanley phase dated at 9000bp -9700bp by Prest (1970);

iv. the top of the glacial varve sequence of Lake Superior cores C-77-3 and C-77-4 dated at 9000bp (Saarnisto 1975).

The character of the oscillation swings of these palaeoinclination curves are similar and the correlation is apparrent. Several methods of identifying the major oscillation swings have been utilized in the various palaeomagnetic studies of the Great Lakes area. The identification method of Creer and Tucholka (1982a) labelling the palaeoinclination oscillations from α to π and the palaeodeclination oscillations from A to K has been employed.

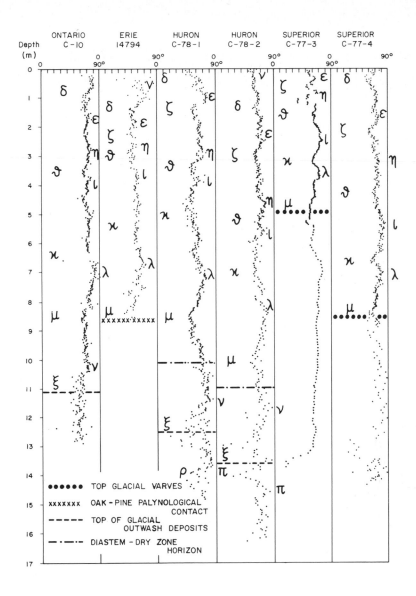

Figure 4-10:   Inclination logs for cores from Lakes Ontario, Erie, Huron and
Superior showing stratigraphic marker horizons.
Maxima and minima marked with Greek letters.

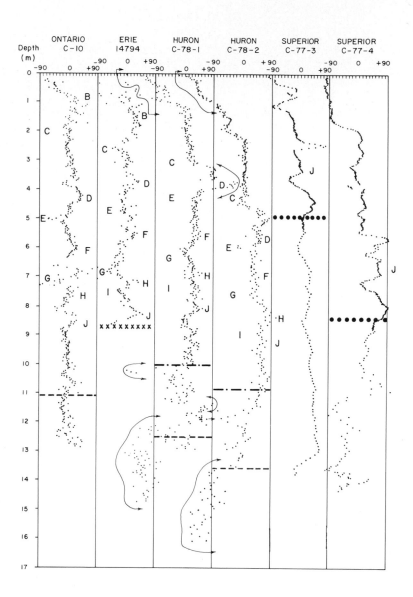

Figure 4-11:   Declination logs for cores from Lakes Ontario, Erie, Huron and
Superior showing stratigraphic marker horizons.
Labels identify westerly and easterly extreme values

4.4.3. Transformation of time scale

Unfortunately radiocarbon has not proved to be effective in dating post–Glacial lacustrine or glacial outwash deposits of the Great Lakes.   Mothersill (1979) suggested an error of approximately 1700yr for the radiocarbon dates of the post glacial sequence of C–77–4. However this should not be utilized as a 'zero' correction for all ages. Creer et al. (1976b) similarly suggested an error of 2000 years or more for Lake Michigan sediments However the cores from Lakes St. Croix and Kylen in Minnesota utilized by Banerjee et al. (1979) (see preceeding section in this chapter) for compiling palaeomagentic logs are rich in organic carbon and the dates determined from radiocarbon would appear to be reasonably accurate (Figure 4–12).   The pattern of oscillation swings for the upper part of the section of the section to about 27m below the ice surface (6000bp) for the Minnesota lakes palaeoinclination curves, show fairly good resolution.

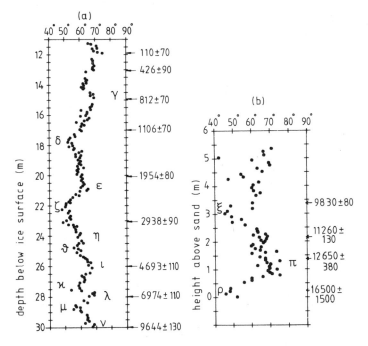

Figure 4–12:   Inclination logs on depth scale for (a) Lake St. Croix core 75B and (b) Kylen Lake core 75A. Data digitized from Banerjee et al. (1979).

Uncorrected radiocarbon dates shown at appropriate depths (after Creer and Tucholka, 1982a).

The oscillation of the composite section below this are not as well defined because of the relatively lower rates of deposition. The character of the palaeoinclination oscillations for the entire post–Glacial sequence is well defined for the cores from Lake Huron, Superior and Ontario because deposition was relatively rapid and the short sampling interval of 3cm. Unfortunately these cores could not be dated accurately using radiocarbon. However, correlation of the palaeoinclination curves show that the age of the stratigraphic and palynological horizons or contacts noted in Figure 4–10 fit reasonably well into the dates compiled from radiocarbon of the Minnesota curves (Figures 4–10 and 4–12). Creer and Tucholka (1982a) combined the Minnesota and Superior–Huron data by first replotting the Minnesota data against time rather then depth. Then the Superior and Huron palaeoinclination curves were plotted as a function of time by correlation with the oscillation swings of the Minnesota palaeoinclination curves. Finally the Minnesota and Superior–Huron palaeoinclination curves were interpolated at equally spaced tiome intervals of 40yr and summed to construct the 'type' curve (Figure 4–13a).

The oscillation maximum labelled λ is the most prominent and consistent feature of the palaeoinclination curve and can be utilized as a reference for correlation purposes. This is fortunate because without a prominent feature such as the λ oscillation, correlation of the palaeoinclination curves would be questionable. Based on absolute dating of the Minnesota cores by radiocarbon, supplemented by the inferred dating of a stratigraphic horizon (top of varve deposition in Lake Superior dated at 9000bp) there would appear to be 3.4 paleoinclination oscillation periods covering the last 9000 years for an average of about 2700 years period. It should be noted that this periodicity is in agreement with the calculations of Mackereth (1971) for the palaeoinclination periodicity for the Late Quaternary of Europe. However, based on the dates assigned to the palaeoinclination swings, it would appear that this periodicity can only be taken as an average and is not consistent.

4.4.4. Applications of type–curves

Creer and Tucholka (1982a) present several versions of the correlation of palaeoinclination curves with the stratigraphy of Lake Michigan (Figure 4–14). This demonstrates the problems and to some extent the limitations of palaeomagnetic curves when a composite curve is compiled from more than one core and radiocarbon cannot be used as a time–check. Additional problems of correlation by palaeomagnetic curves will occur in basins where intermittent or rapidly changing rates of deposition occurred.

The palaeodeclination curves for the cores analyzed for the Great Lakes Basin area also have been plotted as a function of depth (Figure 4–11). The oscillation pattern of these curves are similar, although the pattern is not as well defined or as consistent as the palaeoinclination curves. The declination oscillation swings lettered A to K do show a consistent, relative relationship with the oscillation swings of the palaeoinclination curves labelled α to π for Lakes Ontario, Erie and Huron (Figures 4–10 and 4–11). However the two palaeodeclination logs for Lake Superior although somewhat similar in character

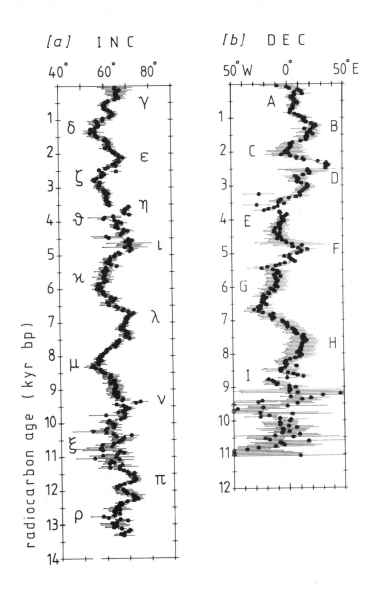

Figure 4-13: Type curves showing (a) inclination variations and (b) declination variations for the Great Lakes area by stacking individual records at 40 year increments (after Creer and Tucholka, 1982a).

do not appear to be correlatable with the palaeodeclination logs of the lower Great Lakes. A 'type' palaeodeclination curve (Figure 4-13b) plotted as a function of time was constructed using the same technique as for the construction of the 'type'

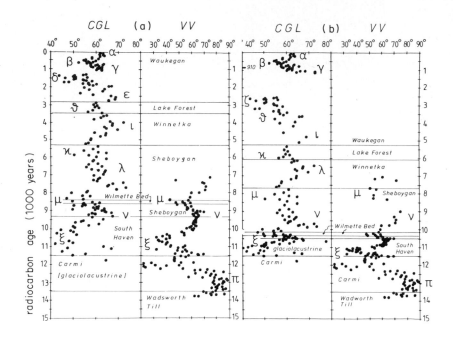

Figure 4-14:   Lake Michigan inclination data transformed to
time-scales using two alternate schemes
(after Creer and Tucholka, 1982a).

palaeoinclination curve by combining the Minnesota and Superior-Huron data (Creer and
Tucholka, 1982a).

These 'type' curves plotted as a function of time would appear to provide the best
control for both time-parallel correlation and absolute dating of late and post-Glacial
sedimentary sequences of the Great Lakes area. The methodology used in compiling
these 'type' curves has undoubtedly led to the smoothing out of some of the finer details
of the oscillation patterns.   In time, as additional data are collected and in particular if
pieces of wood which can be accurately dated are fortuitously found in lake cores,
refinements will be made to the dates assigned to the pattern of palaeomagnetic
oscillations.   Within individual basins additional palaeomagnetic data will result in the
compilation of more detailed palaeoinclination and palaeodeclination curves which may
only be useful for local correlation purposes. The utilization of palaeomagnetic curves
also will be a valuable tool to more accurately determine the late and post-Glacial history
of the Great Lakes area with respect to determination of rate, type and distribution of
sediments during specific lake phases (Mothersill and Brown, in press); in determining
the age of post-Glacial deposits cropping out across southern Ontario; and in accurately
determining the period of time that outlets such as at Fossmill and North Bay acted as
channelways connecting the Great Lakes system to the Champlain Sea.

4.5. RESULTS FROM ARGENTINA

K.M.Creer and P.Tucholka,
University of Edinburgh, UK;
D.A.Valencio, A.M.Sinito and J.F.Vilas,
University of Buenos Aires, Argentina.

4.5.1. Introduction

Cores were collected from three lakes located near 41°S 71.5°W in Rio Negro Province, south-western Argentina, namely: (i) Brazo Campanario, a narrow (~0.5km) arm extending from Lago Nahuel Huapi, (ii) Lago Morenito, a small lake ~500x300m connected to the western basin of Lago Moreno and (iii) Laguna el Trebol, a small almost circular lake of diameter ~600m. Small lakes were chosen in preference to the many large lakes in the region so as to minimize the possibility that the bottom sediments might have been disturbed by turbidity currents.

The bottom sediments were sampled with 6m and 1m pneumatic corers (Mackereth 1958, 1969). The cores were split open and sub-sampled in a field laboratory to avoid the possibility of damage to the physical structure of the sediments and of the implacement of SRM due to vibration and shock during transport to the University laboratories in Buenos Aires.

The methodology and the results are discussed by Creer et al. (1983).

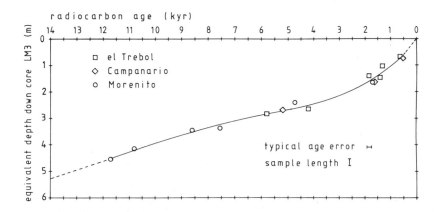

Figure 4-15:   Depth vs. age transformation curve: radiocarbon ages plotted against equivalent (LM3) down-core depth. Detail of the ^{14}C ages and inter-lake transformations are given in Creer et al. (1983).

4.5.2. Sedimentology

The results of grain size analyses showed that the sediments were poorly selected, the sands containing important amounts of clay fraction and the clays containing substantial amounts of sand with some mud fraction present in the sediments from all samples. A definitive within-lake lithological correlation could be made for el Trebol only. The detailed results of sedimentological studies are described in Valencio et al. (1982) for el Trebol, in Mazzoni and Sinito (1982) for Morenito and in Sinito et al. (1982) for Campanario.

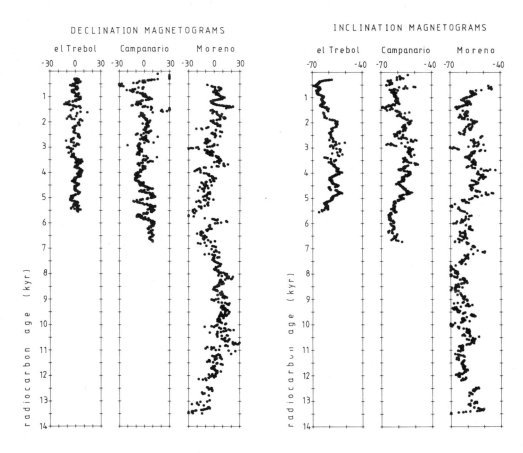

Figure 4-16:   (a) Declination and (b) inclination magnetograms for
the three lakes studied.

4.5.3. Within-lake correlation

The magnetic susceptibility (k) logs constructed from sub-sample measurements showed a sequence of peaks and troughs which could be correlated from one core to another at the within-lake level. The features shown by the NRM intensity (J) logs could also be correlated from core to core within each lake. The susceptibility and NRM intensity based correlations were consistent with one another and, in the case of el Trebol, with the correlation based on lithology.

The depth scales for all cores from the same lake were then adjusted to that of a 'master' core selected for that lake, namely cores LT1, LM3 and LC1 for el Trebol, Morenito and Campanario respectively.

These within-lake correlations could not be extended to the between-lake level.

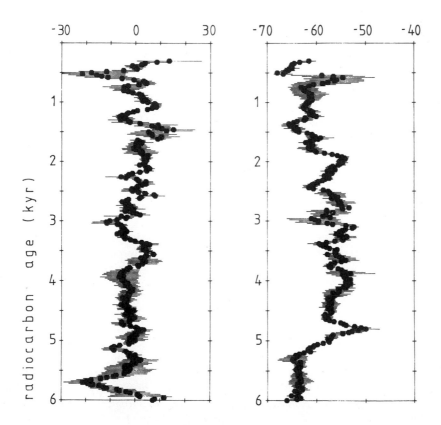

Figure 4-17: Stacked type-curves for the three lakes, 0-6000bp, (a) declination and (b) inclination.

4.5.4. NRM directions

The NRM was measured on fluxgate magnetometers either in Buenos Aires or in Edinburgh. Pilot samples were AF demagnetized progressively up to 600 Oe peak field. Some samples showed evidence of a weak viscous magnetization, probably acquired in the laboratory, which could always be removed by AF demagnetization in 25 Oe or 50 Oe. AF demagnetization in stronger fields produced no further systematic change in NRM directions. Median destructive fields fell in the range 150–350 Oe. Bulk AF demagnetization was then carried out in a peak field of 100 Oe.

Declination, inclination, intensity and susceptibility logs were produced for each core and then the depth scales were adjusted to that of the 'master' core for each lake. Next, these transformed data were put into a common file for each lake and average curves were calculated using running means over a 12cm window with a 2cm step. The resulting declination and inclination curves are shown in Creer et al. (1983).

4.5.5. Radiocarbon dating

Samples consisting of 25cm lengths of core were chosen from depths in different cores corresponding to extreme values of declination or inclination which were considered to be of potential use for between–lake correlation. Reliable dates were obtained from sixteen samples by the NERC Radiocarbon Laboratory at East Kilbride, Scotland. Detailed information is presented in Creer et al. (1983).

4.5.6. Transformation to a time–scale

It was not possible to correlate the cores from one lake with those from another on the basis of lithology or magnetic susceptibility. Also, there were too few radiocarbon ages for them to be used exclusively for between–lake correlation. Therefore this was carried out using the patterns of the declination and inclination records, taking care to satisfy the following two criteria: (i) the same transfer function should apply to both declination and inclination data and (ii) the correlation should be compatible with the suite of radiocarbon ages.

The transfer functions so defined allow the conversion of the down–core depths of samples dated by radiocarbon to a common depth scale; that of Moreno core LM3. The resulting radiocarbon ages and corresponding equivalent depths are plotted in Figure 4–15 and a best fitting cubic spline curve was constructed through the points to define the depth/time transfer function.

4.5.7. Construction of magnetograms

Individual core data were transformed first to the depth scale of the master core for the lake of origin and then to the depth scale of the overall 'master' core LM3 and finally to the time–scale, the object being to maintain the identity of the measured data points right up to the last stage at which interpolation at 20 year intervals was carried out prior to stacking unit vectors for all cores from each lake at each time horizon. The resulting declination and inclination magnetograms are shown in Figure 4–16.   The el Trebol and Campanario records extend back only to about 6000bp while the Morenito records extend back to nearly 14000bp.

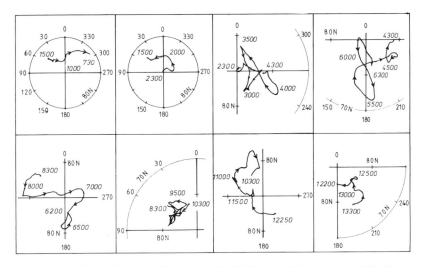

VGP   PATH   700 bp – 13300 bp   W. ARGENTINA   WINDOW = 400 yr   STEP = 20 yr

Figure 4–18:   VGP path corresponding to the declination and inclination magnetograms for L. Morenito, 0–14000bp.

Finally the data at each time horizon were stacked for the whole collection of cores for the interval 0 – 6000bp. Standard errors were computed at each level and the resulting pair of declination and inclination magnetograms are shown in Figure 4–17.

4.5.8. Virtual geomagnetic poles

The VGP curve corresponding to the stacked magnetograms for Lago Morenito are shown in Figure 4–18. The curve has been smoothed over a 100 year window and, to improve clarity, it is shown in seven sections. The sense of geomagnetic drift may be inferred from the sense of looping though such an interpretation is not unambiguous.

Intervals of clockwise and counter-clockwise looping appear to alternate. Prior to ~10000bp it was clockwise after which there were intervals of counter-clockwise and clockwise looping, each of them of about 4000 years duration running through to about 2000bp. From 2000bp to 1000bp there appears to be a short interval of counter-clockwise looping followed by clockwise looping through to the present time.

4.6. RESULTS FROM AUSTRALIA

By C.E. Barton,
University of Rhode Island, USA.

4.6.1. Introduction

Australia is poorly endowed with lakes suitable for palaeosecular variation studies. The best region is in the south-east where explosion-formed volcanic craters (maars) are numerous.   Many of these contain at least 10000 year sequences of undisturbed sediments which preserve a strong palaeomagnetic signal and can be radiocarbon dated.

Volcanic maars are typically flat-bottomed and steep-sided with small catchment areas defined by low profile scoria cones.   Many contain lakes up to 80m deep with very uniform sedimentary sequences.   A palaeomagnetic study has been made on large sets of 6m and 12m length cores from Lake Keilambete and Lake Bullenmerri (Barton and McElhinny, 1981; Barton and Polach, 1980).   Results from Lake Gnotuk were also reported, but these were less detailed and less reliable than those from the other two lakes.

Inter-core correlations of NRM intensity, low field susceptibility and lithology were used to normalize sedimentation rates in different cores so results could be stacked. Independent radiocarbon chronologies for each sequence (19 dates for Keilambete, and 47 dates for Bullenmerri) permitted transformations from depth to age scales (Figure 4-19). After averaging each data set by taking interquartile means at 100 year intervals, the resulting records were sufficiently similar to indicate that directions of remanence were primarily controlled by the geomagnetic field, and that systematic offsets between the three radiocarbon chronologies were no greater than 200 years.

4.6.2. A Revised secular variation curve

Stacking and averaging data from numerous cores provides a more stable estimate of the secular variation, but results in degradation of the primary signal.   Having been through the exercise, one can cast any pretence at objectivity aside and use the averaged record as a guide for selecting which cores, or sections of cores, provide the

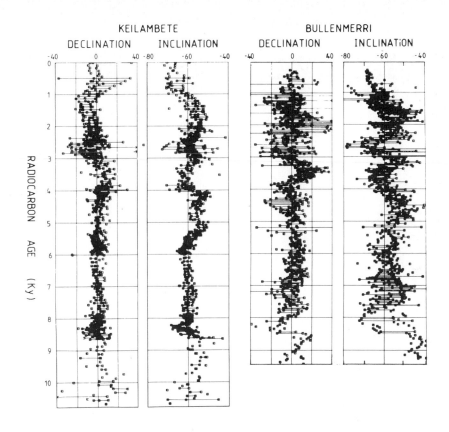

Figure 4-19: Stacked NRM data from sets of cores from Lake Keilambete and Lake Bullenmerri. All results are after AF cleaning and have been scaled to time using independent radiocarbon chronologies. Sedimentation rates were normalized to a common standard prior to stacking. Adapted from Barton and McElhinny (1981).

best estimate of secular variation. The results of doing this for Lake Keilambete are shown in Figure 4-20. The upper part of the record is dictated by historical data and results from numerous short cores (Barton and Barbetti, 1982).

From age intercepts on radiocarbon age – depth plots, it was assumed in previous publications that apparent increases in radiocarbon ages due to incorporation of ancient carbon were limited to less than 200 years, and could probably be ignored. Recently, a detailed comparison of historical observations, archaeomagnetic data, and results from short cores (Barton and Barbetti 1982) indicated that radiocarbon ages of the upper sediments are probably too old by about 450 radiocarbon years (350 calendar years). The revised time scale used for Figure 4-21 (Table 4-5) assumes that all the radiocarbon ages are too old by 450 years, and therefore that there is a large increase in sedimentation rate at the top of the sequence. An adjustment has also been made at the bottom of the sequence in the light of a new radiocarbon date (SUA-ANU2135:

KEILAMBETE COMPOSITE

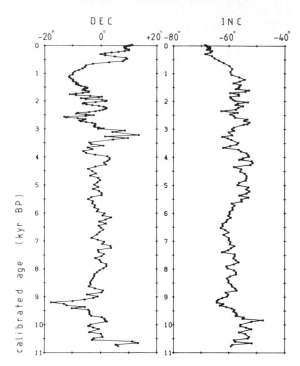

Figure 4-20:   Preferred estimate of the secular variation in south-eastern Aust;alia
based on the results from Lake Keilambete.
The uppermost 122 years of the record is based on observatory data; declinations
from 100 to 3000BP are based on a composite of whole core measurements
(Barton and Barbetti 1982); the remainder of the composite
is based essentially on long core KI and short core K1F scaled to
match the sedimentation rate in master core KF, then transformed to a
time-scale by linear interpolation between the tie-points listed
in Table 4-5.

8560±110bp at 310-320cm depth in Keilambete core KJ).

Figure 4-20 represents our preferred estimate of the geomagnetic secular variation in south-eastern Australia, mean site coordinates 38.2°S, 142.8°E.

4.6.3. Palaeointensity estimation

Laboratory redeposition experiments (Barton and McElhinny, 1979; Barton et al., 1980) show that sediments from Keilambete and Bullenmerri rapidly acquire a stable remanence with no inclination error.   For fields up to 100μT (= 1 Oe) the remanence is

Table 4-5:   Revised chronology for the Keilambete sequence.[*]

Depth in master core KF (cm)	Radiocarbon age (yr bp)	Adjusted[**] radiocarbon age (yr bp)
0	0	0
60	1900	1450
82	2450	2000
160	4000	3550
232	6000	5550
280	7000	6550
324	8500	8050
430	11,000	10,5550

[*] Intermediate ages/depths are obtained by linear interpolation between the above tie-points.

[**] Adjusted by subtracting 450yr bp to correct for the inferred effect of dead carbon.

proportional to the field strength.   However, in the natural sediments this relationship is complicated by the reduction in remanence associated with compaction (Barton et al., 1980).   Normalization of NRM with respect to ARM has been favoured as a means of deriving relative palaeointensity estimates from wet-lake sediments (Levi and Banerjee, 1976).

Within-lake agreement between NRM/ARM ratios for Keilambete and Bullenmerri were tolerable, but the agreement between lakes is poor (Figure 4-21).   Because of this we can only postulate generally higher field strengths prior to about 4500 years ago. Calibration of these relative field strength estimates using whole-core redeposition data, Barton and McElhinny (1979), combined with greater within-lake consistency favours the Keilambete curve.   However, neither curve shows the approximately sinusoidal trend of global archaeointensity data, and both conflict with results from Australia (Barbetti, section 3.15).   We conclude that these simple estimates do not reflect past geomagnetic field strengths.   King et al. (1983) have drawn attention to magnetic mineralogical criteria which should be satisfied before plausible relative intensity estimates can be expected.   We assume that magnetic mineralogy in the Keilambete and Bullenmerri sediments is too heterogenous to provide such estimates.

4.6.4. Palaeo-water depth estimation

Bowler (1970) constructed a water-depth curve for Lake Keilambete for the last 10000 years based essentially on variations of particle size and carbonate chemistry.   The form of the curve is generally consistent with diatom, ostracod and pollen data from the sediments, although there is some controversy over whether the lake level was low

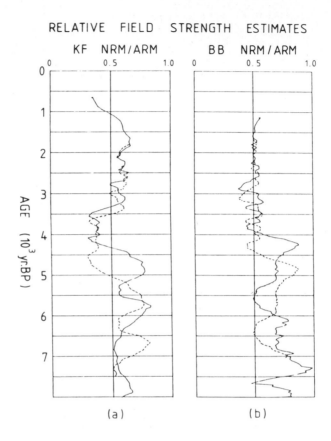

RELATIVE FIELD STRENGTH ESTIMATES

Figure 4-21:   NRM/ARM ratios for Keilambete core KF and Bullenmerri core BB for the
interval 0 to 8000bp. Solid lines are for true age, calibrated according to
Clark (1975).   Both NRM and ARM measurements were made after
AF cleaning at 10mT (100 Oe).

around 9000bp.

The NRM intensity curve for the Keilambete sequence bears a striking resemblance to
the water depth curve (Figure 4-22 a, b).   Bowler used an earlier, less detailed set of
radiocarbon ages (Bowler and Hamada, 1971) which has the effect of contracting the
time scale prior to about 7500bp.   Allowance for this improves the correlation between
the lower parts of the two curves.   Some of the NRM minima are caused by dilution of
the magnetic mineral content during episodes of rapid carbonate precipitation.   If this is
taken into account, there is a strong case for arguing that the NRM curve represents a
high resolution estimate of palaeo-water depths.   In fact, initial magnetic susceptibility is

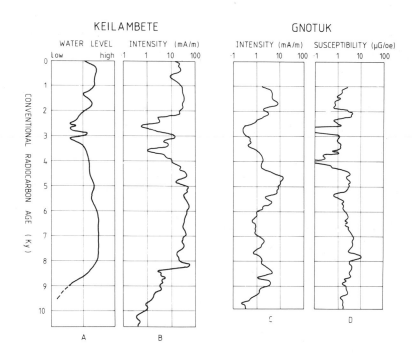

Figure 4-22: (a) Bowler's water depth curve for Lake Keilambete. The curve is intended as a relative measure of water depth: low corresponds to zero depth, the centre line to about 10m, and the peak around 7000bp to about 30m. (b) Smoothed NRM intensity curves for stacked results from 6 long and 2 short cores from Lake Keilambete. (c) Smoothed NRM intensity and (d) susceptibility curves for Lake Gnotuk. Intensity and susceptibility curves are plotted on logarithmic scales.

probably a better estimator. NRM intensity and susceptibility are so highly correlated in the Keilambete sequence that it makes little difference which variable is chosen. However, in nearby Lake Gnotuk, which is hydrologically very similar to Lake Keilambete, the susceptibility curve corresponds most closely to its counterpart for Lake Keilambete (Figure 4-22c, d).

Of the possible mechanisms which could account for the correlation it appears that enhanced input of magnetic material at times of high water level is the most plausible. A detailed discussion of the above points is given by Barton et al. (1981).

4.6.5. Lake Mungo

Lake Mungo in western New South Wales is part of the interconnected Willandra system of which are currently dry. It is apparent from the archaeological record that the lakes were full about 30000 years ago, when it has been suggested that a geomagnetic excursion occurred (Barbetti and McElhinny, 1976). An attempt was therefore made to find evidence of the excursion in sediments from Lake Mungo and nearby Lake Mulurulu and Lake Arumpo.

Lake Mungo and Lake Mulurulu were sampled from 3m deep trenches dug in the lake floors; Lake Arumpo was sampled from a water-cut gulley near the lake shore. The lithology is heterogeneous and difficult to correlate either within or between basins. Based on geomorphology arguments and radiocarbon data from various sites throughout the Willandra lakes system, J. M. Bowler estimates that the sampled sequences span approximately 15000bp to 50000bp.

Pairs of palaeomagnetic samples taken at approximately 3cm intervals were subjected to both AF cleaning and thermal cleaning up to 200°C (the plastic used for impregnation could not be heated to higher temperatures). Median destructive fields were mostly in the range 10 - 20mT for Lakes Mungo and Arumpo, and ranged up to 60mT for Lake Mulurulu. Mean directions before and after cleaning demonstrated the presence of a large secondary component, probably due to a weathering product such as goethite.

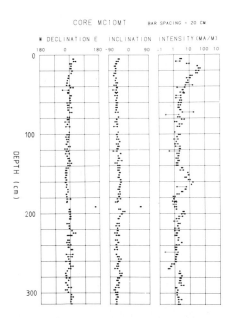

Figure 4-23: NRM data for Lake Mungo trench C after AF cleaning in a peak field of 10mT. Paired data points at the same level are joined by bars. Pairs which differ by more than 30° have been omitted.

Mean angular differences between specimen pairs per section, after AF cleaning at 10mT or 15mT, ranged from 12° to 42°. A filter was applied to eliminate paired directions which differed by more than 30°. Despite this high scatter in the data there was no evidence to support a major excursion of the geomagnetic field away from the axial dipole direction for the region (Figure 4-23). Abnormal directions in some horizons were all either not reproducable, or were associated with disturbed lithologies.

We conclude that no geomagnetic excursions were recorded by lake-floor sediments. We cannot conclude that the Lake Mungo excursion did not occur. Although the age limits of the sequences studied almost certainly encompass 30000bp, it is possible that the sedimentary record of such an excursion was lost, either by ablation of material from the lake floor or by chemical overprinting during subsequent dry phases.

4.7. THE PALAEOMAGNETISM OF CAVE SEDIMENTS

By Stavros Papamarinopoulos and K.M. Creer,
University of Edinburgh, UK.

4.7.1. Introduction

Caves are widely distributed over the terrestrial land surface and many contain sedimentary deposits accumulated over several hundreds to millions of years. Since the cave habitat seems to have been preferred by the hominids at least since the Middle Palaeolithic, relative dating by archaeological and palaeontological controls is often possible.

Two distinct sedimentary regimes have been recognized by Kukla and Lozek (1957). 'Entrance' facies deposits frequently contain datable artefacts and either recent or very old animal and human remains. They show evidence of cave-wide cycles of erosion, clastic deposition, stalagmitic formation, calcite resolution and erosion (Sweeting, 1973) and many of these episodes have been correlated with climatic cycles (Geology Today, 1974). Sedimentation rates including hiatus and removal episodes have varied from 0.25mm/yr to 1.0mm/yr over the past 50000 years in the Mediterranean region (Kopper, 1972).

'Interior' facies deposits are found beyond the reach of surface weathering and accumulate or disperse only under hydrological controls (Reams, 1972). They are characterized by low sedimentation rates, uniform temperatures, high humidities (85% – 100%) and a sparse and unique biota. Karkabi (1965) and Karkabi and Ziade (1973) quoted by Creer and Kopper (1976), have estimated the biomass in one deep cave to be 20 – 30 grams per hectare. Clays predominate and sometimes occur in bands with silts and sands suggesting that sedimentation was controlled by annual or possibly longer period cycles of deposition.

4.7.2. Methods of sampling

Samples are normally taken from the walls of existing or freshly dug trenches. Small ~(20mm x 20mm x 18mm) plastic boxes are pushed directly into the trench wall, taking care to align the edges of the boxes with a plumb-line. The azimuth of the base of each individual box is measured with a hand-held Brunton compass. The inclination of the axis of the base of each individual box with the horizontal is measured with an inclinometer, though normally this is not necessary if care has been taken to push the boxes in horizontally. Before removal, the base of each box is marked with an arrow pointing upwards and labelled (Kopper, 1972). Alternatively, long, open sided boxes ~(1m x 4cm x 4cm) are aligned vertically with a plumb-line and pushed into the exposed face of the trench. The azimuth of the back face of the box is measured with a Brunton compass and any small misalignment of the long axis with the vertical is measured. After removal, the box, full of sediment, is sealed in plastic sheet and on return to the laboratory, the sediment is sub-sampled in small plastic boxes (Kopper, 1975). A third method utilizes a small tubular piston corer which can be driven deep into the trench wall so as to extract fresh sediment. A device for orientating the coring tube, consisting of a hinged table containing a pair of spirit levels and a magnetic compass, is fixed to the top end of the coring tube. The corer is then extracted from the trench wall and the sediment core is extruded into successive (usually 3 to 5) cylindrical specimen holders (Papamarinopoulos, 1978) When this last method is used care must be taken not to give the samples SRM by hammering the tube in too hard (see Games in Chapter 1).

4.7.3. Magnetic properties

Iron compounds are ubiquitous in all cave sediments as the insoluble residues of carbonate rock weathering. They undergo a variety of diagenetic processes depending on their locations within the cave (Broughton, 1972). Actinomycetes, fungi, bacteria and ultra microscopic forms are typical diagenetic agents in the subterranean environment (Caumartin, 1963). Sulphur-oxidizing and nitrogen-fixing bacteria populate the interior zones where organic food supplies are deficient. The bacteria obtain their energy by simple oxidation of inorganic material and they prefer a clay substrate. Ferric hydroxide is thus formed in the host clay beds and this is then transformed to limonite and finally to one of the ferric oxyhydroxides such as lepidocrocite (γ-FeOOH) or goethite (α-FeOOH) (Hedley, 1968). A competing microflora characterized by iron reducers and organic synthesizers, occupies the cave entrances because of the presence of a large proportion of organic material. The net effect is the formation of ferric compounds in the deep interior and ferrous compounds near the entrance. This explains the differences in the magnetic properties of interior and exterior facies sediments. Clays which retain more diagenetic iron than coarser sediments predominate in the interior, but are scarce in the entrance so that interior facies sediments would be expected to register higher susceptibilities and NRM intensities than entrance facies sediments.

The main magnetic properties of cave sediments may be summarized as follows. NRM intensities (J) usually range between 10μG and 30μG and sometimes samples exhibit values up to 100μG, whereas the low-field magnetic susceptibility (k) exhibits values between 10μG/Oe and 50μG/Oe and exceptionally ranges up to 100μG/Oe. The Q-ratios (J/k) steadily exhibit values lower than 0.5.

Although the MDFs are generally low (~100 Oe), high values may occasionally be encountered: for instance samples from a section at Ball's cavern in North America showed MDFs in the range between 450 Oe and 800 Oe. This illustrates the complexity of the magnetic mineralogy.

Optical microscope, X-ray diffraction and Mossbauer studies of magnetic separates have shown that both fine and coarse grained magnetite could be the carrier of the remanence in some cases. However, haematite and goethite have been found occasionally in sediments in small quantities, together with magnetite, e.g. in Canet Cave, Mallorca (Papamarinopoulos, 1978; Papamarinopoulos et al., 1982).

The essential differences between the magnetic properties of interior and exterior facies sediments found in several Mediterranean cave sediments by Creer and Kopper (1976) are: (i) the magnetic low-field susceptibilities of interior facies deposits are between 3 and 60 times higher than those of the exterior facies, (ii) the scatter of NRM directions is greater for exterior facies sediments than for interior facies and (iii) the NRM of interior facies samples is more resistant to AF demagnetization than that of the exterior facies, MDF values being respectively ~120 Oe and 55 Oe to 75 Oe.

4.7.4. Geomagnetic Secular Variations

Palaeomagnetic results from three sections of cave sediments running (a) through Holocene time (Jeita Cave), (b) from about 30000bp to 50000bp (Arbreda Cave) and (c) through the Brunhes, Matuyama, Gauss and Gilbert polarity epochs (Campana Cave) show that vastly different rates of deposition may occur in different cave environments.

Figure 4-24 shows a section through the Jeita Cave complex which is located near Jounie in the Lebanon. Interior facies deposits were sampled from a trench dug into floor of the Upper Dry Cave. The sediments contain black layers which are interpreted as having originated from the burning of surface vegetation. This places a relatively young age on the section since agriculture was practised in the region only from about 8500bp (Creer and Kopper, 1976). Red clay found at the base of the sampled section is thought to have originated from the removal of terra rossa from the surface during exceptionally wet climatic conditions, probably associated with the maximum of the last glaciation some

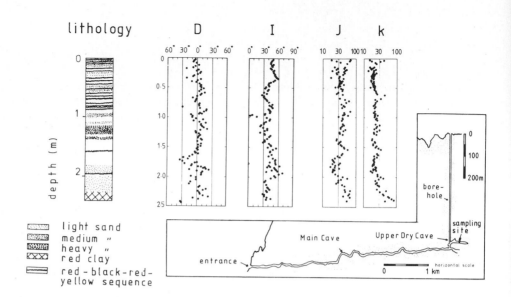

15000 to 20000 years ago.

Declination and inclination logs resulting from the palaeomagnetic measurements after
AF demagnetization in 60 Oe peak field are also shown in Figure 4-24. The MDFs are
mainly in the range 80 Oe to 150 Oe which is lower than the corresponding range of
values typical of lake sediments. Thus cave sediment samples have to be handled very
carefully during sampling, transport and throughout experimental work in the laboratory
because they are particularly prone to acquire secondary magnetization in stray magnetic
fields. The Jeita section was correlated very tentatively, with a sediment core from the
Black Sea by Creer & Kopper (1976) on the basis of the palaeomagnetic inclination logs.
While the correlation they inferred with the results then (and until now) available is not
unique, it is likely that further sampling of these deposits should allow the
palaeomagnetic logs to be firmly dated because the carbon rich layers would appear to
be highly suitable for radiocarbon age determinations. Unfortunately political conditions
in the Lebanon since the original collection of samples was made have prevented any
more field work.

(b) Arbreda Cave, Spain

The section of exterior facies deposits sampled in Arbreda Cave, Catalonia, contains
layers of angular stones (Fig. 4-25) which are interpreted as originating from collapse

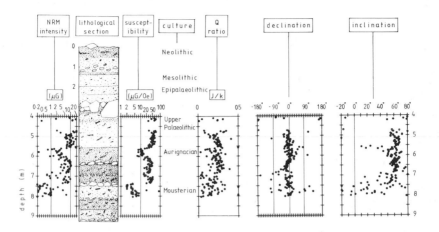

Figure 4-25:   Exterior facies deposits from Arbreda Cave, Calalonia, Spain.
Palaeomagnetic logs after AF demagnetization in 60 Oe. Lithological
log shown between intensity (J – in μG) and susceptibility
(k – in μG/Oe) logs which are both plotted on logarithmic scales.

of the entrance overhang during climatically cold periods. The occurrence of Aurignacian
artefacts (Upper Palaeolithic) and Mousterian tools (Figure   4-25) allow the interval
represented by the sampled part of the section to be placed between 30000bp and
50000bp (Creer & Kopper, 1976). In layers containing large angular stones, samples
were taken from the soft sediment filling the interstitial spaces so that continuous
sampling was possible from 5.5m to the base of the trench at 8.8m.

The patterns of NRM intensity (J) and susceptibility (k) logs appear to have been
strongly influenced by lithology. In particular, periods of heavy occupation (Aurignacian
and Mousterian) are characterized by low values of both J and k. The Q-ratio (J/k) log
bears strong resemblences to the J and k logs: therefore it cannot be accepted as
reflecting geomagnetic intensity variations. There are quite a few reports in the literature
of attempts to use other magnetic quantities such as SIRM (Creer et al. 1976 a, b) and
ARM (Levi and Banerjee, 1976) as a normalizing parameter to allow for the effect of
variations in the concentration of magnetic grains from layer to layer of the sedimentary
structure on NRM intensity.   However, no-one has yet convincingly demonstrated that
secular variations in geomagnetic intensity may be successfully isolated from lake
sediment records of NRM intensity.

The NRM directions recorded in the sediments containing artefacts and tools of these
cultures show unusually high scatter of NRM directions with some declinations strongly
divergent from 0° and some low and even negative inclinations (Figure   4-25).
Therefore, it would not be justified to interpret these aberrant directions as evidence of
geomagnetic excursions.   MDF values are typically less than 100 Oe and the optimal AF
demagnetizating field was 60 Oe in which all samples were cleaned to produce the logs

illustrated in the figure.

(c) Campana Cave, Mallorca.

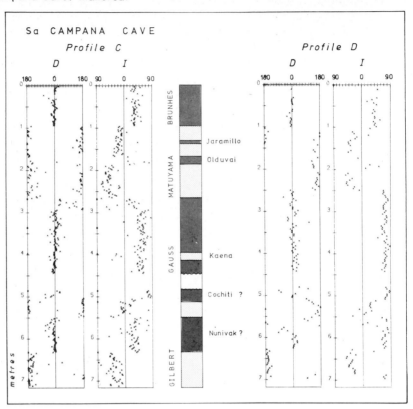

Figure 4-26:   Palaeomagnetic declinations (D) and inclinations (I) recorded down
two parallel sections cut through interior facies sediments deposited in
Sa Campana Cave, Mallorca, (after Creer and Kopper, unpublished).

Declination and inclination profiles obtained from two parallel 7m long sections of
interior facies deposits are shown in Figure 4-26.   Several polarity reversals of the
geomagnetic field are recorded and on this basis, Creer & Kopper (unpublished) have
placed the base of the section within the Gilbert epoch of reversed geomagnetic polarity
which means that it must be about 4.5Myr old.

4.7.5. Discussion

Palaeomagnetic studies of cave sediments carried out to date suggest that the
palaeomagnetic method is potentially capable of contributing to a better understanding of

the cave environment. For example in regions where secular variation type-curves have been constructed for post-Glacial time from lake sediment cores it should be possible to determine rates of deposition in caves by palaeomagnetic correlation. Since sediments are deposited only intermittently in the cave environment, essentially during wet climatic periods, it is to be expected that future studies of cave sediments should hold the potential of providing records of secular variation viewed through many windows of late Glacial and earlier time. A possible advantage of using cave sediments as recorders of the ancient geomagnetic field is that they are sometimes found associated with archaeological remains of early man which can be of some help in dating them. This is of considerable importance because while dating constitutes a major difficulty in constructing post-Glacial secular variation records, the associated problems are compounded many times when attempts are made to extend palaeosecular variation studies into the late Glacial, beyond the optimal range of the radiocarbon method.

From the few palaeomagnetic studies carried out so far on cave sediments, it is clear that rates of deposition vary widely. Therefore cave sediments are likely to provide opportunities of investigating a wide spectral range of secular variations and also polarity reversals of the main field.

4.8. STATISTICAL ANALYSIS OF PALAEOMAGNETIC DATA

By R.M. Clark (Monash University, Australia)

4.8.1. Introduction

Let us consider three basic problems arising in the analysis of palaeomagnetic data such as those obtained from cores of lake sediment, (i) the smoothing of a scalar variable (e.g. susceptibility) as a function of depth down the core, (ii) the smoothing of directional measurements versus depth, and (iii) the alignment or stretching of the depth scale of one core relative to that of another core.

We assume that we have n measurements $z_1, z_2, \ldots, z_n$ at corresponding depths $d_1, d_2, \ldots, d_n$ (not necessarily equally spaced) which satisfy the equations

$$z_i = F(d_i) + e_i \quad i = 1, 2, \ldots, n \tag{1}$$

Here the function F denotes the true or intrinsic functional relationship or regression curve describing how the variable z changes with depth d, while e_i denotes the overall 'error' of measurement in the i-th sample. It is assumed that the e_i's are independent random variables with zero mean and probability distributions of the same general type, although not necessarily identical, e.g. Gaussian but with standard deviation depending on depth. The variable z may be either scalar or vector valued, while any errors in the determination of the depths d are assumed to be negligible. The aim of the statistical analysis is to estimate and draw conclusions about the regression function F.

Equation 1 specifies that each observation z_i may be decomposed into 'signal' $F(d_i)$ plus 'noise' e_i. The methods of statistical analysis discussed below may therefore be applied in any situation in which the observations can be regarded as being of this form. But for consistency we will refer to the independent variable d as 'depth'.

Under the previous assumptions, $F(d_i)$ may be interpreted as the mean or true value of the variable z at depth d_i; the function F defines how this this mean value varies with depth. Thus our task is to estimate not a single mean but a sequence of n means $\{F(d_i)\}$ which must satisfy the constraint that they lie on the presumably smooth function F. If F is slowly varying relative to the magnitude of the error-terms $\{e_i\}$, the actual measurements z_i, z_{i+1} on consecutive samples will be very similar, simply because their corresponding mean values $F(d_i)$, $F(d_{i+1})$ will be very similar. This similarity does not contradict the assumption that the error-terms (or 'de-trended residuals') are independent. This latter assumption is implicit in the specific formulae given below for confidence limits. Although these formulae can be modified to take account of correlation structure in the residuals, the analysis becomes much more complicated by the need then to postulate and estimate a realistic correlation structure.

4.8.2. Scalar Response Variable – Methods

The various alternative methods of analysis can be most easily described in the case where the response variable z in the previous formulation is a scalar, e.g. susceptibility or intensity.

In many situations, the only prior information regarding the unknown function F is that it is 'smooth' in some sense. In such a case, an intuitive approach is to assume that F can be approximated by a low-degree polynomial whose coefficients may then be estimated from the data by the method of least-squares. But it is now well known that polynomials are often unsatisfactory as approximating functions. On the other hand, spline functions are not only preferable to polynomials but are known to be the most successful and adaptable approximating functions for smooth empirical functions (Wold, 1974; Rice, 1969, p. 123; de Boor, 1978). A spline function of degree m may be defined as a piecewise polynomial of degree m whose first (m−1) derivatives are continuous, the join-points between the polynomial pieces being known as knots. Spline functions may be formulated in terms of either truncated power functions or the so-called B-splines, the latter ensuring computational stability. In practice, the most widely used splines are cubic splines (m=3).

For a given set of knots, the parameters of the best-fitting cubic spline can be easily estimated by the method of least-squares, since equation (1) can now be re-formulated as

$$z = X\underline{\beta} + \underline{e} \tag{2}$$

Here, $\underline{z}' = (z_1, z_2, \ldots, z_n)$ (i.e., the n observations written as a vector), $\underline{e}' = (e_1, e_2, \ldots, e_n)$, $\underline{\beta}' = (\beta_1, \beta_2, \ldots, \beta_p)$ a p x 1 vector of parameters to be estimated, and X is an n x p matrix whose (i,j)-th term is $B_j(d_i)$, the j-th B-spline evaluated at depth d_i. The number of parameters, p, equals the number of knots internal to the observed depths $\{d_i\}$ plus 4. Equation (2) defines a linear model or equivalently a multiple linear regression problem. Programs for the solution of such problems may be found in most computer packages such as SPSS or SAS, while special-purpose programs for the evaluation of B- splines and solution of the resulting equations are given in de Boor (1978, Ch. 10 & Ch. 14) and in subroutine packages such as IMSL or NAG.

The choice of the number and location of the knots can be crucial. Wold (1974) gives some practical guidelines, while Jupp (1978) and de Boor (1978, Ch. 12) give algorithms which give, under certain conditions, an optimum choice of knots. Alternatively, if equally-spaced knots are used, the optimum knot-spacing may be determined from the data by cross-validation (see Clark and Thompson (1979) and Section 5.2.3).

Cubic splines arise in an alternative approach to the problem of smoothing the data. Returning to first principles, our aim is to find an estimate of F which is both 'smooth' and a reasonable fit to our data $\{(d_i, z_i)\}$. Clearly, these are conflicting requirements: a smooth function, such as a straight line, may be a poor fit to the data, while at the other extreme, a function which fits the data exactly may be expected not to be smooth. Suppose then, for each potential estimator f of F, we define a smoothness measure $S(f)$ and goodness-of-fit measure $G(f)$ by

$$S(f) = \int (f''(x))^2 dx$$

and

$$G(f) = \sum_{i=1}^{n} w_i (z_i - f(x_i))^2,$$

where $w_i, \ldots, w_n$ are suitable weights. Mathematically, the best compromise between smoothing and goodness-of-fit is achieved by the estimator which minimizes

$$C(f) = G(f) + \lambda S(f) \tag{3}$$

The arbitrary positive parameter λ specifies the importance to be attached to smoothness at the expense of goodness-of-fit. Schoenberg (1964) showed that the unique best estimator in this sense was a cubic spline with knots at every data point $\{d_i\}$ and coefficients depending on the smoothing parameter λ. We denote this spline as an approximating cubic spline to distinguish it from the least-squares cubic spline (obtained by solving (2) by least-squares) which has only $k \ll n$ knots. Algorithms for the evaluation of approximating cubic splines are given by Reinsch (1967), de Boor (1978, Ch. 14) and in some subroutine packages. The parameter λ (controlling the degree of smoothing) may be estimated from the data by cross validation (Wahba and Wold,

1975).

Approximating cubic splines, despite their mathematical and intuitive appeal, have a serious disadvantage in statistical analysis. There is as yet no method for constructing statistical tests of hypotheses based on approximating cubic splines. For example, it is not possible to test whether two different data sets have the same underlying regression function F. In contrast, such a hypothesis can be easily tested with least-squares cubic splines, by invoking the powerful theory of linear models.

Most of the preceding methods involve the specification of F in terms of some parameters which are subsequently estimated from the data. Since our only prior knowledge of F is that it is smooth, it may be more appropriate to use methods which do not require F to be specified parametrically. Examples of such methods include simple moving averages, weighted moving averages and the non-parametric estimators of Priestley & Chao (1972) and Clark (1977). The latter estimator is controlled by a single parameter denoted the **bandwidth** and is shown to be computationally simpler but almost as efficient as the approximating cubic spline.

In all the preceding cases, the resulting estimate $\hat{F}(d)$ of F at depth d is of the form

$$\hat{F}(d) = \sum_{i=1}^{n} a_i(d) z_i \tag{4}$$

where the weights $\{a_i(d)\}$ depend on d, and on the particular smoothing method used. In other words, $\hat{F}(d)$ is simply a weighted **linear** combination of the observations $\{z_i\}$, with the observations 'closest' to d having the greatest weight. This **linearity**, however, means that such estimators are sensitive to **outliers**, i.e. observations which are widely separated from the rest of the data. Since the absence of outliers cannot usually be guaranteed, it will often be preferable to use some of the so-called **robust** methods of smoothing, in which the final smooth curve is insensitive to the presence of outliers.

There is now a variety of related methods of robust non-parametric data smoothing, of which the simplest is the use of running **medians** (Tukey 1977, Ch. 7). Chapters 7 and 14 of Tukey's book contain numerous techniques (e.g. re-roughing, hanning) which are based on running medians and may be used in various combinations to improve the smoothing. The numerical examples in this book clearly demonstrate the stability of these procedures in the presence of gross outliers. A more readable account of Tukey's techniques, together with listings of relevant computer programs, is given by McNeil (1977). Cleveland (1979) gives what is essentially a robust version of the non-parametric estimator of Clark (1977). So far there has been, however, very little research into the statistical properties of these methods, e.g., the construction of associated statistical tests.

There has been much research recently on the robust estimation of the parameter β in the linear model (or multiple linear regression model)

$$z = X\beta + e$$

for an arbitrary matrix X (not necessarily the one defined in (2) for the special case of cubic splines). One approach, outlined by Andrews (1974), is to minimize, with respect to β, the quantity

$$\sum_{i=1}^{n} \psi(r_i/s),\qquad\qquad(5)$$

where $r_i = z_i - (X\beta)_i$ = residual at i-th point, s is a robust estimate of the scatter of the residuals, and $\psi(.)$ is some suitable function. In ordinary least-squares estimation, $\psi(u) = u^2$; this quadratic form tends to magnify the effect of any outliers, by squaring the associated large residuals. Andrews gives some alternative choices of ψ which give relatively little weight to large residuals and which are related to certain well-known robust estimators of location. The scale factor s in (5) may be taken either as proportional to the median of the absolute values of the residuals or the inter-quartile range of the residuals. With these alternative choices of ψ, minimization of (5) is not straightforward and requires some iterative search procedure. Algorithms for the general case of arbitrary 'design matrix' X and 'penalty function' ψ are given, for example, by Dutter (1977), while Lenth (1977) considers the special case (2) of cubic splines with given knots. However, computer programs implementing these algorithms are not yet widely available.

A second approach (Mosteller & Tukey, 1977, Ch. 14) is to use iteratively reweighted least-squares. At each stage in the iteration, a new estimate of β is found by weighted least-squares, in which each observation is given a weight which is some function, w(u) say, of its residual under the current estimate of β. The weight-function is such that observations with large residuals are given little weight, for example, Tukey's bi-square weight function

$$w(u) = (\max(0, 1-u^2))^2.$$

These two approaches are related, since one of the algorithms for minimizing (5) involves interatively-reweighted least-squares but with the weight function w related to the derivative of ψ. Both methods use a robust estimate of the scatter of the residuals, whereas the usual estimate, the sample standard deviation, is extremely sensitive to outliers, as it involves the sum of squares of residuals.

4.8.3. Scalar Response Variable – Recommendations

Least-squares estimation, whether weighted or unweighted, offers a wide selection of statistical techniques such as confidence limits and tests of hypotheses which, under the assumption that the error-terms $\{e_i\}$ have a Gaussian distribution, are equivalent to the highly efficient maximum likelihood methods. The basic theory and computational algorithms may be found in almost any textbook on statistics or computer package for statistical analysis. The main disadvantage of this method is its sensitivity to outlying observations. Robust regression overcomes this problem, while still providing some methods for inference (Andrews, 1974). Since these methods are not as straightforward or as readily available as those of least-squares, it may still be preferable to use the method of least-squares, provided extreme outliers are identified and removed. Case studies by Andrews (1974) and Carroll (1980) confirm that least-squares regression with outliers deleted yields almost the same results as robust regression.

In practice, the **degree** of smoothing is often more important than the method of smoothing. Too little smoothing may produce spurious kinks or undulations in the fitted regression curve, while too much smoothing may destroy some of the fine detail of F. In some of the methods discussed, the degree of smoothing is controlled by a single parameter, e.g., the bandwidth of Clark's non-parametric estimator, the smoothing parameter λ of the approximating cubic spline or the knot-spacing of the least-squares cubic spline (if equally-spaced knots are used). In such cases, the optimum value of this parameter can be determined objectively from the data themselves by the method of cross-validation. In its simplest form, the method involves the computation of the **cross-validation** mean-square error

$$\text{CVMSE} = (1/n) \sum_{i=1}^{n} (z_i - f_i(d_i))^2,$$

where $f_i(d_i)$ is the predicted value of z at depth d_i based on the estimator f_i derived from the data set with the i-th point (d_i, z_i) deleted. In general, the estimator f_i must be recomputed as each successive point is deleted. The optimum value of the smoothing parameter is taken as that for which the CVMSE is a minimum. Variations on this procedure (in which several points are are deleted at each step) are used by Wahba & Wold (1975), Clark & Thompson (1978) and Cleveland (1979).

Any smoothing procedure will produce a 'smooth' curve from any data set, even a set of random numbers. We therefore recommend using a method whereby the reality or otherwise of any kinks or wiggles in the fitted curve can be assessed by confidence limits. If the error terms $\{e_i\}$ in (1) are independent and Gaussian distributed with constant variance, then for linear estimators of the form (4), 95% confidence limits for the true mean response F(d) at arbitrary depth d are given by

$$\hat{F}(d) \pm t_\nu [s^2 \sum a_i^2(d)]^{1/2} \tag{6}$$

where $s^2 =$ Residual Mean-Square on, say, ν degrees of freedom (d.f.), and t_ν denotes the appropriate percentile of Student's t-distribution on ν d.f.. In the special case of cubic splines with k interior knots fitted by weighted least-squares,

$$\sum a_i^2(d) = \underline{x}'(X'W^{-1}X)^{-1}\underline{x}$$

where the matrix X is as in equation 2, the j-th element of the p x 1 vector $\underline{x}$ is $B_j(d)$ (the j-th B-spline evaluated at d), and $\nu = n - k - 4$. The weights are specified in the n x n matrix W which is assumed to be of full rank and in practice is usually of diagonal form (see Draper and Smith, 1981). In the special case of unweighted least-squares, W is simply the n x n identity matrix. For $\nu > 20$, it is usually sufficiently accurate to set t_ν

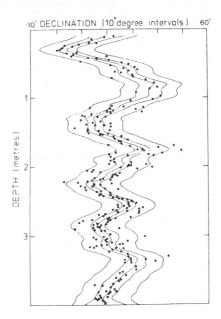

Figure 4-27:   Declination versus depth in a core from
Lake Windermere, England. Central curve is smooth estimate of true function F
relating declination to depth; two curves immediately adjacent
to this curve are 95% confidence limits for F. Dots represent
the original observations. (The two outer curves are 95%
prediction limits for future observations of declination.)

From Clark & Thompson (1978).

= 2.

Figure 4-27 shows the corresponding confidence band for mean declination versus depth in a core from Lake Windermere, based on the non-parametric estimator of Clark (1977) with bandwidth determined by cross-validation. The confidence limits (formulae 6) are for the **mean** response $F(d)$, **not** an individual observation z_i. In the case of least-squares regression with Gaussian errors, the proportion of original observations lying within the 95% confidence band for the mean response will be considerably less than 95%, depending on the ratio p/n, as shown in Table 4-6.

When, as in Figure 4-27, the confidence band is constructed as a sequence of pointwise 95% confidence intervals at successive depths, it should be noted that the probability that all of these pointwise intervals are simultaneously 'correct' is not necessarily 0.95. However, in cases where the fitted curve $\hat{F}$ is obtained by least-squares multiple regression, an exact simultaneous 95% confidence band for the entire response curve F can be obtained by replacing t_ν in formulae 6 by $(pF^*_{0.05})^{1/2}$, where $F^*_{0.05}$ is the upper 5% point of the F-distribution on p, n-p d.f. (Seber, 1977, Section 5.2).

The choice of smoothing method must be subjective to some extent, depending on the

Table 4-6:   Approximate proportions of observations lying within
95% confidence limits for a mean curve when estimated by
fitting p linearly-independent parameters by least-
squares to n observations whose 'errors' of measurement
follow a Gaussian distribution.

p/n	Proportion
0.02	0.22
0.04	0.31
0.06	0.38
0.08	0.43
0.10	0.47
0.15	0.56
0.20	0.63

aim of the analysis and on one's prior knowledge. We recommend using cubic splines with knots chosen by a combination of Wold's guidelines, cross-validation or one of the knot-placement algorithms, with the parameters estimated by iteratively-reweighted least-squares with a weight-function such as Tukey's bisquare weight (Mosteller & Tukey, 1977, Ch. 14). This method utilizes the flexibility and adaptability of cubic splines, while providing straight forward formulae for confidence limits and tests of hypotheses, and offering good resistance to outliers.

Whatever method is chosen, the statistical analysis should include a thorough examination of the residuals $r_i = z_i - \hat{F}(d_i)$ from the fitted curve $\hat{F}$. As discussed by Andrews (1974), Draper and Smith (1981, Ch. 3), Seber (1977, Section 6.6) and Daniel & Wood (1971), plots of Studentized residuals against (i) the independent variable (depth, in this case), (ii) the predicted response F and (iii) any other related factor are a most effective way of detecting outliers or patterns in the residuals, the latter indicating a possibly incorrect model or over-smoothing of the data. Probability plots can be invaluable in checking the probability distribution of the error terms (e_i) and in detecting outliers. These graphical procedures may be supplemented by statistical tests such as the Runs Test (Draper & Smith, 1981, Ch. 3) and D'Agostino's test for normality (D'Agostino, 1972). It should be noted that outliers do not necessarily have large associated residuals (see, for example, Figure 1 of Yale & Forsythe (1976)), while sometimes apparent outliers arise from the use of an inappropriate smoothing method, for example, an incorrect parameterization of F. In addition, some outliers are more important than others, depending on the values of associated variables (Andrews & Pregibon, 1978). One should not accept the fitted curve unless these plots and tests show that the pattern of residuals is reasonable.

4.8.4. Directional data

The direction of natural remanent magnetization (NRM) in a sample of lake sediment may be specified in terms of either angular coordinates (declination and inclination) or the direction cosines of a unit vector. A series of directional measurements at various depths $\{d_i\}$ down a core may therefore be represented as a series of n unit vectors $z_1, z_2, \ldots, z_n$. A natural extension of model (1) to this situation is that each unit vector z_i follows the Fisher(1953) distribution with mean vector $F(d_i)$, where $F(.)$ is some smooth unit-vector-valued function of depth d. Geometrically, we may think of $F(.)$ as tracing out a smooth curve on the surface of a unit sphere, while the individual observations $\{z_i\}$ are scattered about this curve. As in the scalar case, it is not unreasonable to assume that the z_i's are independently distributed, and that their scatter about the curve F, as specified by the dispersion parameter κ may vary with depth.

The direction cosines or angular coordinates must be defined relative to some set of coordinate axes - XYZ. Any change of coordinate axes may be viewed either as a rotation of axes or alternatively a rotation of the data. Since the choice of coordinate axes is arbitrary, it is essential that the operations of smoothing and rotation be **commutative**: we should get the same answer whether we rotate the data and then smooth, or smooth first and then rotate.

This property of 'rotational invariance' is not necessarily satisfied when smoothing is based on angular coordinates, as the following example shows. Suppose that the path traced out by $F(.)$ on the unit sphere is a great circle. If the coordinate axes are chosen so that this great circle intersects the Z-axis, then all points on this curve have the same declination. The data may then be smoothed simply by estimating this common declination by fitting a horizontal line to the graph of observed declinations versus depth. But for any other choice of axes, the declination of points on the great circle will not be constant, and clearly it would not be sensible to fit a horizontal line to the observed declinations under this choice of axes. This difficulty with smoothing angular coordinates is of theoretical rather than practical importance. Provided the smoothing procedure is not too restrictive, any differences due to interchanging the operations of smoothing and rotation should be small.

Rotational invariance does apply if smoothing is achieved by smoothing the three sets of direction cosines using any **linear** estimator of the form (4), and then rescaling so that the smoothed vector is a unit vector. The **same** estimator (i.e., the same set of coefficients $\{a_i(d)\}$ must be used for the X, Y and Z coordinates. This class of linear estimators includes least-squares splines, approximating splines and moving-average estimators. The sample mean direction, the familiar estimate (Fisher, 1953) used when all n observations have the same theoretical mean direction, corresponds to the special case of the method in which $a_i(d) \equiv 1/n$. Unfortunately, there is at present no theory available for the construction of confidence limits or tests of hypotheses in the general case of arbitrary weights $\{a_i(d)\}$.

In many cases, the observed vectors $\{z_i\}$ are clustered within 10° to 15° of the overall mean direction. In such cases, it is convenient to convert the observations to stereographic equi-angular coordinates $\{(X_i, Y_i)\}$ relative to the sample mean direction. This reduces the problem to that of smoothing in the plane rather than on the sphere. Furthermore, by a property of the Fisher distribution (Mardia, 1972, Section 8.7.2b), these stereographic coordinates are, for large κ, approximately distributed as bivariate Gaussian with zero correlation and equal variances (given by $1/\kappa$). Thus if the function F is expressed parametrically, maximum likelihood estimation of the corresponding parameters reduces to least-squares estimation based on the X and Y coordinates separately.

Since these stereographic coordinates are defined relative to an arbitrary set of axes, any method of smoothing based on these coordinates should also be rotationally invariant. If least-squares estimation is used, rotational invariance can be assured by fitting the same form of regression equation to both the X and Y coordinates, e.g., cubic splines with the same knots for X and Y. The problem then becomes one of bivariate multiple linear regression, for which the theory and computations are well known (see, for example, Finn (1974)).

In particular, pointwise confidence ellipses for the true stereographic coordinates at any arbitrary depth d can be computed from well-known formulae (e.g., Finn, 1974, equation 5.2.40). Figure 4-28 shows the resulting confidence bands, in terms of stereographic coordinates relative to the overall mean direction, for the mean direction of NRM over a depth range of 120-170 cm. in a core from Loch Lomond, Scotland. This confidence band was derived by first fitting least-squares cubic splines to the X and Y coordinates as functions of depth, with knots at the same depths for both coordinates. Further details of this analysis are given in Clark & Thompson (1982).

In many respects, bivariate multiple regression corresponds to univariate multiple regression on the X and Y coordinates separately. Hence the suggestions in section 4.2.3 above on the choice of smoothing, detection of outliers and examination of residuals still apply.

4.8.5. Alignment of depth scales

So far we have considered only the problem of smoothing data in a single palaeomagnetic record. But in many situations the ultimate objective of any such smoothing is the comparison of the signals in several related palaeomagnetic records or cores.

The basic assumption in such comparisons is that, apart from random errors or noise in measurement, the variation of the response variable with **time** is identical in all cores. However, since the response can be measured only as a function of depth, it is

LOCH LOMOND CORE 1

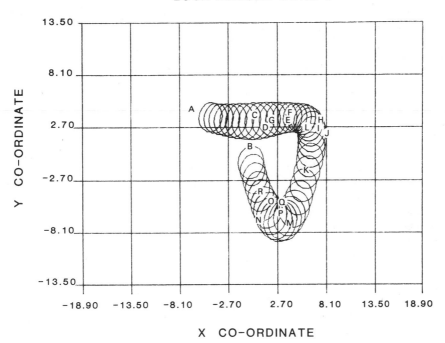

Figure 4-28: Confidence band for function F defining
variation of direction of NRM with depth in a core from Loch Lomond,
Scotland. Band comprises pointwise 95% confidence ellipses at
40 equally-spaced depth-points over the range 120-170cm.
Coordinate axes define stereographic equi-angular coordinates
(scaled in degrees) relative to mean direction for the entire core.
Confidence ellipses are derived from least-squares cubic splines
fitted to stereographic coordinates over the depth-range
100-300 cm. Letters A, B, C, ... represent original observations
in order of increasing depth over the range 120-170cm.

necessary to adjust or stretch the depth scales of the various cores relative to one another, to account for different sedimentation rates within cores. If, in the case of two cores, F_1 and F_2 denote the mean response as functions of depth in cores 1 and 2 respectively, Clark & Thompson (1979) show that this adjustment of depth scales reduces mathematically to the problem of estimating the <u>stretching</u> function g defined implicitly by the identity

$$F_2(g(d)) \equiv F_1(d) \quad \text{for all depths d in core 1} \tag{7}$$

For any depth d in core 1, g(d) is the depth in core 2 corresponding to the same time as depth d in core 1. Our only prior information about this unknown function g is that it must be monotonic. The additional complication is that both F_1 and F_2 are unknown, and must be estimated from observations which are subject to random errors of measurement.

Several alternative methods of aligning two cores (or equivalently estimating the function g have been proposed, namely (i) visual matching), (ii) cross-correlation, (iii) curve-matching and (iv) sequence-slotting using a distance measure.

In the first method, the plots of response versus depth for each core are scanned visually, and key features common to both plots are identified. The function g may then be estimated directly from its definition by plotting against one another the estimated depths in each core corresponding to successive features (e.g. Stober & Thompson (1977)). This method, being subjective, does not lead to any statistical tests but is very simple and can provide a useful check on more complicated methods.

Cross-correlation techniques are based on the observation that, if there were no errors of measurement, the responses at corresponding depths in the two cores would, when plotted against one another, lie exactly on the straight line Y = X. Thus the function g may be estimated by that function (from some suitable class of functions) which maximizes the correlation coefficient between the corresponding paired responses. Since the observed depths in each core do not necessarily correspond to equal time-intervals, some preliminary smoothing of the responses in one or both cores is necessary (Rudman & Lankston, 1973). However, Clark and Thompson (1979) point out several serious deficiencies of such cross-correlation methods. In particular, a high correlation between paired comparisons need not necessarily imply that the corresponding function g is correct.

Clark & Thompson (1979) proposed an alternative method derived directly from identity (7). This identity implies that if the correct stretching function g is used, the plot of response versus **adjusted** depth should be identical in both cores, apart from noise in the data. If, then, the response curve is represented as a multiple linear regression model with parameters estimated by least-squares, this hyphothesis of identical response curves may be tested by a standard statistical test. This will be the case if, for example, the response curve is estimated by a least-squares cubic spline. In principle, the compatibility with the data of any arbitrary stretching function g may be assessed by this straightforward F-test. In practice, if g is assumed to be **linear**, the method easily yields best estimates and confidence limits for the parameters of this linear function. This assumption of linearity, although arbitrary, may not be unreasonable since it implies that sedimentation rates in the two cores have remained **proportional** to one another throughout the time-interval considered.

Gordon (1973, 1980) and Gordon & Reyment (1979) considered the slightly different problem of 'slotting' the two sequences together, that is, of constructing the best possible ordering on a common depth scale of the combined observations from both cores. The algorithm is based on a measure of the 'distance' between each pair of observations, and in particular it can take account of the simultaneous variations in several response variables. The slotting automatically preserves the correct ordering of depths within each core and can be forced to satisfy other constraints specified by the user. The degree of concordance between the two cores is quantified by two criteria ψ and κ defined by Gordon.

Gordon's method makes only the very weak but indisputable assumption that the function g is monotonic, but in turn produces only an ordering of the combined observations. Methods (such as that of Clark & Thompson (1979)) which produce an actual estimated stretching function must naturally make stronger assumptions concerning that function. Gordon's method does not have any associated statistical tests to assess the validity of the final slotting, and has the minor disadvantage that if the response curves are too flat, the optimum slotting may contain long blocks of observations from the same core which must then be un-blocked by hand. On the other hand, if there are one or more gaps or missing segments in either core, the function g in 7 will necessarily be discontinuous, and hence non-linear, but still monotonic. In such situations, Gordon's method should be used because of its flexibility and realistic assumptions. Each gap would be expected to result in a block of observations, enabling the location of gaps to be more easily identified. The adjustment of the depth scale could then be refined by subsequently applying the Clark & Thompson method to only those segments of the cores between gaps, resulting in a piecewise linear stretching function.

4.8.6. Summary

With the widespread availability of computers and the large amounts of numerical experimental data being collected by palaeomagnetists, statistical analysis is now an essential part of most palaeomagnetic investigations. This contribution has reviewed a variety of statistical methods, all based on the concept of 'smoothing' the data. Most of these methods apply not only for palaeomagnetic data but for any data of the form **signal plus noise**. In practice, the choice of statistical method must be to some extent subjective, as it will depend both on the prior information and aims of the analysis, e.g., whether simple exploratory analysis or more formal statistical inference such as testing of hypotheses or construction of models. Preference should be given to methods which are resistant to outlying observations, which enable underlying assumptions to be tested, and which do provide for statistical inference (such as confidence limits) if required.

4.9. SPECTRUM ANALYSIS

By C.E.Barton,
University of Rhode Island, USA.

4.9.1. Introduction

Much of our effort to analyse the past behaviour of the geomagnetic field involves searching for repetitive behaviour, e.g. westward drift of the non-dipole field, oscillations in dipole moment. Spectral analysis provides a powerful tool to do this, albeit one which generally requires strong assumptions about the time-series being analysed (Figure 4-29). The heart of the problem is that the frequency spectrum is a statement of the nature of an infinite time-series, whereas the information at our disposal comes from a short, noisy and distorted sample.

Figure 4-29: Spectrum analysis: a powerful tool.

This contribution is a synopsis of a review published recently (Barton, 1983). The reader should also consult a review on techniques by Kay and Marple (1981).

4.9.2. Methods

The two traditional methods of spectral analysis are the best understood and remain the backbone of the most analyses. The periodogram, or discrete Fourier transform method (Schuster, 1898; Toman, 1965; and for example, Bloomfield, 1976) requires the assumption that the time-series is infinitely repetitive outside the observed window An estimate of the spectrum is derived from smoothing the discrete Fourier transform of the series. The auto- covariance method (Blackman and Tukey, 1959) assumes that the time-series is zero outside the observed window. An estimate of the spectrum is obtained from the Fourier transform of the autocovariance functions. (The latter are derived from the product of adjacent terms when the series is compared with itself displaced in time by a discrete number of lags, i.e. sample intervals).

An alternative approach is to model the series as an autoregressive process (Yule, 1927). Each term in the series, $X(t)$, is treated as a linear sum of a finite number of preceeding terms together with a random noise component:

$$X(t) = a_1 X(t-1) + a_2 X(t-2) + a_3 X(t-3) + \ldots \ldots + a_m X(t-m) + B(t)$$

The number of terms, m, is the order of the autoregressive (AR) process, $B(t)$ is the noise component, and the constants a_1 to a_m are known as the prediction error coefficients. The spectrum of the autoregressive process can readily be computed from the set of prediction error coefficients, thus the problem boils down to estimating the coefficients of the model which best fits the data.

Yule and Walker (see Smylie et al., 1973) did this by assuming that the unknown values of $X(t)$ were all zero. An innovation by Burg (1967, 1968, 1975) was to select the coefficients according to a certain criterion for minimizing the variance in the prediction error series $B(t)$. This does not avoid making assumptions about the unknown values of $X(t)$, but does have the attraction that the assumed values introduce the least amount of "information" – hence the name maximum entropy method (MEM), or maximum entropy spectral analysis (MESA). The method is data-adaptive (Lacosse 1971) in the sense that the assumed values of $X(t)$ are determined from the known values. Introductions to the method have been published by McGee (1969), Smylie, et al. (1973), Anderson (1971), Ulrych and Bishop (1975), Kanasewich (1975), Kay and Marple (1981) and many others.

A feature of parametric modelling techniques such a Burg's MEM is that they offer very high resolutions, even for time series which are short compared with the periodicities of interest. Because of this, combined with its computational simplicity, the MEM has become popular for analysing geophysical time-series, e.g. the Chandler Wobble (Brillinger, 1973) and variations in geomagnetic polarity bias (Ulrych, 1972).

A serious limitation is imposed by the propensity of the method towards peak shifting and splitting. The user has the difficult task (impossible sometimes for series which cannot be modelled as AR processes) of selecting the appropriate order of the AR model. Various methods and criteria for selecting the order are referenced by Barton (1983). No one criterion is always appropriate, and furthemore the optimum choice of order appears to be frequency dependent. The peak shifting problem has led to a revaluation of traditional methods. Despite its low resolution, it turns out that the periodogram method can be superior to the maximum entropy method even at very low frequencies (Swingler, 1980).

More sophisticated parametric modelling techniques have been described (Pisarenko, 1972; Ulrych and Clayton, 1976; Swingler, 1979) which purport to circumvent the peak shifting and splitting problems of the MEM. At the moment these methods are largely untested.

4.9.3. Scalar time-series

There is some justification for treating declination and inclination as independent scalars: uncertainties created by palaeomagnetic recording processes (e.g. DRM inclination errors), sampling processes (e.g. core twisting and non-vertical penetration), and measuring processes (e.g. proximity of the magnetic vector to one of the specimen coordinate axes) do not affect declination and inclination measurements equally. Furthermore, certain non-dipole source geometries will influence either declination or inclination preferentially (see section 4.10).

Historically, declination and inclination time-series were analysed independently (e.g. Yukutake 1962). One of the more surprising results to emerge was that the spectra of the separate time series of the same secular variation record are often, but not always, markedly different. Barton has argued (Barton, 1983a; Barton and McElhinny, 1982) that these differences can arise as a natural consequence of the drift of non-dipole sources in the outer part of the core.

4.9.4. Vector time-series

Spectral analysis is commonly performed using complex number series because of notational simplicity (see for example Bloomfield, 1976). For real valued data the complex parts are equated to zero. Spectral estimates are then obtained at both positive and negative frequencies, (within the range of ± the Nyquist frequency). For real data, the positive and negative frequency halves of the spectrum are always mirror images. It is then normal practice to express all the variance (i.e. power) in terms of positive frequencies only by doubling the amplitude of the spectral peaks.

One approach to expressing vector time series in an analytically tractable form is to map each vector, V(t), on to a complex plane surface and then express it as a complex number, X(t) + iY(t), where X and Y are the real and imaginary coordinates. Spectral analysis is then performed on the complex number time series. However, if the imaginary part is non-zero, positive and negative frequency halves of the power spectrum are not symmetrical. Furthermore, the degree and sense of asymmetry at a given frequency is determined by the degree of ellipticity and sense of rotation of the path traced out by the vector as a function a time at that frequency. For directional data the situation is simplified as all the vectors are unit vectors. Circular precession of the vector at a given frequency will result in a positive spectral peak with a zero negative counterpart for rotation in one sense (e.g. clockwise, depending on the mapping rotation) and the reverse for rotation in the opposite sense. The case of no rotation, i.e. oscillation along a straight line path in the complex plane, results in equal spectral power at both positive and negative frequencies.

The technique has been used, for example, by Brillinger (1973) to analyse the Chandler wobble and by Gonella (1972) to analyse wind and ocean current data. Denham (1975) introduced the technique for analysing palaeomagnetic time series and drew attention to the elegant relationship between the asymmetry in the complex number power spectrum, and westward or eastward drift of the non-dipole field, as given by Runcorn's rule (Runcorn, 1959; Skiles, 1970) – see also Dodson (1979) and Creer and Tucholka (1982b) regarding limitations of this rule. Denham's lead has now been followed by a number of authors, e.g. Oberg and Evans (1977), Turner and Thompson (1981), Barton and McElhinny (1982).

4.9.5. The geomagnetic spectrum

Time series analyses have now been performed on a large number of secular variation records from around the world. The general forms of the spectra are similar, with most of the variance concentrated at the low frequency end of the spectrum and a fairly rapid fall off towards higher frequency, with a small number of 'well defined' peaks. Few authors give confidence limits so the significance of the peaks is difficult to assess. A wide range of discrete periodicities have been cited in the literature, from a few hundred years (Dubois, 1982) to greater than 25000 years (Negrini et al., 1981). The current lack of a consensus regarding which discrete periodicities characterize the global geomagnetic secular variation must be taken as evidence that they do not exist. Peaks in spectra of individual records tend to cluster within three broad bands (Barton, 1982, 1983; Lund and Banerjee, 1983):

   i. a very long period, between 3000 and 9000 years which accounts for most
      of the variance and is generally attributed to dipole fluctuations,

   ii. and intermediate period, between 2000 and 3000 years generally
       attributed to non-dipole fluctuations and drift,

iii. a short period, between 600 and 1000 years, at relatively low power which
      may, or may not be multiples of stronger low frequency peaks.

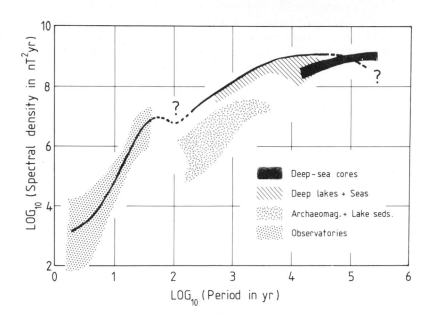

Figure 4-30:   A preliminary assessment of the geomagnetic power spectrum derived
                from observatory and palaeomagnetic data. Individual spectra from the
                four classes of records fall within the shaded zones. The solid line is a
                guess as to where the actual spectrum lies. Reasons why the lacustrine
                records underestimate the spectrum are discussed by Barton (1982, 1983).
                Reproduced from Barton (1982).

Compilations of spectra from secular variation time series from sites around the globe
(Figure 4-30) show a smooth increase in power in the geomagnetic spectrum form
periods of 100 years to 10000 years (Barton 1983a, 1983b).   This is contrary to earlier
speculations, e.g. Doell and Cox (1972), that dipole and non-dipole effects would be
associated with recognizable frequency bands.   Interchange of energy between the dipole
and non-dipole fields (Verosub and Cox 1971) would account for this, though from an
analysis of global paleointensity data, McElhinny and Senanayake (1982) argue that the
power in the dipole and non-dipole fields vary directly.

REFERENCES

Aitken, M.J. Dating by archaeomagnetic and thermoluminescent methods. *Phil. Trans. R. Soc.*, 1970, *269A*, 77–88.

Andersen, N. On the calculation of filter coefficients for maximum entropy spectral analysis. *Geophysics*, 1974, *39*, 69–72.

Andrews, D.F. A robust method for multiple linear regression. *Technometrics*, 1974, *16*, 523–531.

Andrews, D.F. and Pregibon, D. Finding the outliers that matter. *J. Roy. Statist. Soc. B.*, 1978, *40*, 85–93.

Baag, C. and Helsley, C.E. Shape analysis of paleosecular variation. *J. geophys. Res*, 1974, *79*, 4923–4932.

Banerjee, S,K., Lund, S.P. and Levi, S. Geomagnetic record in Minnesota Lake sediments – absence of the Gothernburg and Erieau excursions. *Geology*, 1979, *7*, 588–591.

Barbetti, M.F. and McElhinny, M.W. The Lake Mungo geomagnetic excursion. *Phil. Trans. R. Soc.*, 1976, *A281*, 515–542.

Barraclough, D.R. Spherical harmonic analysis of the geomagnetic field for eight epochs between 1600 and 1910. *Geophys. J. R. astr. Soc.*, 1974, *36*, 497–513.

Barton, C.E. Spectral analysis of palaeomagnetic time series techniques and applications. *PHTR*, 1982, *A306*, 203–209.

Barton, C.E. Analysis of palaeomagnetic time series techniques and applications. *Geophysical Surveys*, 1983. in press.

Barton, C.E. and McElhinny, M.W. Time series analysis of the 10000 yr geomagnetic secular variation record from SE Australia. *Geophys. J. R. astr. Soc.*, 1982, *68*, 709–724.

Barton, C.E. and Barbetti, M. Recent geomagnetic secular variation from lake sediments, ancient fireplaces and historical measurements in southeastern Australia. *Earth Planet. Sci. Lett.*, 1982, *59*, 375–387.

Barton, C.E., McElhinny, M.W. and Edwards, D.J. Laboratory studies of depositional DRM. *Geophys. J. R. astr. Soc.*, 1980, *61*, 355–377.

Barton, C.E. and McElhinny, M.W. Detrital remanent magnetization in five slowly redeposited long cores of sediment, *Geophys. Res. Lett.*, 1979, *6*, 229–232.

Barton, C.E. and McElhinny, M.W. A 10000 yr geomagnetic secular variation record from three Australian maars. *Geophys. J. R. astr. Soc.*, 1980, *67*, 465–485.

Barton, C.E. and Polach, H.A. ^{14}C ages and magnetic stratigraphy in 3 Australian maars, *Radiocarbon*, 1980, *22*, 728–739.

Blackman, R.B. and Tukey, J.W. *The Measurement of Power Spectra from the point of view of Communications Engineering.* Dover, New York, 1959.

Bloomfield, P. *Fourier Analysis of Time Series: an introduction.* Wiley, New York, 1976.

Bowler, J.M. and Hamada, T. Late Quaternary stratigraphy and radiocarbon chronology of water level fluctuations in Lake Keilambete, Victoria. *Nature*, 1971, *232*, 330–332.

Brillinger, D.R. An empirical investigation of the Chandler Wobble and two proposed excitation processes. *Bull. Int. Stat. Inst.*, 1973, *45*, 413–434.

Broughton, P.L. Secondary mineralization in the cave environment. *Studies in Speleology*, 1972, *2*, 191–207.

Burg, J.P. *Maximum Entropy Spectral Analysis.* Thirty-Seventh Ann. In. Meeting Soc. Explor. Geophys., Oklahoma City, 1967.

Burg, J.P. *A new analysis for time series data.* Adv. Study Inst. on Signal Processing, NATO, Enschede, 1968.

Burg, J.P. *Maximum Entropy Spectral Analysis.* Doctoral dissertation, Stanford University, 1975.

Carroll, R.J. Robust methods for factorial experiments with outliers. *Appl. Statist.*, 1980, *29*, 246–251.

Caumartin, V. Review of the microbiology of the underground environment. *Bull. National Speleological Society*, 1963, *25*, 1–14.

Champion, D.E. *Holocene geomagnetic secular variation in the western United States: implications for the global geomagetic field* (Open file report 80-824). U.S. Geological Survey, 1980. 314 pp.

Clark, R.M. Non parametric estimation of a smooth regression function. *J. Roy. Statist. Soc. B.*, 1977, *39*, 107–113.

Clark, R.M. and Thompson, R. A new statistical approach to the alignment of time series. *Geophys. J. R. astr. Soc.*, 1979, *58*, 593–607.

Clark, R.M. and Thompson, R. Smoothing and comparison of palaeomagnetic records in lake sediments. , 1982.

Clark, R.M. and Thompson, R. An objective method for smoothing palaeomagnetic data. *Geophys. J. R. astr. Soc.*, 1978, *52*, 205–213.

Cleveland, W.S. Robust locally weighted regeression and smoothing scatterplots. *J. Am. Stat. Ass.*, 1979, *74*, 829–836.

Cox, A. Latitude dependence of the angular dispersion of the geomagnetic field. *Geophys. J. R. astr. Soc.*, 1970, *20*, 253–269.

Cox, A. The frequency of geomagnetic reversals and the symmetry of the non- dipole field. *Rev. Geophys. Space Phys.*, 1975, *13*, 35–51.

Creer, K.M. Geomagnetic variations for the interval 7000–25000 BP as recorded in a core of sediment from station 1474 of the Black Sea cruise of 'Atlantis II'. *Earth Planet. Sci. Lett.*, 1974, *23*, 34–42.

Creer, K.M., Valencio, D.A., Sinito, A.M., Tucholka, P. and Vilas, J.F. Geomagnetic secular variations 0–14000 years before present as recorded by lake sediments from Argentina. *Geophys. J. R. astr. Soc.*, 1983, Vol. *in press*.

Creer, K.M., Anderson, T.W. and Lewis, C.M.F. Late Quaternary geomagnetic stratigraphy recorded in Lake Erie sediments. *Earth Planet. Sci. Lett.*, 1976a, *31*, 37–47.

Creer, K.M., Gross, D.L. and Lineback, J.A. Origin of regional geomagnetic variations recorded by Wisconsinan and Holocene sediments from Lake Michigan U.S.A. and Lake Windermere, England. *Geological Society of America Bulletin*, 1976b, *87*, 531–540.

Creer, K.M., Hogg, T.E., Malkowski, Z., Mojski, J.E., Niedziolka-Krol, E., Readman, P.W. and Tucholka, P. Palaeomagnetism of Holocene lake sediments from north Poland. *Geophys. J. R. astr. Soc.*, 1979, *59*, 287-314.

Creer, K.M., Hogg, T.E., Readman, P.W. and Reynaud, C. Palaeomagnetic secular variation curves extending back to 13400 years BP recorded by sediments deposited in Lac de Joux, Switzerland. *J. Geophys.*, 1980, *48*, 139–147.

Creer, K.M. and Kopper, J.S. Secular oscillations of the geomagnetic field recorded by sediments deposited in caves in the Mediterranean region. *Geophys. J. R. astr. Soc.*, 1976, *45*, 35–58.

Creer, K.M., Readman, P.W. and Papamarinopoulos, S. Geomagnetic secular variations in Greece through the last 6000 years obtained from lake sediment studies. *Geophys. J. R. astr. Soc.*, 1981, *66*, 193–219.

Creer, K.M., Thompson, R., Molyneux, L and Mackereth, F.J.H. Geomagnetic secular variation recorded in the stable magnetic remanence of recent sediments. *Earth Planet. Sci. Lett.*, 1972, *14*, 115–127.

Creer, K.M. and Tucholka, P. Secular variation as recorded in lake sediments: a discussion of North American and European results. *Phil. Trans. R. Soc.*, 1982b, *A306*, 87–102.

Creer, K.M. and Tucholka, P. Construction of type curves of geomagnetic secular variation for dating lake sediments from east central North America. *Can. J. Earth Sci.*, 1982a, *19*, 1106–1115.

Daniel, C. and Wood, F.S. *Fitting Equations to Data*. Wiley, New York, 1971.

De Boor, C. *A Practical Guide to Splines*. Springer Verlag, New York, 1978.

Denham, C.R. Spectral analysis of paleomagnetic time series. *J. geophys. Res*, 1975, *80*, 1891–1901.

Dodson, R.E. Counterclockwise precession of the geomagnetic field vector and westward drift of the non-dipole field. *J. geophys. Res*, 1979, *84*, 637–644.

Dodson, R.E., Day, R., Dunn, J.R. Fuller, M. Henyey, T. and Kean, W. Continuous records of secular variation. *EOS, Trans. Amer. Geophys. Un.*, 1977, *58*, 308.

Dodson, R., Fuller, M. and Pilant, W. On the measurement of the remanent magnetism of long cores. *Geophys, Res. Lett.*, 1974, *1*, 185–188.

Dodson, R.E., Fuller, M.D. and Kean, W.F. Paleomagnetic secular variations from Lake Michigan sediment cores. *Earth Planet. Sci. Lett.*, 1977, *34*, 387–395.

Doell, R.R., and Cox, A. The Pacific geomagnetic secular variation anomaly and the question of lateral uniformity in the lower mantle. In Robertson, E.C. (Ed.), *The Nature of the Solid Earth*. MaGraw–Hill, New York, 1972.

Draper, N.R. and Smith, H. *Applied Regression Analysis*. Wiley, New York, 1981. (2nd edition).

Dubois, R.L. Some archeomagnetic results and their analysis: southwestern US and Meso–America. In *Geomagnetic Workshop Abstracts*. USGS, Golden, Colorado, 1982.

Dutter, R. Numerical solution of robust regression problems: computational aspects, a comparison. *J. Statist. Comput. Simul.*, 1977, *5*, 207–238.

Finn, J.D. *A General Model for Multivariate Analysis*. Holt, Rinehart and Winston, New York, 1974.

Fisher, R.A. Dispersion on a sphere. *Proc. Roy. Soc. London*, 1953, *A217*, 295–305.

Geology Today. Communications Machines Research Inc. Del Mar, Cal. 529, 447–450.

Gonella, J. A rotary component method for analyzing meteorological and oceanographic vector time series. *Deep–Sea Res.*, 1972, *19*, 833–846.

Gordon, A.D. A sequence–comparison statistics and algorithm. *Biometrika*, 1973, *60*, 190–200.

Gordon, A.D. SLOTSEQ: A Fortran IV program for comparing two sequences of observations. *Computers & Geosciences*, 1980, *6*, 7–20.

Gordon, A.D. and Reyment, R.A. Slotting of borehole sequences. *J. Math. Geology*, 1978, *11*, 309–327.

Griffiths, D.H. The remanent magnetism of varved clays from Sweden. *Mon. Not. R. astr. Soc. Geophys. Suppl.*, 1955, *7*, 103–114.

Hedley, I.G. Chemical remanent magnetization of the $FeOOH-Fe_2O_3$ system. *Phys. Earth. Planet. Inter.*, 1968, *1*, 103–121.

Hough, J.L. *Geology of the Great Lakes*. University of Illinois Press, Urbana, 1958. 313 pp.

Hough, J.L. Lake Stanley, a low stage of Lake Huron indicated by bottom sediments. *Geological Society of America Bulletin*, 1962, *73*, 613–620.

Hough, J.L. Correlation of glacial lake stages in the Huron–Erie and Michigan Basins. *J. Geol.*, 1966, *74*, 62–77.

Ising, G. On the magnetic properties of varved clay. *Ark. Mat. Astron. Fys.*, 1943, *29A*, 1–37.

Johnson, G.L., Murphy, T. and Torreson, O.W. Pre-history of the Earth's magnetic field. *Terr. Magn. Atmos. Elec.*, 1948, *53*, 349–372.

Jupp, D.L. Approximation to data by splines with free knots. *S.I.A.M. J. Numer. Anal.*, 1978, *15*, 328–343.

Kanasewich, E.R. *Time Sequence Analysis in Geophysics*. University of Alberta Press, Edmonton, 1975. 364 pp.

Karkabi, S. L'exploration des eaux du karst en Liban. 4th International Speleological Congress, Postojna, Ljubljana, Dubrovnik.

Kay, S.M. and Marple, S.L.Jr. Spectrum analysis in geophysics. *IEEE*, 1981, *69*, 1380–1419.

King, R.F. Remanent magnetism of artificially–deposited sediments. *Mon. Not. R. astr. Soc. Geophys. Suppl.*, 1955, *7*, 115–134.

Kopper, J.S. Geophysical surveying of cave sites. *Pyrenae*, 1972, *8*, 7–18.

Kovacheva, M. Summarised results of the archaeomagnetic investigation of the geomagnetic field variation for the last 8000 years in south–eastern Europe. *Geophys. J. R. astr. Soc.*, 1980, *61*, 57–64.

Kukla, J. and Lozek, V. The problems of investigation of cave sediments. *Ceskoslovensky Kras*, 1957, *10*, 19–83.

Kullenberg, B. The piston core sampler. *Svenska hydrogr-biol. Komm. Skr. Tredge. serien hydrografi*,

1947, pp. 1-16.

Lenth,R.V. Robust Splines. *Comm. Stat. Theor. Method.* , 1977, *A6*, 847-854.

Levi,S and Banerjee, S.K. On the possibility of obtaining relative paleointensities from lake sediments. *Earth Planet. Sci. Lett.* , 1976, *29*, 219-226.

Livingston,D.A. A lightweight piston sampler for lake deposits. *Ecology*, 1955, *36*, 137-139.

Lund, S.P. *Late Quaternary Secular Variation of the Earth's Magnetic Field as recorded in the Wet Sediments of three North American lakes.* Doctoral dissertation, University of Minnesota, Minneapolis, 1981. 300 pp.

Mackereth, F.J.H. A portable core sampler for lake deposits. *Limnol. Oceanog.* , 1958, *3*, 181-191.

Mackereth, F.J.H. A short core sampler for subaqueous deposits, *Limnol. Oceanogr*, 1969, *14*, 145-151.

Mackereth, F.J.H. On the variation in the direction of the horizontal component of the magnetization in lake sediments. *Earth Planet. Sci. Lett.* , 1971, *12*, 332-338.

Mardia, K.V.   *Statistics of Directional Data.* Academic Press, New York, 1972.

Mazzoni, M.M. and Sinito, A.M. Estudio paleomagnetico y sedimentologico de ambientes lacustres – II Lago Moreno. *Rev. Asoc. Geol. Arg.* , 1982, Vol. *in press.*

McElhinny, M.W. and Senanayake, W. Variations in the geomagnetic dipole 1: the past 50000 years. *J. Geomag. Geoelectr.* , 1982, *34*, 39-52.

McElhinny, M.W. and Merrill, R.T. Geomagnetic secular variation over the last 5 million years. *Rev. Geophys. Space Phys.* , 1975, *13*, 687-708.

McGee, T.M.   On Burg's method of spectral analysis. 1969. unpublished manuscript, Vening Meinesz Laboratory, Utrecht.

McNeil, D.R.   *Interactive Data Analysis.* Wiley, New York, 1977.

McNish, A.G. and Johnson, E.A. Magnetization of unmetamorphosed varves and marine sediments. *Terr. Magn. Atmos. Electr.* , 1938, *43*, 401-407.

Molyneux, L. and Thompson, R. Rapid measurements of the magnetic susceptibility of long cores of sediment. *Geophys.J. R. astr. Soc.* , 1973, *32*, 479-481.

Molyneux,L., Thompson, R., Oldfield, F. and McCallan, M.E. Rapid measurement of the remanent magnetization of long cores of sediment, *Nature*, 1972, *237*, 42-43.

Morner, N-A., Lanser, J.P. and Hospers, J. Late Weichselian palaeomagnetic reversal. *Nature, Phys. Sci.* , 1971, *234*, 173-174.

Mosteller, F. and Tukey, J.W.   *Data Analysis and Regression.* Addison-Wesley, Reading, Mass., 1977.

Mothersill, J.S. The paleomagnetic record of the late Quaternary sediments of Thunder Bay. *Can. J. Earth Sci.* , 1979, *16*, 1016-1023.

Mothersill, J.S. Late Quaternary paleomagetic record of the Goderich Basin, Lake Huron. *Can. J. Earth Sci.* , 1981, *18*, 448-456.

Negrini, R., Davis, J.O. and Verosub, K.L. Secular variation during a 250000 year interval in the Middle Pleistocene as recorded by lake sediments in northern Nevada. *EOS*, 1981, *62*, 851.

Noel, M. The palaeomagnetism of varved clays from Blekinge, southern Sweden. *Geol. Foren. Stockh. Fohr.* , 1975, *97*, 357-367.

Noel, M. and Tarling, D.H. The Laschamp geomagnetic 'event'. *Nature Phys. Sci.* , 1975, *253*, 705-706.

Oberg, C.J. and Evans, M.E. Spectral analysis of Quaternary palaeomagnetic data from British Columbia and its bearing on geomagnetic secular variation. *Geophys.J. R. astr. Soc.* , 1977, *51*, 691-699.

Opdyke, N.D., Ninkovich, D., Lowrie, W. and Hayes, J.D. The palaeomagnetism of two Aegean deep-sea cores. *Earth Planet. Sci. Lett.* , 1972, *14*, 145-159.

Papamarinopoulos, P.St. *Limnomagnetic studies on Greek sedimants.* PhD thesis, University of Edinburgh, 1978.

Papamarinopoulos, P.St., Readman, P.W., Maniatis, Y.   and Simopoulos, A. Magnetic characterization and Mossbauer spectroscopy of magnetic concentrates from Greek lake sediments. *Earth Planet. Sci. Lett.* , 1982, , 173-181.

Pisarenko, V.F. On the estimation of spectra by means of non-linear functions of the covariance matrix. *Geophys.J. R. astr. Soc.*, 1972, *28*, 511–513.

Prest, V.K. Quaternary geology of Canada. In R.J.W. Douglas (Ed.), *Geology and economic minerals of Canada*. Geological Survey of Canada, 1970.

Priestley,M.B. and Chao,M.T. Non-parametric function fitting. *J, Roy. Statist. Soc. B*, 1972, *34*, 385–392.

Reams, M.W. Deposition of calcite, aragonite and clastic sediments in a Missouri cave during four and one-half years. *Bull. Nat. Speleological Society*, 1972, *34*, 137–141.

Reinsch, C.H. Smoothing by spline functions. *Numer. Math.,*, 1967, *10*, 177–183.

Rice, J.R. The Approximation of Functions. *Addison-Wesley, Reading Mass*, 1969, Vol. 2.

Rudman, A.J. and Lankston, R.W. . Stratigraphic correlation of well logs by computer techniques. *Amer Assoc. Petroleum Geol. Bull.*, 1973, *57*, 577–588.

Runcorn, S.K. On the theory of the geomagnetic secular variation. *Ann. de Geophys.*, 1959, *15*, 87–92.

Saarnisto, M. Stratigraphical studies on the shoreline displacement of Lake Superior. *Can. J. Earth Sci.*, 1975, *12*, 300–319.

Schuster, A. On the investigation of hidden periodicities with application to a supposed 26 day period of meteorological phenomena. *Terr. Mag.*, 1898, *3*, 13–41.

Seber, G.A.F. *Linear Regression Analysis*. Wiley, New York, 1977.

Skiles, D.D. A method of inferring the direction of drift of the geomagnetic field from paleomagnetic data. *J. Geomag. Geoelectr.*, 1970, *22*, 441–461.

Smylie, D.E, Clarke, G.K.C. and Ulrych, T.J. Analysis of irregularities in the Earth's rotation. In *Methods in Computational Physics*. Academic Press, New York, 1973.

Stober,J.C. and Thompson, R. Palaeomagnetic secular variation studies of Finnish lake sediments and the carriers of remanence. *Earth Plan. Sci Letters*, 1977, *37*, 139–149.

Sweeting, M.M. *Karst Landforms*. Columbia University Press, New York, 1973. 362 pp.

Swingler, D.N. A comparison between Burg's maximum entropy method and non-recursive technique for the spectral analysis of deterministic signals. *J. geophys. Res*, 1979, *84*, 679–685.

Swingler, D.N. Burg's maximum entropy algorithm versus the discrete Fourier transform as a frequency estimator for trunctated real sinusoids. *J. geophys. Res*, 1980, *85*, 1435–1438.

Terasme, J., Kerrow, P.F. and Dreimanis, A. Stratigraphical studies on the shoreline displacement of Lake Quaternary stratigraphy and geomorphology of the eastern Great Lakes region of southern Ontario. In . 24th International Geological Congress, Montreal, 1972. Field excursion A42.

Thellier, E. Sur la direction du champ magnetique terrestre en France durant les deux derniers millenaires. *Phys. Earth. Planet. Inter.*, 1981, *24*, 89–132.

Thompson, R. Palaeolimnology and palaeomagnetism. *Nature*, 1973, *242*, 182–184.

Thompson, R. Long period European secular variation confirmed. *Geophys.J. R. astr. Soc.*, 1975, *43*, 847–859.

Thompson, R. The paleomagnetism of varved clays from Blekinge, southern Sweden: a comment. *Geol. Foren. Stockh. Fohr.*, 1976, *98*, 283–284.

Thompson, R., Oldfield, F., Dearing, J. and Garrett-Jones, S.E. Some magnetic properties of lake sediments and their links with erosion rates. *Pol. Arch. Hydrobiol.*, 1978, *25*, 321–331.

Thompson, R. and Barraclough, D.R. Cross validation, cubic splines and historical secular variation. In *IAGA Bulletin 45.* , 1981.

Thompson, R.and Berglund, B. Late Weichselian geomagnetic 'reversal' as a possible example of the reinforcement syndrome. *Nature*, 1976, Vol. *263*.

Thompson,R., Bloemendal,J., Dearing,J.A., Oldfield,F., Rummery,T.A., Stober,J.C. and Turner, G.M. Environmental applications of magnetic measurements. *Science,*, 1980, *207*, 481–486.

Thompson, R. and Kelts, K. Holocene sediments and magnetic stratigraphy from Lakes Zug and Zurich, Switzerland. *Sedimentology*, 1974, *21*, 577–596.

Thompson, R., Longworth, G., Becker, L.W., Oldfield, F., Dearing, J.A. and Rummery, T.A. Mossbauer effect and magnetic studies of secondary iron oxides in soils. *J. Soil Sci.*, 1979, *30*, 93–110.

Thompson, R. and Morton, D.J. Magnetic susceptibility and particular–size distribution in recent sediments of the Loch Lomond drainage basin. *J. Sed. Pet.*, 1979, *49*, 801–812.

Thompson, R, and Turner G.M. British geomagnetic master curve 10000–0 yr. BP for dating European sediments. *GRL*, 1979, *6*, 249–252.

Toman, K. Spectral shifts of trunctated sinusoids. *J. geophys. Res*, 1965, *70*, 1749–1750.

Tukey, J.W. *Exploratory Data Analysis*. Addison–Wesley, Reading, Mass., 1977.

Turner, G.M. and Thompson, R. The behaviour of the Earth's magnetic field as recorded in the sediments of Loch Lomond. *Earth Planet. Sci. Lett.*, 1979, *42*, 412–426.

Turner, G.M. and Thompson, R. Lake sediment record of the geomagnetic secular variation in Britain during Holocene times. *Geophys. J. R. astr. Soc.*, 1981, *65*, 703–725.

Turner, G.M. and Thompson, R. Detransformation of the British geomagnetic secular variation record for Holocene times. *Geophys. J. R. astr. Soc.*, 1982, *70*, 789–792.

Ulrych, T.J. Maximum entropy power spectrum of long period geomagnetic reversals. *Nature*, 1972, *235*, 218–219.

Ulrych, T.J. and Bishop, T.N. Maximum entropy spectral analysis and autoregressive decomposition. *Rev. Geophys. Space Phys.*, 1975, *13*, 183–200.

Ulrych, T.J. and Clayton, R.W. Time series modelling and maximum entropy. *Phys. Earth. Planet. Inter.*, 1976, *12*, 188–200.

Valencio, D.A., Creer, K.M., Sinito, A.M., Vilas, J.F.A., Mazzoni, M.M., Spalletti, L.A., Romero, E.J. and Fernandez, C.A. Estudio paleomagnetico, sedimentologico y palinologico de ambientes lacustres. *Asociacion Geologica Argentina, Revista*, 1982, *37*, 183–204.

Verosub, K.L. and Cox, A. Changes in the total magnetic energy external to Earth's core. *J. Geomag. Geoelectr.*, 1971, *23*, 235–242.

Verosub, K.L., Davis, J.D. and Valastro, S.Jr. A paleomagnetic record for Pyramid Lake, Nevada, and its implications for proposed geomagnetic excursions. *Earth Planet. Sci. Lett.*, 1980, *49*, 141–148.

Vitorello, I. and Van der Voo, R. Magnetic stratigraphy of Lake Michigan sediments obtained from cores of lacustrine clay. *Quaternary Research*, 1977, *7*, 398–412.

Wahba, G. and Wold, S. A completely automatic french curve fitting spline function by cross–validation. *Comm. Statist.*, 1975, *4*, 1–17.

Wold, S. Spline functions in data analysis. *Technometrics*, 1974, *16*, 1–11.

Yale, C. and Forsythe, A.B. Winsorized regression. *Technometrics*, 1976, *18*, 291–300.

Yukutake, T. Free decay of non–dipole components of the geomagnetic field. *Phys. Earth. Planet. Inter.*, 1968, *1*, 93–96.

Yukutake, T. Review of the geomagnetic secular variations on the historical time scale. *Phys. Earth. Planet. Inter.*, 1979, *20*, 83–95.

Yule, G.U. On the method of investigating periodicities in disturbed series, with special reference to Wolfer's sunspot numbers. *Phil. Trans. R. Soc.*, 1927, *A226*, 267–298.

5.
Epilogue

By K.M. Creer and P. Tucholka,
University of Edinburgh, UK.

5.1. Respective roles of lake sediments and baked clays as recorders of the ancient field direction

Lake sediments are capable of providing virtually continuous records of the ancient geomagnetic field and since lakes are to be found in several regions of most continents, good geographical coverage is possible. However, these palaeomagnetic records are always somewhat smoothed and attenuated due to the processes by which sediments acquire their magnetization. Nor is it easy to obtain true zeros for either declination or inclination since sediment cores are not usually orientated. On the other hand, archaeological materials provide 'spot' values of the ancient field and, with care, fully orientated samples can be recovered. The snag here is that it is possible to obtain only sporadic coverage in both space and time, since civilizations flourished within strictly limited geographical regions and for limited periods of time. Even at those sites for which there is evidence of occupation over extended periods, there may be some uncertainty as to whether this was continuous or broken. Thus, palaeomagnetic data from lake sediments and archaeological data from human artefacts are, in many respects, complementary to one another and they may be knitted together to produce a more complete picture of the behaviour of the ancient field over recent millennia than can be obtained from either source alone. In particular, archaeomagnetic results may be used to correct the amplitudes and also to establish the lines of zero declination and ADF inclination of the SV recorded by lake sediments. Similarly if, in future, a method is ever devised to allow the recovery of reliable palaeomagnitude records, it is unlikely to yield absolute values of the

COMPARISON OF ARCHAEOMAGNETIC WITH *PALAEOMAGNETIC* DATA

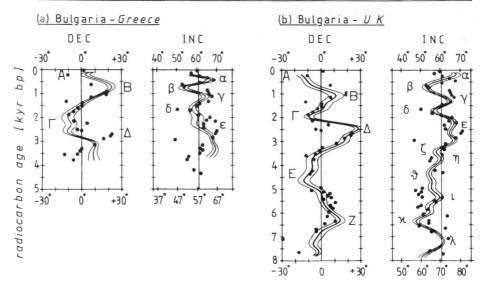

Figure 5-1: Calibration of palaeomagnetic records: (a) stacked curves for Lake Trikhonis, Greece (Creer et al. 1981), 0–4500bp; (b) UK type-curves (see section 4.2 of this volume and Turner and Thompson [1981]), 0–9000bp. The points in both (a) and (b) represent archaeomagnetic data for SE Europe obtained by Kovacheva(1980) and described in section 3.6 of this book. The palaeomagnetic records have been (i) amplified to make the average deflection equal to that of the corresponding archaeomagnetic curve over the same time range and (ii) shifted laterally so as to adjust the mean declination and inclination to the corresponding archaeological values (see Table 5-1). The scale at the top (base) of the inclination plots applies to the archaeomagnetic (palaeomagnetic) results.

ancient intensity, and calibration against archaeomagnetically determined values will again be necessary.

Let us first compare the palaeomagnetic records for Lake Trikhonis, Greece (Creer et al. 1981) with archaeomagnetic results for SE Europe (Kovacheva [1980] and section 3.6 of this book). The size of the region from which the archaeomagnetic and palaeomagnetic samples were taken is sufficiently small (3° of longitude by 5° of latitude approximately) that we should expect the long period secular variations (with which we are concerned in the context of this book), to have occurred virtually synchronously everywhere within it). The stacked declination and inclination curves for Lake Trikhonis, constructed by fitting cubic splines with 21 knots, are shown in

Figure 5-1a and may be compared with the archaeomagnetic results which are represented by dots. (Note that the choice of the optimum number of knots is not critical over quite a wide range – see section 5.4). Ages are given in radiocarbon years. The declination and inclination scales are absolute, the mean palaeomagnetic values having been shifted to the corresponding mean values defined by the archaeomagnetic results over the same time range. The scale at the top of the inclination plot applies to the archaeomagnetic data and that at the base to the palaeomagnetic data and the vertical line shows the ADF inclination (61° for Bulgaria, 58° for Greece). Also the amplitudes of the palaeomagnetic curves have been adjusted by increasing the average deflection to equal the average archaeomagnetic deflection. Declination and inclination were treated separately, using Gaussian statistics because it was found that different amplification factors were required – see Table 5-1. The lateral shifts applied to the Greek palaeomagnetic data correspond to the mean declination and inclination calculated for the archaeomagnetic data over the last 4000 years for which alpha 95 = 2.2° for N = 27.

Table 5-1:   Calibration of Greek and UK palaeomagnetic curves.

parameter	correction to palaeomagnetic data	
	Trikhonis (0 – 4000 bp)	Windermere/Lomond (0 – 8000 bp)
Dec. factor[+]	1.5	3.8
Inc. factor	1.9	1.6
Dec. shift[*]	2.7°	1.8°
Inc. shift	0.0°	-2.2°

[+] Factors to normalize palaeomagnetic data to the archaeomagnetic (Kovacheva, 1980) over the dated time span

[*] Shift applied to the mean palaeomagnetic direction after amplitude normalization. Latitude difference of sites has been allowed for.

We have carried out a similar comparison between the UK palaeomagnetic declination and inclination curves with the SE Europe archaeomagnetic data. In this case we must take account of the fact that the ADF inclination in UK is ~71°, some 10° steeper than in the region where the archaeomagnetic studies were made and that UK is situated some 20° to the west so we should expect secular variations arising from westward (eastward) drifting sources to occur later (earlier) in UK than in SE Europe (by some 130 yr for a drift rate of 0.15 deg/yr which is approximately the value recorded historically). The UK palaeomagnetic declination and inclination type-curves are plotted in Figure 5-1b and the shifts and amplification factors applied are listed in Table 5-1. Note that throughout the last 8000 years, declination has been biased towards the east by some 1.8° and that inclination has shown a bias to shallow values by some 2.2° these values again being defined by the mean of the population of archaeomagnetic directions for which alpha 95 = 2.2° and N =44.

There appears to be no systematic phase difference between the times at which the maxima and minima appear along the SE European and UK records. At about 1000bp the SE European declination peak labelled `B` and inclination peak labelled `γ` in Figure 5-1b lag behind the corresponding UK peaks which is consistent with eastward drifting sources. However at about 2500bp the SE European easterly maximum `Δ` occurs before the corresponding UK maximum and this is consistent with westward drifting sources - on the other hand, the inclination maximum `ε` occurs at about the same time on the records for both geographical locations. An earlier easterly maximum on declination `Z` is recorded at both locations between about 5000 and 7000bp, with the peak for SE Europe appearing later than for UK. It is possible that these discrepancies arise from inaccuracies in the respective time scales which were derived quite independently for the archaeomagnetic and palaeomagnetic data sets.

5.2. Geomagnetic objectives of palaeomagnetic and archaeomagnetic research

There are two main reasons why the objectives set for any interpretative study of lake sediment or baked clay magnetic data must necessarily be limited to an investigation of the more basic and fundamental properties of the geomagnetic field and its longer period time variations. First, the number of regions for which type-curves have, as yet, been constructed (or for which they are likely to be constructed in the near future) is extremely small as compared with the substantial number of geomagnetic observatories where the recent historic field has been recorded. Furthermore, MAGSAT data, which have become available over the last decade, are of higher quality and provide much better global coverage even than observatory data, resulting in a marked improvement in the definition of the detailed topography of the field for the last decade or so. Thus we can never expect to reach the stage where we will be able to carry out sophisticated analyses on pre-historic data such as those routinely applied to recent data. Next, the time control of lake sediment derived SV type-curves is based, at best, on radiocarbon ages which are subject to substantial random and systematic errors, rather than on reliable and accurate mechanical or electronic clocks. Thus, we cannot expect to do more than contribute to answering some very basic questions such as the following.

    i. Are the spatial and temporal variations of the geomagnetic field, as observed in detail for the last few centuries, typical of its behaviour over many millennia?

    ii. What has been the relative importance of the dipole as compared with the non-dipole parts of the field as sources of long-period secular variations?

iii. Noting that it has been found useful to separate the historic NDF into 'standing' and 'drifting' parts, the question arises as to whether we can identify similar 'standing' and 'drifting' NDF foci for the pre-historic

field. Realizing that all foci of the NDF show evidence of both movement and of waxing or waning of their magnitudes, let us define a 'standing' source as one for which the associated SV is dominated by the contribution due to changes in its magnitude and let us define a 'drifting' source as one for which the associated SV is dominated by the contribution from its motion.

iv. What is the typical lifetime of any given NDF source? In particular, have any individual sources persisted long enough to have passed beneath a fixed point on the surface several times as they continued to drift to the west (or east)?

v. How uniform has the rate of drift been (a) over the global surface at any given time and (b) through time as recorded at any given place? Are the historically observed drift velocities typical samples from the overall range of values through extended time intervals?

vi. Have there been 'epochs' of predominantly westward (or eastward) drift? If such 'epochs' of drift have, in fact, occurred, was the sense of drift the same for all the NDF sources which existed at that time?

vii. Are the characteristic times of drift of NDF foci shorter, longer or about the same as those of their fluctuations in magnitude?

viii. Is there any reliable evidence that the field strength or direction changed abnormally rapidly in the past (as compared to historically observed rates of change), noting that only data from the same stratigraphic or occupational level should be used to estimate rates of change?

ix. Do 'excursions' constitute real geomagnetic phenomena? If so are they to be considered as 'aborted' reversals or as large amplitude SV? To answer these questions satisfactorily, it is necessary to show (a) that any observed aberrant palaeomagnetic directions have a geomagnetic origin and (b) to determine the geographical extent over which any proposed 'excursion' is to be observed. Thus, tighter dating control than is currently available is an essential requirement.

5.3. Basic characteristics of synthetic SV produced by some simple models

One approach to finding answers to some of these questions is to examine the characteristic properties of secular variations synthesized from simple computer models. Throughout, it will be assumed that the time varying magnetic fields originating from the postulated sources are not attenuated or distorted as they pass through the liquid core because it is not clear how these effects should be allowed for.

5.3.1. Precession of Main Dipole

Precession of the main dipole may be simulated by allowing its equatorial component (GED) to rotate with uniform angular velocity about its axial component (GAD). The resulting time variations of declination, inclination and intensity are shown in Figure 5-2a. The time-scale is marked in units of T, the period of precession of the resultant inclined dipole. These particular curves correspond to the SV which would be recorded by an observer located at latitude 45°N for a model with m/M = 0.50 where m and M respectively are the moments of the GED and GAD. The curves exhibit the following characteristics.

    i. The declination and inclination oscillations are out of phase by T/4.

    ii. The form of the declination oscillations is not symmetrical between the zeros and the extreme values.

    iii. The amplitudes of the inclination minima (as measured from the ADF value) are larger than those of the maxima.

    iv. The VGP path is circular and centred on the geographic pole (see Figure 5-3a).

    v. The sense of motion around the VGP path, with advancing time, is clockwise for westward rotation and anti-clockwise for eastward rotation of the GED.

    vi. The relative amplitudes of the declination and inclination anomalies depend on the latitude of the observer.

    vii. Using <u>unit</u> vectors, the inclination, I_u, averaged over a whole number of rotations of the GED, is always shallower than the ADF inclination, I_o, for the site. Similarly, the time averaged intensity, f_u, is too great. It is necessary to use <u>total</u> vectors to obtain the <u>true</u> average inclination, I_t, and intensity, f_t, (Creer, 1983), as demonstrated in Table 5-2. Although the VGP path is circular the time averaged VGP, calculated using unit vectors, will always be 'far-sided' because of the phase relationship between intensity and inclination.

5.3.2. Drifting radial dipoles of fixed moment

The drifting part of the non-dipole field, as computed for example by Yukutake and Tachinaka (1968), has an approximate quadrupolar symmetry (see Creer, 1981). It may thus be modelled for any given epoch, again very approximately, by four radial dipoles (RD) separated by 180° of longitude and located at mid-latitudes. Two of these RD would be situated under the northern hemisphere, one pointing up and the other pointing down. The other two would be located in the southern hemisphere and each one would be of opposite sign to its companion at the same longitude in the

Table 5-2: Unit and total vector statistical parameters for Secular Variation
Models
(Inclination and Field Magnitude)

Source		m/M	I_u	I_t	ΔI	f_u	f_t	Δf
dr	GED	0.50	60.2°	63.4°	-3.2°	0.4560	0.4278	+0.0282
dr	GED	0.14	62.8°	63.4°	-0.6°	0.4332	0.4278	+0.0054
dr	RD(up)	-0.20	60.7°	63.5°	-2.8°	0.3665	0.3530	+0.0135
dr	RD(down)	+0.20	62.1°	63.2°	-1.1°	0.5093	0.5025	+0.0068
dr	RD(pair)	±0.20	57.2°	63.4°	-6.2°	0.4610	0.4278	+0.0332
dr	RD(72)	±0.001	58.7°	63.4°	-4.7°	0.4447	0.4278	+0.0169
osc	RD(0°E)	±0.20	57.4°	63.4°	-6.0°	0.4440	0.4278	+0.0162
osc	RD(30°E)	±0.20	60.5°	63.4°	-2.9°	0.4376	0.4278	+0.0098
osc	RD(30°W)	±0.20	60.5°	63.4°	-2.9°	0.4376	0.4278	+0.0098
osc	RD(ph90°)	±0.20	56.9°	63.4°	-6.5°	0.4487	0.4278	+0.0209
osc	RD(ph135°)	±0.20	61.8°	63.4°	-1.6°	0.4461	0.4278	+0.0183

dr	=	drifting sources of fixed moment
osc	=	standing, oscillating sources
GED	=	geocentric equatorial dipole, moment m
RD	=	radial dipole, moment m, at r = 1750 km
RD72	=	line of radial dipoles, at r = 3400 km,
		moment per degree of longitude m = 0.001 M at r = 3400 km
I_o, I_u, I_t	=	inclination (ADF value, unit vector mean,
		total vector mean respectively);
f_o, f_u, f_t	=	corresponding intensities
		where I_o = 63.4° and f_o = 0.4278
ΔI	=	$I_u - I_u$; $\Delta f = f_t - f_u$
M	=	Moment of geocentric axial dipole (GAD)

northern hemisphere.   The SV in declination, inclination and intensity produced by
westward drift of a single RD pointing upwards is illustrated in Figure 5-2b and by a
single westward drifting RD pointing downwards in Figure 5-2c. The particular curves
illustrated are for both RDs located at latitude 45°N with m/M = 0.20 (m, M being
respectively the moments of each RD and of the GAD) and with r = 1750km: the RDs
are located deep in the outer core, like those with which Alldredge and Hurwitz (1964)
modelled the geomagnetic field for epochs 1945AD and 1955AD. The time-scale is
marked in units of the period of one complete revolution of RD drift relative to the
surface: for example, if the drift rate were 0.072deg/yr, each RD would pass beneath
any given meridian periodically at intervals of T = 5000 years, provided its life-time
exceeded T and if the drift rate remained constant.   Figure 5-2d shows SV curves for
the combined effect of both RDs. Note that in all examples, attenuation due to the
conducting core has been ignored.   The curves have the following characteristics.

i. The shape of the declination anomalies due to a single RD is not
symmetrical between the peak values and the zeros on either side.

ii. The shape of declination anomalies due to the combined effect of both

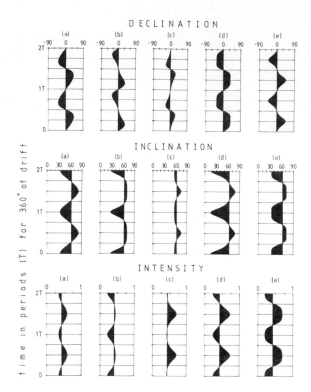

Figure 5-2:   Drifting RD sources: computer simulated secular variations (SV)
in declination, inclination and intensity resulting from westward drift of
(a) a geocentric equatorial dipole (GED), (b) a single radial dipole (RD)
pointing upwards, (c) a single RD pointing downwards. The observer and the RDs
are situated at latitude 45°N and the RDs are located on opposite meridians.
The ratios of the moments of the GED and of the RDs respectively to the
moment M of the GAD are m/M = 0.50 and 0.20. The curves in columns (d) and (e)
illustrate how the SV patterns produced by point sources differ from those
produced by distributed sources: model (d) consists of the combination of the
pair of RDs of columns (b) and (c); model (e) consists of a line of RDs in
two blocks each spanning 180° of longitude around 45°N latitude,
one block pointing up and the other down, located near the core surface at
r = 3400km. The curves are calculated for
m/M = 0.00086 per degree of longitude.

RDs is flat-topped.

iii. The magnitudes of the inclination anomalies produced by the upward-
pointing RD are larger than those produced by the downward pointing
RD.

iv. The shapes of the inclination maxima and minima due to the pair of RD
are pointed, and the minima are of larger amplitude than the maxima.
These unequal magnitudes are due to the form of the tangent function.

v. There is a phase difference of T/4 between the times at which the
extreme values of declination and inclination occur.

vi. The VGP loop for the single upward pointing RD is 'far-sided' and that for the downward pointing RD is 'near-sided' as one would expect (Figures 5-3b and c). [Note that the opposite is true for southern hemisphere sites and RD: see <viii>].

vii. The VGP loop for the pair of RDs is also 'far-sided (see Figure 5-3d).

viii. For the corresponding situation in the southern hemisphere, an observer would compute 'near-sided' VGP loops (for the N pole) for the pair of drifting RDs and also for the pair of lines of RDs located under 45z[o]S or any southern latitude.

ix. The sense of motion around the VGP paths is again clockwise for westward drift and anti-clockwise for eastward drift.

x. For any axial symmetrical distribution of sources, it is necessary to use total vectors rather than unit vectors to calculate the time averaged field inclination and intensity (see Table 5-2).

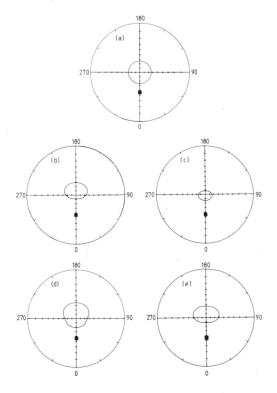

Figure 5-3: Drifting sources: VGP paths corresponding to SV producedby models described in caption to Figure 5-2 a-e. Location of observer indicated by dot at 45°N 0°E.

5.3.3. Drifting current–loop sources of fixed moment

Alldredge and Hurwitz (1964) found that they had to locate the eight RDs with which they modelled the 1945AD and 1955AD geomagnetic fields deep in the outer core to obtain the best fit between the model field and the observed field. It has often been suggested, on the basis of skin–depth arguments, that the sources of non–dipole field and its SV must be located within the top few hundred kilometres of the outer core. Although such arguments cannot strictly be valid since they apply to solid conductors while the core is fluid, the conclusion that the sources of SV must lie near the core surface has persisted largely because it is not clear whether even low frequency ($\sim 10^{-2}$ to 10^{-3} yr^{-1}) electro–magnetic signals might be propagated over long distances in the liquid core and still remain recognizible. We can avoid the problem of attenuation, at least for sources on the same side of the core as the observer, by replacing each RD by an equivalent electric current–loop which, by virtue of its geometry, would have to be located at rather shallower depth beneath the outer–core surface to produce a comparable effect (Peddie, 1979).

The patterns of SV of D and I produced by westward drift of two distributed sources, simulated in this case by a line of radial dipoles consisting of two blocks, arranged end to end, each extending through 180° of longitude around a line of latitude (45°N), are shown in Figure 5–2e. These curves exhibit the following features which distinguish them from the corresponding curves due to deep–seated RD sources.

    i. The shape of the declination anomalies is pointed .

    ii. The inclination maxima and minima are flat–shaped with the minima having rather larger amplitudes than the maxima.

    iii. The phase shift of T/4 between the extreme values of declination and inclination (which is characteristic of drifting sources) again results in a looped VGP path with the sense of looping (clockwise or anti–clockwise) depends on the direction of drift.

    iv. Distributed sources can be distinguished from point sources by the shapes of the declination and inclination ancmalies. These diffrences in the shapes of the D,I and J curves are diminished however, when the signals are smoothed (as palaeomagnetic signals always must be).

    v. Again, use of unit vectors rather than total vectors will cause errors in computations of time–averaged field inclinations and intensities (see Table 5–2).

5.3.4. Oscillating or pulsating radial dipoles fixed in position

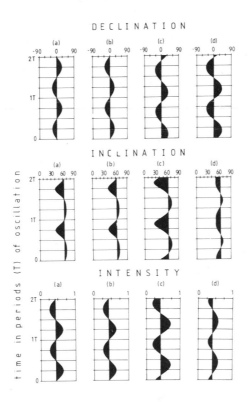

Figure 5-4: Standing sources: SV in declination and inclination produced at 45°N, 0°E
(a) by a single RD (m/M = 0.20) located under 45°N 30°W; (b) by the same RD
under 45°N, 30°E. Both are at r=1750km and oscillate with a period of 5000yr.
Curves (c) and (d) are due to the combined effect of this pair of RD
with the source at 30°E lagging behind the 30°W source by 90°
and 135° respectively.

In the context of "standing' sources, secular variations may be introduced, most simply, by regular oscillations, with period T, of the moment m , and in this case the sign of the associated NDF anomaly will change periodically. Anomalies which simply grow and decay in magnitude, maintaining their positive or negative polarity, may be represented by pulsating sources which may be modelled by RDs, the moments of which have an oscillating part m_o and a steady (bias) part, m_b with $m_b \geqslant m_o$.

The SV produced by single standing oscillating RD sources is shown in Figure 5-4a and b. The observer is located at 45° N and longitude 0° E. The parameters specific to the models are given in the caption to the figure. The combined effect of two identical RD sources located on either side of an observer, oscillating out of phase, is illustrated in Figure 5-4 c and d. In case (c) the oscillations of RD(b) lag behind those of RD(a) by 90°, and in case (d) by 135°. The curves exhibit the following characteristics.

i. The declination anomalies are rounded in shape.

ii. The inclination maxima are flat-shaped but the minima are pointed, and of larger amplitude than the maxima.

iii. Declination and inclination oscillate exactly in-phase or out-of-phase with one another, depending on source-observer geometry (counting east declinations and high inclinations positive and west declinations and low inclinations negative).

iv. These properties, computed for an oscillating source, would still hold for any single standing source which pulsated, i.e. which grew and decayed maintaining the same polarity, either repeatedly or once only.

v. The VGP paths resulting from oscillations of single sources will always consist of straight lines as shown in Figure 5-5 a and b: one might draw an analogy with plane polarized light.

vi. The time averaged directions for a whole number of complete oscillations computed using unit vectors, will deviate from ADF values: the mean declination, D_u, will not equal zero and the mean inclination, I_u, will be too shallow. Using total vectors, the mean directions correspond to the ADF values (over a whole number of oscillations).

vii. The combined effects of two sources, oscillating with the same period will produce open, loop-shaped VGP plots as illustrated in Figure 5-5 c and d, corresponding respectively to the cases of circularly and elliptically polarized light.

viii. Again, if time averaged directions are computed using unit vectors, I_u will be shallower than the ADF value, I_o, but D_u will be found to equal zero if the amplitudes of the moments of the two sources are equal. Using total vectors, the mean directions correspond to the ADF values (over a whole number of oscillations) (see Table 5-2 and Creer (1983)).

ix. In the more general case, with two oscillating sources one might expect to obtain VGP paths having the appearance of Lissajous figures.

x. The sense of motion along such VGP paths may be either clockwise (right elliptically polarized) or anti-clockwise (left elliptically polarized) depending on the sign of the phase difference between the oscillations of the sources.

xi. If the sources pulsate rather than oscillate, and this may be a more realistic assumption geomagnetically, the VGP paths should look like short lengths of Lissajous figures. This conclusion will also hold for the case of a source which grows and decays once only.

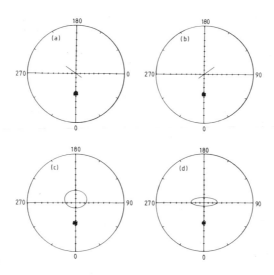

Figure 5-5:   Standing sources: VGP paths corresponding to SV
illustrated in Figures 5-4 a-d.

Thus, interference between the growth and decay patterns of two or more standing sources must contribute to rotation of the geomagnetic vector (which is usually detected by curvature of VGP paths). The currently popular interpretation of the sense of curvature (see many of the contributions in Chapters 3 and 4 of this volume, for example) in terms of drifting geomagnetic sources is not unique and it is conceivable that circumstances might exist, not uncommonly, for two or more sources which grow and then decay as they drift west, to yield VGP paths which exhibit counter-clockwise loops, if the influence of the changing variations in their intensity were to outweigh the influence due to that of their movement.

5.3.5. Spatial variation of SV due to a standing oscillating or pulsating source

The relative amplitudes of the declination and inclination variations, produced by pulsations or oscillations of a standing RD source, will clearly depend on source-observer geometry. Figure 5-6 provides a particular illustration of this point, showing sets of declination and inclination curves which would be recorded around 45°N latitude due to an oscillating RD located under 45°N, 0°E.

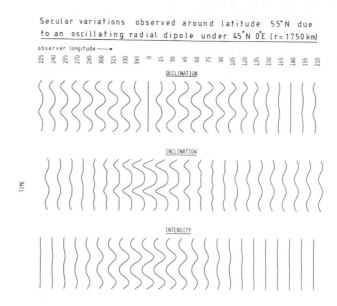

Secular variations observed around latitude 55°N due
to an oscillating radial dipole under 45°N 0°E (r = 1750 km)

observer longitude ——→

Figure 5-6: Standing sources: differences in the form of the SV produced
by a single oscillating RD source located at different longitudes around
45°N latitude. Observer is located at 45°N, 0°E. Other parameters
as given in the caption to Figure 5-4.

Note that observers located on either side of the source meridian will record
variations in declination which are apparently out-of-phase. Maximum amplitudes in
declination (for the model under discussion where the source is located near the
bottom of the liquid core) will be observed to occur at about 50° to the east and west
of the source (assuming no attenuation in the core or mantle) while maximum
inclination amplitudes occur when the source occupies the same meridian as the
observer. Note that, depending on distance from the source, some observers will see
large amplitude inclination variations and small amplitude declination variations while
others will see large declination variations and small inclination variations. Creer and
Tucholka (1982c) discuss this point in their treatment of European and North
American results.

5.3.6. Drifting sources of variable moment

Having examined, in the forgoing sections (5.3.1 - 5.3.5), the properties of the
two simplest types of source of SV with only one time variable parameter, i.e. either
moment or position, let us now examine the general properties of RD sources which
vary in magnitude while they drift, again restricting the discussion to the case of
purely azimuthal drift and neglecting attenuation by the liquid core and by the mantle.

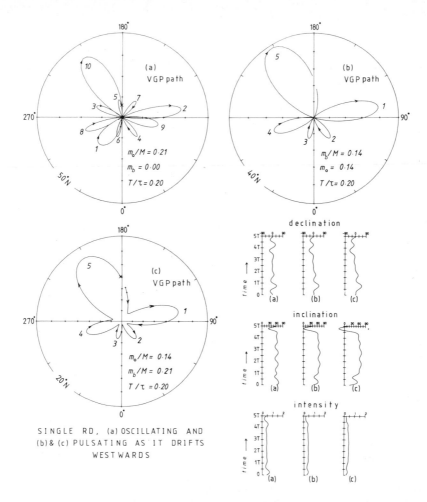

Figure 5–7: Drifting and Fluctuating Sources with $T/\tau < 1$:
SV in direction and field intensity and corresponding VGP paths for:
(a) an oscillating and westward drifting RD source with $m_o/M = 0.21$, $m_b = 0.00$, $T/\tau = 0.20$;
(b) a pulsating and westward drifting RD source with $m_o/M = m_b/M = 0.14$, $T/\tau = 0.20$;
(c) a pulsating and westward drifting RD source with $m_o/M = 0.14$, $m_b/M = 0.21$, $T/\tau = 0.20$.
The records cover a time τ, i.e. the time taken for one
complete revolution of drift relative to the surface.
m_o and m_b are respectively the oscillating and the bias parts of the moment of the RD,
M = moment of the GAD, T = period of oscillations of m_o. The time scale along the direction
and intensity plots is in units of T.
Note that 5T = τ and one complete revolution of drift is illustrated.

The SV pattern produced will depend essentially on the relationship between the characteristic time of variation of the strength of the magnetic moment of the RD, i. e. the period T of oscillation or pulsation, and τ, the time taken for one complete revolution of drift of the RD source relative to the surface. The case, $\tau \gg T$ will approximate to that of a drifting source of fixed strength and the case $\tau \ll T$ will approximate to that of a standing source which oscillates or pulsates.

Let us first consider the case $\tau > T$ in which the waxing and waning of moment occurs more rapidly than drift. Let the RD moment m consist of two parts, an oscillating component m_o and a steady (bias) component m_b. The case of an oscillating source would then be modelled by setting $m_b = 0$. If τ/T is an integer (=n), the VGP path will consist of 2n petals and a particular case for n=5 is shown in Figure 5-7a. The amplitude of each petal will depend on the position of the source relative to the observer when it attains a maximum (positive or negative) intensity. For a pulsating source, we set $m_b > m_o$ and in this case the pattern will exhibit n petals i.e. half as many as for the corresponding oscillating source, as shown in Figure 5-7, (b) for the special case of $m_o = m_b$, and (c) for $m_b = 1.5 \, m_o$. In both cases illustrated, n = 5. Obviously, in the general case, when the ratio τ/T is not an integer, the pattern will not repeat itself exactly on successive revolutions of drift. The sense of curvature around the loops follows the Runcorn(1959) rule, but the order in which the petals appear is complex and may even be the opposite of that predicted for some oscillating source parameters (NB: an interesting analogy is the apparent retrograde rotation of the wheels of vehicles in the early (silent) cinema movies). The order in which the petals occur is shown by the italicized numbers in Figure 5-7.

In Figure 5-7, the SV in declination, inclination and intensity for the three specific sets of parameters are illustrated. The particular values used for the adjustable parameters of the model were chosen so that the amplitudes of the variations should be typical of the actual geomagnetic SV, as observed historically. This will be seen to be so, but it will also be observed that there is a sharp 'excursion' of inclination to negative values near the top of the record. Such 'excursions' should be expected to occur whenever a source attains maximum moment at about the time when it passes beneath the longitude of the observer. The duration of such an 'excursion' will be of the order of the 'time constant', T, associated with its pulsations. Setting $\tau = 2000$yr (which is comparable with the historically observed rate of westward drift), and noting that $\tau/T = 5.0$ in our particular models, we find that T = 400 yr which is of the same order as the times characterizing the growth and decay of NDF foci as observed for recent geomagnetic epochs. Thus, the models presented here reproduce some of the basic features exhibited by the secular variations of the historic field and furthermore, they predict that the occasional occurrence of so-called geomagnetic 'excursions' should not be regarded as unexpected although none have been observed during the last few centuries for which instrumental records are available. For the case of $\tau \leqslant$ T, the VGP path traces out a rolling motion, as shown in Figure 5-8.

Although, in this section, radial dipoles and electric current loops have been used to model geomagnetic secular variations, they cannot be taken as providing a correct physical interpretation of the processes which actually occur in the liquid core since the geomagnetic field and its secular variations must originate from

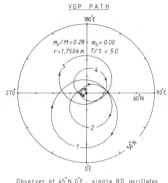

VGP PATH

Observer at 45°N 0°E ; single RD oscillates
with period, T, as it drifts west ('period'
= τ), in the equatorial plane

Figure 5-8:   Drifting and Fluctuating Sources with T > τ :
oscillating and westward drifting RD source located in the equatorial plane
with m_o/M = 0.28, m_b = 0.00, T/τ = 5.0. The length of the illustrated
record = τ which is the time taken for one revolution of drift relative to
the surface. m_o and m_b are respectively the oscillating
and bias parts of the RD moment and M is the moment of the GAD.
T is the period of oscillation of m_o.

magnetohydrodynamic processes. However, we cannot 'see' inside the core: we can
only observe the field which exsists outside the core. Conventional geomagnetists had
used geocentric multipole and eccentric dipole models to describe the historic field as
observed over the last century and a half, and they continue to do so even though it is
universally accepted that such sources have no physical reality. Thus we think it is
useful also, to use radial dipoles to model the palaeofield and its secular variations.

5.4. Palaeomagnetic SV type-curves

5.4.1. Construction of standardized curves

In Chapter 4, the main results of palaeomagnetic studies on sediment cores from
four regions, well separated geographically, are outlined. The resulting data have
been presented in the form of type-curves depicting SV in declination and inclination,
through post-Glacial time, for the four respective regions, W. Europe – Turner and
Thompson (1981), N. America – Creer and Tucholka (1982a), Australia – Barton and

McElhinny (1981) and W. Argentina – Creer et al. (1983). Although these type-curves were constructed using the same general procedures (inter-core correlation, smoothing, filtering and stacking of data), these steps were not necessarily carried out in the same order, nor were the same analytical methods used by the different research groups that made the original studies. Thus any comparison of the statistical reliabilities of the respective data sets as represented, for example, by error-bars at each time horizon or by lines drawn on either side of the type-curves at some specified distance (e.g. one or two standard-errors) is not a straightforward matter. In section 8 of Chapter 4, Clark describes a rigorous approach to the problem of curve fitting. There is clearly some advantage in adopting, as standard, a procedure which employs the best of modern statistical methods. Therefore we have recalculated the type-curves for all four regions following Clark's approach. Since the quality of declination records is usually much lower than that of inclination records, (for purely technical reasons associated with taking sediment cores), final stacked curves have been produced separately for declination and inclination.

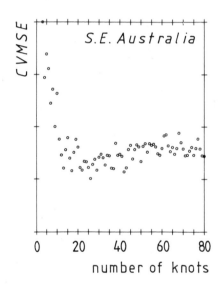

Figure 5-9:   CVMSE parameter of Clark plotted against the number of knots used to construct cubic spline fits to the inclination data for S.E. Australia.

We have preserved the original measured data points through the whole process of time-scale alignment and stacking, in order to avoid unnecessary interpolation. Thus the final curve fitting and statistics were calculated on the data set containing original declination and inclination values for individual samples transferred to the common time scale. The stepwise procedure of time-scale alignment is shown in Figure 4-1 in the introduction to Chapter 4. Following Clark's advice, equally spaced cubic splines were used to estimate the representative curves. There are two obvious advantages in using cubic splines rather than running means for formally defining functions to represent serial arrays of data points: cubic spline fits are less affected by occasional outlying data points and 'robust' methods can be applied for estimating the statistical reliability of the fitted curves. In order to select the optimum degree of smoothing, we attempted the cross validation technique. However, with the large number of points and the high density (1000 to 1600 points in 1000 years) the CVMSE parameter of Clark did not provide a sharp minimum because the fitted curves were found to be very similar for quite a wide range of knot spacings (see Figure 5-9). Therefore we used the pattern of the residuals to estimate the optimum degree of smoothing.

It is important to ensure that none of the signal is lost in the process of curve fitting. This means that the residuals should be randomly distributed about the chosen curve and show no regular pattern. As mentioned above, it turned out that the choice of the number of knots was not critical over a wide range (from about n = 20 to 30), and this can be seen from an inspection of the patterns of residuals having fitted cubic splines with an increasing number of knots (n = 5, 10, 15, 20, 25 and 30) for the Australian inclination and declination data (see Figure 5-10). The data for the other three regions gave a similar result. Our final type-curves , illustrated in Figure 5-11, have been constructed using 29 equally spaced knots (i. e. at ~350 year intervals). The data used to construct these curves are summarized in Table 5-3. It is interesting to note that the patterns of these curves are not noticibly different from those of the stacked data points illustrated in Chapter 4 and previously published in Creer and Tucholka (1982 a, b and c), so the advantage of adopting Clark's approach is not so much that it reveals more information about the pattern of any given type-curve but rather that, being a rigorous method, the error limits are statistically more reliable.

These type-curves, constructed with 29 knots, are considered to present the highest degree of detail obtainable from the data being analysed. As we shall show in the following section, it is, nevertheless, instructive to subject the data to increasing degrees of smoothing with the object of investigating possible correlations between different parts of the spectra of the palaeomagnetic SV records. Thus, for all the results from the four observation regions, we have fitted cubic spline curves with a range of equally spaced knots, from 5 (i. e. at intervals of about 2000yr) to 30 knots (i. e. at intervals of about 350yr). Surfaces have been constructed illustrating the relationship between the curves comprising each family are shown in Figure 5-12. These figures complement Figures 4 and 5 of Creer and Tucholka(1983) in that the surfaces are viewed here from the opposite (n = 29) side.

SE AUSTRALIA

(a) RESIDUALS TO FITTED <u>INCLINATION</u> CURVES

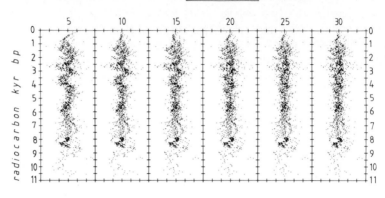

(b) RESIDUALS TO FITTED <u>DECLINATION</u> CURVES

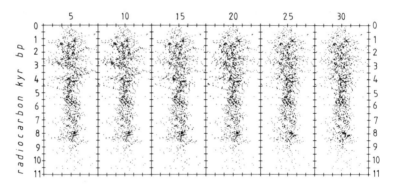

Figure 5-10:   Residuals of stacked Australian (a) inclination and (b) declination data
fitted with cubic splines with an increasing number of equally spaced knots
(n=5, 10, 15, 20, 25 and 30). The horizontal scale is marked at 10° intervals
and the number of knots is shown at the top of each curve.

5.4.2. Correlations between the type-curves

A. Northern Hemisphere

The relationship between the N. American and W. European curves has been
discussed by Creer(1981) and by Creer and Tucholka (1982 b, c).   The inclination

Table 5-3:   Summary of Data used for construction of Type-curves

region	lake	core label	no. cores	no. samples per core	total	Reference
W. Europe (UK)	Lomond	1	4	217	1328	
		2		188		Turner and Thompson (1981)
		3		224		supported by earlier work
		4		199		in UK and extensive
	Windermere	1	3	124		studies in Continental
		2		132		Europe
		3		244		
N. America	Superior	3	2	192	1605	
		4		312		Mothersill (1979, 1981)
	Huron	1	2	379		Banerjee et al., 1979
		2		307		supported by earlier
	St. Croix	B	2	155		studies of Lake Erie and
		C		141		Michigan
	Kylen	A	2	68		
		B		51		
S.E. Australia	Keilembete*			798	1676	
	Bullenmerri*			878		Barton and McElhinny (1982)
W. Argentina	Moreno	2	4	120	603	
		3		165		
		4		160		Creer et al. (1983)
		5		158		

* Records stacked by Barton were used

curves in particular, exhibit a number of similar features in that, when the long wavelength trends have been subtracted, the same number of oscillations are observed along the curves from both sides of the N. Atlantic through post-Glacial time: the peaks of the swings along the N. American curve, are labelled with lower case Roman letters a- n and the corresponding swing peaks along the W. European curve are labelled with lower case Greek letters, $\alpha - \nu$. In general, corresponding swings occur earlier along the W. European curve than along the N. American and Creer and Tucholka (1982c) have shown that the best correlation between the curves is obtained when a phase shift of ~650yr is introduced. This is suggestive of westward drift at an overall average rate of ~0.13deg/yr (equivalent to τ = 2750yr in the models discussed in section 5.3). However, the drift rate has not been uniform because the time lag between successive pairs of peaks (which gives the time taken for a source to drift some 90° across the N. Atlantic), is not constant.

The relationship between the declination type-curves is more complex: while there

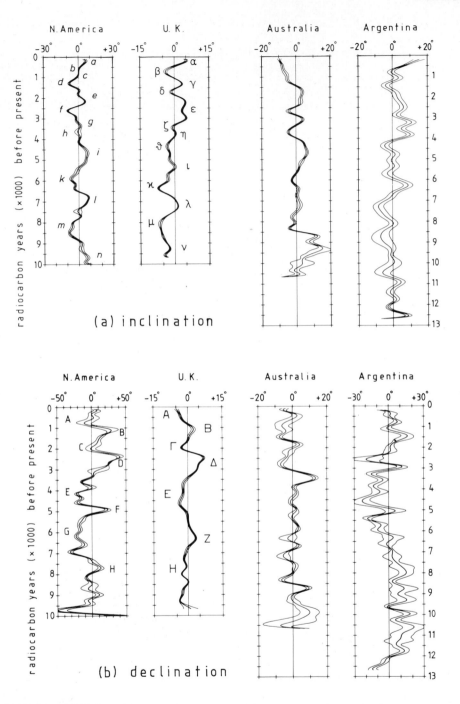

Figure 5-11: (a) Inclination and (b) declination type-curves, constructed by fitting cubic splines with 29 equally spaced knots (corresponding to intervals of ~ 350yr) for the data from N. America, W. Europe (UK), S.E. Australia, W. Argentina.

INCLINATION

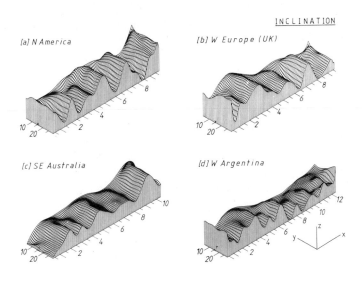

DECLINATION

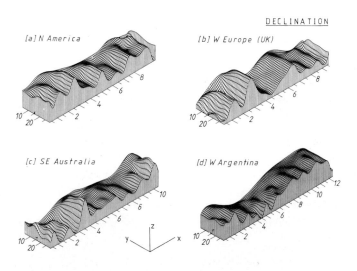

Figure 5-12:   Surfaces formed by a series of type-curves constructed by fitting cubic splines with from 5 to 29 equally spaced knots to the palaeomagnetically recorded SV.

appears to be a very good correlation between swings A, B, C, D and E with A, B, Γ, Δ, and E for zero phase shift for the post- 4750bp data, there is a clear suggestion of an anti-correlation prior to ~4750bp and these observations have been interpreted in terms of standing sources which fluctuate in intensity (Creer and Tucholka, 1982b, c).

This apparent paradox, that a dominance of westward drifting sources is suggested by the inclination results while a dominance of standing sources which fluctuated in intensity is suggested by the declination results, has been discussed by Creer and Tucholka(1982c) who suggested that the explanation could be associated with the source-observer geometry; the paradox will only be resolved satisfactorily with additional data from other northern hemisphere sites which occupy a range of longitudes. Provisionally, five sources, two of them drifting and three standing, were proposed to model the declination and inclination results (considered as a pair) post- 4570bp and four sources, two drifting and two standing were proposed to model the pre- 4750bp SV (see Tables 4 and 5 of Creer and Tucholka 1982c).

Considering the complex character of the historic geomagnetic SV, we should not be surprised to have found that it is necessary to consider fairly complex models to reproduce even the grosser behaviour of the post-Glacial SV. In fact, at first sight, the problem of gaining an understanding of the pre-historic field and its secular variations would appear to be a singularly intractible one, bearing in mind the paucity of data now available (or likely to become available in the near future). However, there is some justification for taking an optimistic view because, intuitively, one might expect to find that the longer period variations (~10^3yr) which are recorded by lake sediments would show coherence over wider geographical areas than the shorter period variations ~10^2yr recorded with instruments through the last few centuries.

It is unlikely that an increased volume of data from either N. America or W. Europe will appreciably modify the form of the existing type-curves as reproduced here, though some improvement to the N. American declination curve should be possible if records from cores taken with Mackereth type corers could be obtained. Improved and more precise dating control will undoubtedly become available, but this will not affect the interrelationships between the declination and inclination curves from a given region and the paradox mentioned above is unlikely to be resolved in this way. Thus, high priority shoul be given to the construction of type-curves of similar quality from other regions which occupy mid-latitudes in the northern hemisphere, that is across Siberia and China and from Japan. This would allow the models provisionally proposed by us to be tested and improved, or if necessary, replaced by better ones.

B. Southern Hemisphere

Type-curves for S. E. Australia and W. Argentina are also shown in Figure 5-11.

No correlatable features are to be observed along the declination nor inclination curves and this is somewhat surprising, considering the evidence from the N. hemisphere curves. A possible reason is that the curves are based on relatively small data sets, as compared with those on which the N. American and W. European curves are based, bearing in mind that the latter curves are supported by results from other studies not included in the data banks used to construct our type-curves. It is also possible that the transformations to a time scale may require modification for one or both of the S. hemisphere regions. Such errors could only be identified by broadening the range of sediment types and environments included in the studies.

Thus, in our opinion, the apparent absence of any correlatable features along the S. hemisphere type-curves should, at present, not be regarded as established. The question as to whether the S. hemisphere SV has been significantly different in character from the N. hemisphere SV will only be answered by further studies at a larger number of sites.

5.5. VGP paths and curvature plots

5.5.1. Description

Virtual geomagnetic pole paths have been computed for the pairs of declination and inclination type-curves, illustrated in Figure 5-12, and constructed by fitting cubic splines with increasing numbers of knots (n = 5, 10, 15 and 20). These VGP plots are shown in Figure 5-13 for (A) N. America, (B) W. Europe, (C) S.E. Australia and (D) W. Argentina.

The VGP paths exhibit increasing degrees of complexity as the number of knots along the respective declination and inclination curves is increased from 5 to 20. Ages, in radiocarbon years, are indicated along the VGP paths to facilitate their comparison. However, beyond n = 15 knots, it becomes difficult to trace the detailed pattern of the paths and we have found that it is easier to compare them by inspecting plots of their curvature. These have been constructed by computing the second derivative of the VGP curves as projected stereographically. (Strictly, the curvatures should be calculated on the surface of the unit sphere, but since we are concerned with the qualitative rather than the quantitative features of the curvature plots, the errors involved in our method of computation are not important.) The curvature plots are shown in Figure 5-14 a-c.

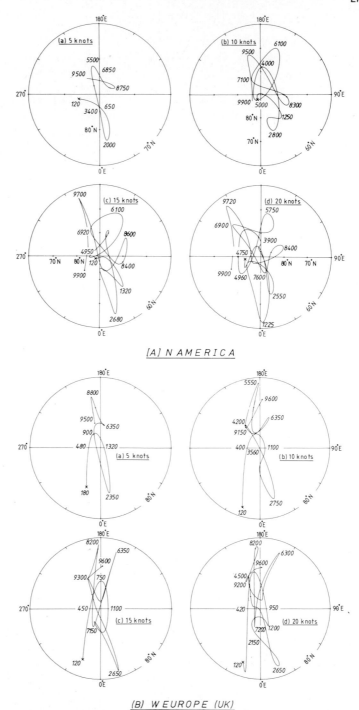

Figure 5–13:   VGP paths corresponding to the declination and inclination curves constructed
by fitting cubic splines with (a) 5 knots, (b) 10 knots, (c) 15 knots,
(d) 20 knots (see Figure 5–12).

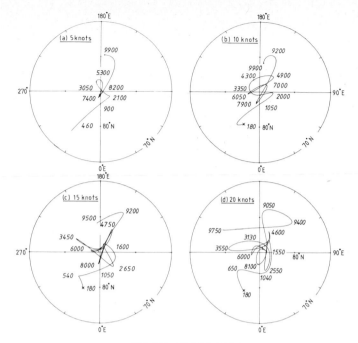

[C] SE AUSTRALIA

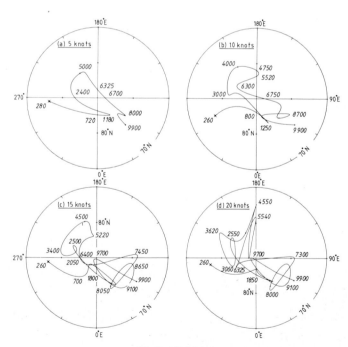

[D] W ARGENTINA

Figure 5-13:   (continued)
Polar stereographic projections: latitudes marked at 10° intervals;
longitudes marked at 30° around the primitive, the latitude of which is shown.

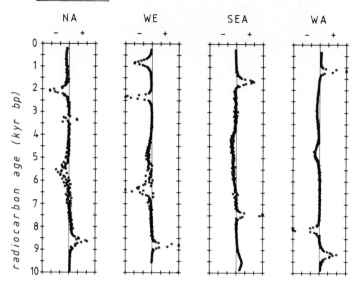

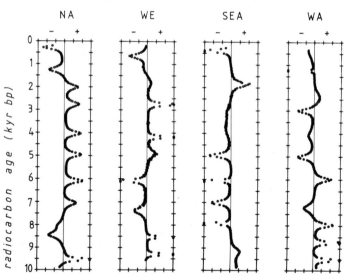

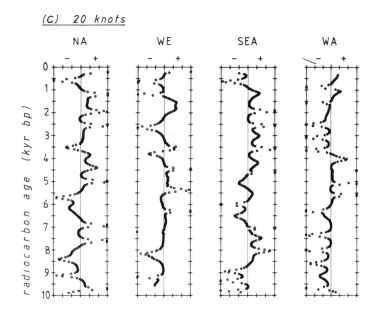

Figure 5-14:

Plots of curvature vs. time for VGP plots corresponding to (a) the 5 knot,

(b) the 10 knot and (c) the 20 knot cubic spline declination and inclination curves

for N. America (NA), W. Europe (WE), S.E. Australia (SEA) and W. Argentina (WA).

Note that negative values correspond to anti-clockwise rotation of the

geomagnetic vector and positive values to clockwise rotation.

5.5.2. Discussion

One of the objectives of studying palaeomagnetic and archaeomagnetic secular variations, is to deduce the relative importance of changes in the orientation and magnitude of the main dipole as against changes in strength and configuration of the NDF in contributing to the whole range of SV which the geomagnetic field undergoes while it is taking on its average axial dipolar configuration. We may do this because the effects of wobble and/or precession of the main dipole should be observed synchronously at all geographical sites, whereas drifting sources should influence the SV observed within a broad latitudinal band within which the phase of the signal would

depend on longitude while fluctuations in intensity of NDF sources would be expected
to have a regional effect only. Intuitively, we should expect the 5 knot curves (Figures
5-13 and 5-14) to provide us with information about variations of the form of the main
dipole field, and the other sets of curves to provide progressively more information
about the non-dipole field variations as the number of knots is increased up to the
optimum value of n = 20 to 30 (see section 5.4).

A. 5 knot curvature plots

The 5 knot curves for N. America and for W. Europe look remarkably similar,
especially from the present back to about 5500bp: the patterns for both regions,
which are separated by some 90° of longitude are rather elongated along an axis
running about 10°E-190°E.   Note that the orientation of one of the standing sources
proposed by Creer and Tucholka (1982c) is aligned along this axis.   The sense of
curvature is almost exclusively counter-clockwise back to about 7000bp. The curvature
plots show a number of sharp maxima which correspond to 'hairpin' bends along the
VGP paths. In section 5.3.6 it is shown that these sharp bends can be produced by
the intensity of a source of the NDF growing, passing through a maximum and then
decaying as it drifts beneath an observer for the case when $\tau > T$, i.e. when the rate
of change of intensity is greater than the rate of drift.   However, there are some
difficulties about such an explanation because the peaks along the curvature plots
occur earlier for W. Europe than for N. America which would suggest westward drift
while the negative curvatures observed since about 7000bp, are suggestive of eastward
drift.   These apparently contradictory conclusions point to the possibility that
interference between at least two fluctuating sources (see section 5.3.3) has made
an important contribution to the observed curvature. Prior to about 7000bp, both lines
of reasoning suggest westward drift.

Although the similarity in the patterns of the N. American and W. European curves
is most striking, there would appear to be one obvious difference in that the cusp at

the top of the W. European curve does not appear on the N. American. Could it be
that it will appear in the near future?

The patterns of the curvature plots for S.E. Australia (SEA) and W. Argentina
(WA) are also similar to one another, though there would appear to be some
discrepancy in the relative ages assigned to these data sets because of the large
apparent differences in time at which corresponding cusps appear.

The patterns of the pair of N. hemisphere curves seems to be quite different to
those of the S. hemisphere pair.

B. 10 knot curvature plots

These are shown in Figure 5-14b. Comparing, first, these 10 knot curves with the corresponding 5 knot curves described in the last section, we note one striking difference: this is the appearance of cusps locating loops with positive curvature, suggestive of a new generation of westward drifting sources. It is reasonable to suppose that these features are associated with sources which have shorter time constants (the interval between knots for these plots is ~1000yr while for the 5 knot curves it is ~2000yr).

There are 10 cusps along the N. American curvature plot and 9 along the W. European. During the last 1500yr, the curvature for both plots is negative (eastward drifting components discussed by several authors in Chapters 3 and 4 ?). There are two cusps for N. America and one for W. Europe. Between 1500bp and 5000bp, there are four positive cusps along both plots (westward drifting sources observed at both sides of the N. Atlantic?) but between 5500bp and 7500bp, two negative cusps occur along the W. European curve while two positive cusps appear along the N. American curve. Creer and Tucholka (1982 b and c) have referred to a 'turning point' in the distribution of geomagnetic SV sources at ~5000bp, before which they proposed that the dominant standing and fluctuating source was orientated along a longitude somewhere in the N. Atlantic (i.e. between the two regions) and after which another similar source appeared, aligned along an azimuth to the same side of both regions.

The S. hemisphere curves both show about equal proportions of positive and negative curvature.

C. 20 knot curvature plots

At 20 knots, the CVMSE parameter of Clark has entered its broad minimum (see Figure 5-9) so these curves can be taken as representing the greatest amount of detail that can be extracted from the data.

The curvature plots for all four regions are more complex and show more cusps than the corresponding plots for 5 or 10 knots. Hence we shall make only some very general remarks. All four curvature plots show intervals of positive and negative curvature, but it would appear that no single sense of curvature occurs at any given time. This indicates that, at least for this part of the spectrum, (one knot every 500yr approximately), the observations cannot be explained simply in terms of sources which are homogeneously drifting in one direction. In fact, interference between standing sources which have waxed and waned in intensity, possibly drifting slowly as they did so, would appear to provide the most plausible explanation of tha available results.

5.6. The Future

In our epilogue, we have concerned ourselves more with the lake sediment side of the investigation of ancient geomagnetic field secular variations than with the archaeological side. This can partly be attributed to the fact that lake sediment work constitutes our own major interest, but there is also some justification for our approach in that the archaeomagnetic aspects have been summed up in our closing discussion to Chapter 3 and by McElhinny's contribution on global analysis of intensities also in Chapter 3.

If we are to point to any single conclusion which can be drawn from the results of the research presented in this book, it must be that a monumental effort will be required to obtain a satisfactory understanding of the long term behaviour of the geomagnetic field and its secular variations. We note that very substantial efforts have, in fact, been applied to the elucidation of some geophysical problems in the recent past. In the geomagnetic context, it would have been impossible to arrive at an understanding of the global pattern of marine magnetic anomalies without the widespread participation of the major oceanographic and geological research institutes to which (by our standards, at least) huge funds were allocated and large numbers of scientists and auxilliary workers were employed. The recovery of a comparably good understanding of the structure of the geomagnetic field through space and time would require an even greater effort and the chances of this coming about are absolutely nil because there are no economic or military reasons for any government funding its scientific institutions to become engaged in such an activity. Thus, we shall have to be content to make the best use of very limited and inadequate resources. This being so, it is important that future work be effectively planned with full international collaboration, concentrating work in the more critical geographical areas, good global coverage being a prime consideration. Also, standardization of methods and techniques is essential so that results from different laboratories can be combined without difficulty.

Another important conclusion is the need to combine archaeomagnetic and palaeomagnetic data whenever it is possible to do so, first because palaeomagnetic data require calibration both in amplitude and azimuth and second because it is highly desirable that we should determine the complete geomagnetic vector, i.e. we need to know both the direction and magnitude of the field if we are to correctly deduce past patterns of the non-dipole field and the symmetry of the ancient field. Since archaeomagnetism is uniquely suited to the determination of ancient magnitudes while, as yet, reliable magnitudes cannot be extracted from sedimentary records, it is of prime importance that archaeomagnetists and limnomagnetists work together in closest harmony. Finally, it should be remarked that, if any single advance were to be made that would improve the effectiveness of our research, it would be in the field of absolute dating, for it is here that some of the biggest uncertainities in the establishment of secular variation type-curves lie.

There is no shortage of problems waiting to be tackled: not only in the areas covered by the contributions to this book which have been mainly concerned with geomagnetic objectives, but also in the application of the results of our research to regional geological correlation and to the resolution of archaeological problems.

REFERENCES

Alldredge, L.R. and Hurwitz, L. Radial dipoles as the source of the Earth's main magnetic field. *J. geophys. Res*, 1964, *29*, 2631-2640.

Banerjee, S,K., Lund, S.P. and Levi, S. Geomagnetic record in Minnesota Lake sediments – absence of the Gothernburg and Erieau excursions. *Geology*, 1979, *7*, 588-591.

Barton, C.E. and McElhinny, M.W. Time series analysis of the 10000 yr geomagnetic secular variation record from SE Australia. *Geophys. J. R. astr. Soc.*, 1982, *68*, 709-724.

Barton, C.E. and McElhinny, M.W. A 10000 yr geomagnetic secular variation record from three Australian maars. *Geophys. J. R. astr. Soc.*, 1980, *67*, 465-485.

Creer, K.M. Long period geomagnetic secular variations since 12000 yr B.P. *Nature*, 1981, *292*, 208-212.

Creer, K.M. Lessons resulting from some simple computer models synthesizing palaeosecular variations of the geomagnetic field. *Nature*, 1983, Vol. *in press*.

Creer, K.M. and Tucholka, P. On the current state of lake sediment research. *Geophys. J. R. astr. Soc.*, 1983, Vol. *in press*.

Creer, K.M. and Tucholka, P. The shape of the geomagnetic field through the last 8,500 years over part of the Northern Hemisphere. *J. Geophys.*, 1982c, *51*, 188-198.

Creer, K.M., Valencio, D.A., Sinito, A.M., Tucholka, P. and Vilas, J.F. Geomagnetic secular variations 0-14000 years before present as recorded by lake sediments from Argentina. *Geophys. J. R. astr. Soc.*, 1983, Vol. *in press*.

Creer, K.M., Readman, P.W. and Papamarinopoulos, S. Geomagnetic secular variations in Greece through the last 6000 years obtained from lake sediment studies. *Geophys. J. R. astr. Soc.*, 1981, *66*, 193-219.

Creer, K.M. and Tucholka, P. Secular variation as recorded in lake sediments: a discussion of North American and European results. *Phil. Trans. R. Soc.*, 1982b, *A306*, 87-102.

Creer, K.M. and Tucholka, P. Construction of type curves of geomagnetic secular variation for dating lake sediments from east central North America. *Can. J. Earth Sci.*, 1982a, *19*, 1106-1115.

Kovacheva, M. Summarised results of the archaeomagnetic investigation of the geomagnetic field variation for the last 8000 years in south-eastern Europe. *Geophys. J. R. astr. Soc.*, 1980, *61*, 57-64.

Mothersill, J.S. The paleomagnetic record of the late Quaternary sediments of Thunder Bay. *Can. J. Earth Sci.*, 1979, *16*, 1016-1023.

Mothersill, J.S. Late Quaternary paleomagetic record of the Goderich Basin, Lake Huron. *Can. J. Earth Sci.*, 1981, *18*, 448-456.

Peddie, N.W. Current loop models of the Earth's magnetic field. *J. geophys. Res*, 1979, *84*, 4517-4523.

Runcorn, S.K. On the theory of the geomagnetic secular variation. *Ann. de Geophys.*, 1959, *15*, 87-92.

Turner, G.M. and Thompson, R. Lake sediment record of the geomagnetic secular variation in Britain during Holocene times. *Geophys. J. R. astr. Soc.*, 1981, *65*, 703-725.

Yukutake, T. and Tachinaka, H. The westward drift of the geomagnetic secular variation. *Bull. Earthquake Res. Inst. Tokyo*, 1968, *46*, 1027.

Index

Author Index